Threatened species legislation

is it just an Act?

The National Library of Australia Cataloguing-in-Publication entry:

Threatened Species legislation: is it just an Act?

ISBN 0 95856085 9 8

I. Hutchings, P. A., II. Lunney, D., III. Dickman, C. R.
IV. Royal Zoological Society of New South Wales.
V. Fauna; Environment and conservation; Threatened species; Environmental legislation

Published October 2004

Published by

Royal Zoological Society of New South Wales

P.O. Box 20, Mosman NSW 2088
Phone: 9969 7336
office@rzsnsw.org.au
www.rzsnsw.org.au ACN 000 007 518

Design and layout by P. & J. Sherratt, Printed in Australia by Southwood Press, Marrickville NSW

Threatened species legislation

is it just an Act?

Edited by Pat Hutchings, Daniel Lunney and Chris Dickman

Royal Zoological Society of New South Wales
Mosman, New South Wales, Australia

Foreword

A foreword is a rare opportunity to be slightly outrageous and perhaps provocative without the normal discipline of modern publication. The Royal Zoological Society of NSW Forum 2003 was entitled "Threatened species legislation: is it just an Act?" The very title invites such an approach. This foreword draws on the experience of chairing a threatened species scientific committee whose task is to recommend listings and whose decisions are expected to show "*Solomonic*" judgement. This foreword also draws on the experience of working with students who challenge the very "Lilies of the Field" Matthew 6 to 28, let alone the robes of Solomon[1].

Rare flora and fauna species occur naturally in all natural systems. In fact, all extinct species were rare just before they went extinct. However, some rare species have unique adaptations that allow them to exist at low abundance and these species receive considerable conservation attention. The problem is determining the difference. Some rare species may or may not have attributes requiring explicit conservation management measures and the associated allocation of valuable resources. Other species are rare because of human action and the consequential limitation of range and density or both; in such cases, abating a threat could be positive. Other species, while not rare at a moment, may be on the brink of precipitous decline. In this case, the impact may have reached its limits or it may be continuing episodically in an unknown fashion. What is certain is that often frequently-unknown or poorly-understood interactions are occurring between human-driven factors and the natural variability of climate, season and landscape geometry to create an unknown outcome. Such difficulties are not restricted to the "nature" of the actual problems. Often the rates of change and the relationships between the multivariate factors involved are reflected in the expert opinion that at times is a confusing mix of science, opinion, a best guess and advocacy.

Consequently, it is not surprising that the relationship between rare species, threatening processes and listing criteria used in many jurisdictions has become a source of confusion and debate. Threatened species listings are made after applying criteria that involve combinations of both the extent and nature of the threat. Such criteria often vary between jurisdictions. Into this *legally constructed net*, species and communities that are naturally rare within secure but circumscribed distributions or wide and variable distributions are caught in addition to those that have become rare through human actions. Species and communities that are not rare, but are threatened, often fail to be detected.

The reality is that rare species can be, by nature, ecologically rare, or have become threatened rare species because of the impact of a threatening process. The problem is to know which is which and how to transcend this dichotomy to find the *sleeper threatened species*. Success in such cases rests on whether the problem could be detected and, if so, would result in a decidedly difficult political outcome. The papers in this forum reflect the difficulty, confusion and the challenges that confront making judgments in listing.

In Australia, threatened rare species may be afforded protection under the Commonwealth's *Environmental Protection and Biodiversity Conservation Act 1999* (EPBC Act 1999) based on national extent, relevant State or Territory legislation based on jurisdictional extent or both. The occasional result is that, on one side of the border a species or community may be listed, and on the other side, not. In addition, the entity's national status may be different again. This confusion is a fundamental problem for the setting of conservation priorities and it needs to be rectified.

The situation is more complex with ecological communities. A community might be a unique assemblage of common elements or it can be a common structural form made up of unique species. The question in this case is where to draw the line. A community may change over time adding to the problems of listing and mapping a moving biological expression. This change can be truly natural, a natural response to a change in some limiting factor by human action or a change directly induced by human action. In such cases, knowing when a community is threatened, and if threatened, what parts of the residual are, in fact, the ecological community is inordinately difficult. Adding to these difficulties is the natural propensity of humans to alter the functional characteristics of communities to suit the purpose to which they put the community over time.

At the foundations of these problems is the dimensionality of our thinking. This can be represented as a cube (Figure 1) that is often greatly distorted into a twisted polygon by the exercise of power. The dimensions of the figure are defined by what the authors in these proceedings variously refer to, explicitly or implicitly, as the three dimensions of judgement. These are the quantitative to the qualitative; the objective to the subjective; and the continuum from "scientific knowledge" to "community knowledge".

The problem of legislation being *just an Act* is that it is just that: it is an Act of a parliament that attempts to reconcile the dimensions of the models that in turn reflect community thinking. As such, the interplay of power that is peculiar to jurisdictions will always give a particular local flavour to the Local Act, its interpretation and application. Expert opinion will suggest solutions, clever biological and biophysical models, clever legal models and even clever political models called compromises or visionary acts, depending on one's perspective, will influence how the Act is framed. However, in the end, they are like the figures of Lascaux, an action by society to represent our relationship with nature. They can be more or less objective, more or less scientific and more or less reflective of community views. Ultimately, they are a reflection of our relationship with the natural world.

---

[1] I particularly acknowledge the work of Ms Danaleigh Irvine who was a postgraduate in the School and Ms Nancy Dahl Tocconi a PhD scholar in the School.

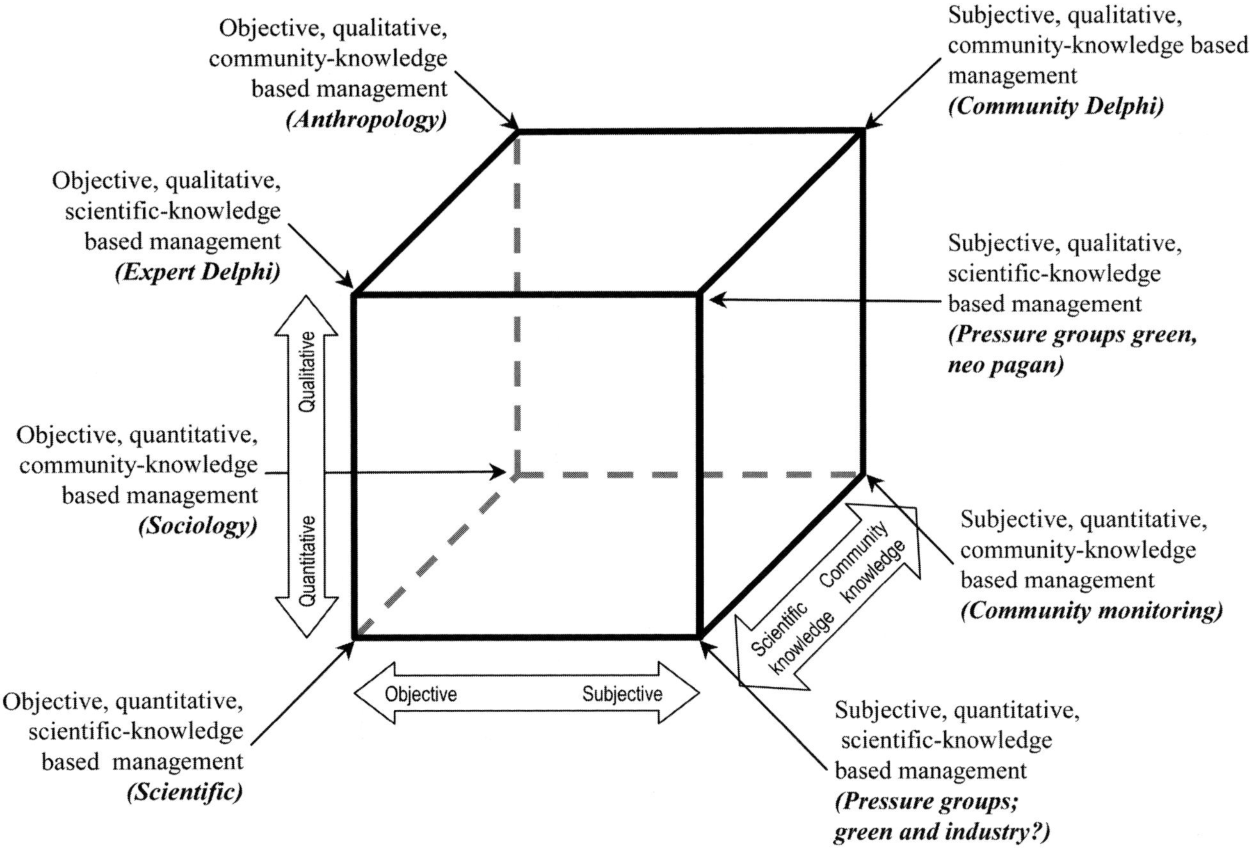

Figure 1. The relationship between ways of knowing about nature that inform thinking about management of nature.

For Robert Frost's neighbour, "Good fences, make good neighbours"; this symposium is part of trying to design a fence which may or may not be needed. The problem is that time is short, and the need for a good fence, a good neighbour or both is great. Consequently, committees will have to make decisions based on judgement. The dimensions of the cube (Figure 1) suggest that many views could be seen as valid but only some can prevail. In such circumstances, the only defence against error is appeal. Appeal can only occur if reasons are given and open to challenge. In such circumstances, those of us who try to be wise cannot become seduced by our power or apparent sophistication. In making judgements, equitable weighting to differing points of view is not always possible because the law does not reflect this weighting. We must be accepting of fallibility and not commit the error of believing in the infallibility of our science and that it will ensure our judgments are free of assumptions, subtly gained in the context in which we work, with all its consequences. The problem for the few who read these words is that in the modern world, fallibility is seen as error and not as the basis of learning. Consequently, politicians fear fallibility as the pillory, if not the guillotine, of the press, rather than something to be applauded when discovered in learned forums such as this one. We must always recognise the one truth of science, namely that science only progresses through the failure of the fences of hypotheses and the process of proposing new ones. We must be glad to be shown to be wrong, because those who prove us so give to science a great service. That service is to advance all of our thinking and achieving better conservation outcomes.

R.J.S. (Bob) Beeton

Bob Beeton Chairs The Australian Threatened Species Scientific Committee and The Australian State of Environment Advisory Committee. He works at The School of Natural and Rural Systems Management, The University of Queensland.

CONTENTS

Preface

There is nothing like retirement to enable one to reflect on a hectic period filled with conflict, choice and the possibility of doing something positive to conserve biodiversity. In January 2003, two of us (Chris Dickman and Pat Hutchings) found ourselves in this position. We had just served seven years on the NSW Scientific Committee convened under the NSW *Threatened Species Conservation Act 1995*. Further, one of us (Chris Dickman) had been the chair of this committee for these formative years. This position had been continuous from being one of a three-person scientific committee under the preceding Act, the NSW *Endangered Fauna (Interim Protection) Act 1991*, which expired at the end of 1995. That committee (Hal Cogger, Chris Dickman and Dan Lunney) had written up the procedures that had been adopted, but discussions and debates on the matter of how to conserve biodiversity and the role of threatened species legislation in that endeavour, had increased since 1995. Thus any reflections on the decision-making process of the early 1990s were of more than historical curiosity. Fellow council members of the Royal Zoological Society of NSW, who had heard us hold forth on various aspects of this subject, proposed in early 2003 that the time had arrived to stop the in-house discussions and run the 2003 RZS forum on this theme. Further, since the subject was increasing in sharpness in the public debate, and legislative changes were in the wind, there was much to be lost by postponing the holding of any public forum. The idea was floated for only a brief period before it became apparent that 2003 was a suitable time to evaluate not only the effectiveness of the NSW *Threatened Species Conservation Act 1995*, but also the broader philosophy underlying threatened species legislation more generally.

While this forum tended to concentrate on the NSW *Threatened Species Conservation Act 1995*, and the complementary NSW *Fisheries Management Act 1994*, we actively sought papers on threatened species legislation in general. Further, the speakers addressed quite different aspects of threatened species legislation, from the question of assessing the certainty of the scientists on the committees which decide which species should be on the lists, to the impact on various affected parties, including government agencies, consultants and community groups. The themes also varied biologically by addressing such subjects as invertebrates, ecological communities and threat abatement plans. A centrepiece was the view from those who look at the effectiveness of the law, whether from the legal angle or from the biologist coming to grips with the law as an instrument to conserve biodiversity.

Since hosting the forum it has become clear that the NSW legislation will be changed. Thus these proceedings are a timely documentation of this 1990s legislation and its legacy. Given the depth of thought that lies behind each of the papers in this set of proceedings, we sense that there will be more changes yet and that other states and the Commonwealth will keep the matter under regular review. These proceedings should help not only in producing more effective formulations of threatened species legislation, but also in presenting new ways of tackling the complex task of conserving biodiversity across a wider legislative and ecological landscape.

The forum on the day (25 October 2003) was intellectually stimulating, it raised more questions than it answered, and it is evident that much has been learnt by the various players over the last decade or so. All this is manifest in the papers in this forum. That the day was stimulating can be seen in the transcript (edited for clarity) of the plenary session, which is included in this published forum.

We remain indebted to each of the authors for their original insights and skilled presentation of their ideas. We are also indebted to the many referees from whom we sought an independent assessment of the papers. Each contribution had at least two referees. The procedure was to referee for the facts, the logic of the presentation and the clarity of expression. Individual ideas, conclusions and conjectures were allowed to stand to keep the diversity of ideas flowing.

The Council of the Royal Zoological Society of NSW is most appreciative of a financial contribution from the Director-General of what was the NSW National Parks and Wildlife Service to help with production costs of the forum and these proceedings. The forum was held just after the NPWS became part of the new Department of Environment and Conservation (NSW).

The Council of the Royal Zoological Society of NSW has remained the key support group for this exercise in testing these difficult ideas in a public forum. Council members not only contributed in detail, but also in sustaining the principle of keeping ideas (and manuscripts) moving. Committed councils keep all parties alive to the needs and interests of the wider community in this critical matter of how to understand and conserve Australia's biological heritage, especially those elements that are fading in the face of the relentless loss of land to what is too often called "development".

Pat Hutchings, Daniel Lunney and Chris Dickman
Editors
October 2004.

The evolving intent and coverage of legislation to protect biodiversity in New South Wales

Peter J Jarman# and Margaret A Brock*#

#Ecosystem Management, University of New England Armidale NSW 2351

*Centre for Natural Resources DIPNR PO Box U 245 Armidale NSW 2351

Email addresses:pjarman2@pobox.une.edu.au, margaret.brock@dipnr.nsw.gov.au

ABSTRACT

In the mid to late 19th century, New South Wales passed laws protecting from over-hunting just a few native bird and mammal species, despite scientists having documented the disappearance and decline of once-common species. As public concern grew over the decline of useful or commercially exploited species, and as scientific societies lobbied for their increased protection, so more bird and mammal species, and from 1927 some native plant species, gained protection. Realisation of the detriment of broad-scale agriculture led to laws setting aside reserves where land-uses that threatened species could be excluded. Mid-20th century laws introduced professional administration of reserves and of biodiversity laws, and increased the range of protected taxa. From the 1990s, with conservation politically potent, modern laws sought to conserve all biodiversity, not just individual species, but also ecosystems and their intrinsic processes, and to manage threatening processes. Such laws affect other planning and land-use laws, at all jurisdictional levels; they are embedded in a nationally integrated set of laws and reflect international agreements to conserve biodiversity. Biodiversity laws have grown in complexity, coverage and importance. Science and society will contribute to their continuing change.

Key words: threatened species legislation, flora, fauna, laws

Introduction

Legislation to protect biodiversity is a major expression of community attitudes towards, and hopes for, the conservation of biodiversity in general and of threatened species in particular. The primary aim of this paper is to review the history of the intent and taxonomic coverage of legislation in New South Wales to protect native species, and to identify some of the influences upon legislation, including the sources of community concern. We have not tried to review the effectiveness of the laws or how they were applied, except where those affected the emergence of amended or additional laws.

Our secondary aim is to suggest that legislation reacts to changes in circumstances (e.g. species really do become rarer), in scientific and management understanding (as we learn more about species, or about our intentional or unintentional effects upon them), and in public perceptions (e.g. the balance of attitudes within the community may change). However, legislation may also lead the way, inducing changes in community attitudes.

Legislation to protect biodiversity in NSW has been diverse (Table 1). Since colonial days, there have been changes in the objectives of legislative protection (what the laws set out to achieve), the objects being protected by legislation (what taxa or ecological or human-influenced processes the laws control), and the kinds, and integration, of legislative protection. In parallel with these changes, our understanding of the need for conservation has changed, and the social pressures for laws affecting biota have grown and changed direction. Through a review of the history of the legislation, we explore each of these themes.

Historical analysis shows that conservation legislation has never been perfect, nor has it been static. Legislation will and should change, and scientists and managers alike can contribute directly or indirectly to those changes. If we are to contribute to future legislation, we should have some grasp of how legislation has tracked and responded to community attitudes and environmental realities in the past. We begin our review in the colonial era.

The Colonial Era: 19th Century

Governors' environmental edicts

Attempts to control the environmental impacts of European settlers in colonial Australia pre-date specific laws. The settlement at Sydney, trying to become quickly self-sufficient in materials and agricultural produce, was bound to change the environment. Yet some changes were avoidable and seen at the time to be detrimental. Cattle escaped and became feral within weeks of the arrival of the First Fleet in 1788. Over-cutting of trees and inappropriate clearing occurred so early that, in 1795, Governor Hunter forbad the destruction, on Crown or private land, of trees fit for ship-building. Within five years of settlement, regulations were promulgated to prevent the felling of trees along the Tank Stream, Sydney's water supply. However, the dumping of rubbish and effluent into the stream led Governor King, in 1802, to impose a fine of five pounds on anyone found polluting the stream. Particularly prescient was Governor King's edict of 1803 to halt the clearing of riparian vegetation along the Hawkesbury River, which was already seen as leading to bank erosion, silting of the stream and consequent flooding (Powell 1976).

Pp 1 - 19 in *Threatened species legislation: is it just an Act?* 2004, edited by Pat Hutchings, Daniel Lunney and Chris Dickman. Royal Zoological Society of New South Wales, Mosman, NSW, Australia.

Table 1. Some NSW Acts relating to the hunting, control, protection or conservation of biodiversity, listed chronologically. Not all Acts are shown, most amending Acts being omitted.

Year	Reference	Title
1852	16 Victoria No. 44	*Native Dog Destruction Act*
1866	29 Victoria No. 22	*Game Protection Act*
1879	42 Victoria No. 10	*Animals Protection Act*
1880	44 Victoria No. 11	*Pastures and Stock Protection Act*
1881	45 Victoria No. 29	*Birds Protection Act*
1883	46 Victoria No. 14	*Rabbit Nuisance Act*
1890	54 Victoria No. 29	*Rabbit Act*
1893	56 Victoria No. 18	*Birds Protection Act*
1901	No. 26, 1901	*Birds Protection Act*
1903	No. 18, 1903	*Native Animals Protection Act*
1918	No. 21, 1918	*Birds and Animals Protection Act*
1922	No. 37, 1922	*Birds and Animals Protection (Amendment) Act*
1927	No. 2, 1927	*Wild Flowers and Native Plants Protection Act*
1930	No. 12, 1930	*Birds and Animals Protection Amendment Act*
1948	No. 47, 1948	*Fauna Protection Act*
1967	No. 35, 1967	*National Parks and Wildlife Act*
1974	No. 80, 1974	*National Parks and Wildlife Act*
1991	No. 66, 1991	*Endangered Fauna (Interim Protection) Act*
1994	No 38, 1994	*Fisheries Management Act*
1995	No. 101, 1995	*Threatened Species Conservation Act*
1997	No. 133, 1997	*Native Vegetation Conservation Act*
2002	No. 78, 2002	*Threatened Species Conservation Amendment Act*

These early edicts were local, often site-specific, in effect. Yet such early insight into the local environmental harm of uncontrolled land-uses did not forewarn all later governors and governments and was not generalised to other areas as settlement expanded. Instead, Australia has experienced 200 more years of environmental degradation (including the clearing of riparian vegetation) as part of the unresolved conflict between individual land-holders' rights to develop and longer-term community interest in sustainable land management, fully conserved biodiversity and functioning ecosystems.

The need to legislate for the management of biodiversity emerged piecemeal. Two kinds of laws relating to vertebrates were passed in the second half of the 19th century in New South Wales: laws that protected individuals of some species, usually so that there would be enough of them to hunt ('game laws'), and laws to encourage the killing of individuals of species seen as pests of agriculture.

Game laws

From about 1830 to 1860, pastoral settlement spread rapidly across the tablelands and onto the plains of NSW. Squatters took up huge stations (just 11 stations spanned the 200 km from Armidale to the present Queensland border); but the pastoral development of those stations was not intense, depending, for example, on natural watering points and on shepherding rather than fences to control stock. Many pioneering graziers quickly became rich from the productivity of cattle and sheep grazing native vegetation communities that had never had stock on them. With wealth and ownership of land came attitudes akin to those of contemporary landed gentry in England, including possessive attitudes towards the right to hunt (for sport) either native or introduced animals on their land. Some landowners introduced game birds and mammals from Europe and America to improve the sport to be had on their properties.

The first pieces of colonial legislation to protect biota were game laws, modelled on then-current English hunting laws and common law. The first in NSW, the *Game Protection Act 1866* (the next, the *Animals Protection Act 1879*, was similar) protected individuals of exotic and native game species by banning their hunting totally for an initial five years, and then imposing a 'closed season' (a non-hunting period), meant to match the breeding season. Game species were listed on schedules to the Act (see Appendix 1). The scheduled exotic game were pheasants, partridges, grouse, swans, Brown Hare *Lepus capensis*, deer and antelopes; but the schedule ends with the phrase "All birds not indigenous to Australia and their produce", so the Act was intended to be used to protect acclimatised birds in general. However, the Act was clearly premonitory in that very few exotic birds had been deliberately introduced to and released in NSW at the time of these Acts. The scheduled native game species predominantly were the equivalents of British game bird species: ducks, Black Swan *Cygnus atratus*, quail, pigeons, plovers, Australian Bustard *Ardeotis australis*, Malleefowl *Leipoa ocellata*, Australian Brush-turkey *Alectura lathami*, Bush Stone-curlew *Burhinus grallarius* and rail species. The Emu *Dromaius novaehollandiae* was listed, as it had been hunted for food and sport since the First Fleet, often to the point of local extirpation. However, the Laughing Kookaburra *Dacelo novaeguineae* was included, apparently out of sentiment for its song. All other native species remained unprotected, i.e. they could be taken or killed at will.

The assumptions behind this form of legislated protection were that:

- Uncontrolled hunting would threaten game populations and reduce sport in future;
- Populations would survive as long as hunting did not target the breeding stock or occur during the breeding season.
- People who had invested to increase the public good by importing game species (in particular) should have their investment protected.

Thus direct human killing of too many individual animals, exotic or native, was seen as the only threat (conservation risk) to be controlled by legislation; without that killing, everything would be fine. The acts reflected the value the community placed upon these species, and protected the opportunity for people to hunt them for sport, except for the kookaburra, which was the first native animal species to be legally protected for aesthetic pleasure alone.

Game laws developed in complexity, limiting who could hunt the species listed in the Schedules of the various Acts, when they could be hunted, by what methods (permissible size of gun, no use of nets, dogs, snares, bird-lime, etc), where they could be hunted, and, eventually, how many individuals could be taken (bag limits). The *Birds Protection Act 1881* again encouraged the importation of exotic birds (not just for hunting, and most of the named species had not at that time been released or become established in NSW, even if they had been in Victoria (Long 1981)), but laid equal emphasis on protecting listed native birds in the breeding season. The listed native birds (see Appendix 1) included several "songbirds" (and included kookaburras), mostly seen as useful to agriculture because they ate insects. The Act also made possible the declaration of reserves for "preserving any of the scheduled birds". Thus began the reservation of areas where protected species could safely breed and find refuge from killing; the examples of reserved areas known to us were wetlands, apparently dedicated for breeding waterfowl.

The *Birds Protection Act 1893* gave similar protection to listed exotic birds (including many that had not been introduced to NSW Long 1981), and was the last NSW act specifically to protect introduced vertebrate species other than game fish (e.g. trout species). The list of non-game native birds protected under the act was greatly extended (see Appendix 1 for a full list) to include many more insectivores (whose eating of insect pests was seen to help horticulture) and also several birds that had been hunted for their plumage, such as Superb Lyrebird *Menura novaehollandiae*, Nankeen Night Herons *Nycticorax caledonicus*, Paradise Riflebird *Ptilorus paradiseus*, Dollarbird *Eurystomus orientalis* and kingfishers, for whom protection was seen as urgently needed: "if we wait...there will be nothing done. Birds, like the rifle and lyre birds, which are shot for their plumage for the decoration of dresses, will be destroyed wholesale, and snared right and left" (W.R. Campbell, moving the Birds Protection Act, NSW Parliamentary Debates Session 1893, 1st series, Volume 63, p.4563). Thus this Act extended protection in two ways: addressing threats from killing for trade rather than sport, and protecting birds thought to be useful for agriculture.

The *Bird Protection Act 1901* consolidated the series of 'game laws' of the 19th century and was more prescriptive of the weapons and methods allowable in hunting; it relied mostly on closed seasons for protection. The attitude still prevailed that over-kill was the only substantial threat, but the Act recognised a wider range of reasons why birds were being killed and hence a longer list of species needing legal protection.

These Acts stirred considerable passion, some legislators fearing that they would "reproduce in this colony the accursed game laws of England. I know that those laws have caused a vast amount of bloodshed. Do not issue licences for shooting game, but protect the birds in any other way you like" (Poole, debate on the second reading of the Animals Protection Bill 25th November 1881, NSW Parliamentary Debates, Session 1881, 1st Series Volume 6, p. 2199; see also Melville p. 2200 and Poole and Jacob, p 2201). Legislators were wary of protecting bird species merely because they had been introduced at someone's personal expense, citing, as examples of protected birds that had become pests, House Sparrow *Passer domesticus* and Indian Mynah *Acridotheres tristis* (Norton, NSW Parliamentary Debates, Session 1893, Volume 63, p. 4568).

'Pest' control

Pastoralists, especially those raising sheep, were beset by mammals and birds that, they believed, ate their stock or their pasture and crops.

Some native and increasing numbers of introduced animal species were seen as threats to pastoral and cropping agriculture. Normally, individual landholders dealt with pests, but landholders called for more organised action against pests on uncontrolled properties. The first NSW attempt at legislation to coordinate vertebrate pest control was the *Native Dog Destruction Act 1852* (see Appendix 1), which tried to ensure that neighbours at least contributed to the cost of strychnine baiting of Dingoes *Canis lupus dingo* and wild Dogs *Canis lupus familiaris*.

Similar Acts aimed to co-ordinate and consolidate destruction of the European Rabbit *Oryctolagus cuniculus*. Under the first, the *Rabbit Nuisance Act 1883*, any animal could be declared and then protected as "a natural enemy of the rabbit"; and possession of a live rabbit or interstate importation of rabbit scalps could attract a fine of up to one hundred pounds or 6-months imprisonment (severe penalties for the time). The *Rabbit Act 1890* facilitated the co-ordinated building of rabbit-proof fences to stop the spread of rabbits.

From the 1860s to the early 20th century, public pressure for 'closer settlement' saw the original vast stations subdivided into smaller holdings, usually fenced and with well-distributed artificial watering points. Graziers were trying to support themselves on more intensively stocked acreages, with native pastures depleted by several decades of the impact of stock. Dingoes, Wedge-tailed Eagles *Aquila audax* and grazing macropods were seen as threats to livelihood.

In 1878, graziers, faced with what they saw as a plague of kangaroos, petitioned the NSW Legislative Assembly for special legislation to "check the rapid increase of the marsupials and so prevent threatened ruin to pastures and crops" (NSW Votes & Proceedings of the Legislative Assembly 1878-79, Vol.VII, pp. 889, 891 and 893). A Marsupials Destruction Bill, tabled but lapsed in 1879, led to the *Pastures and Stock Protection Act 1880*, which initiated 50 years of organised, co-ordinated scalp bounties on pests. Under the Act, landholders, through Pastures Protection Boards, could apply to the Governor to have any species declared a pest, with a defined bounty on its scalp. The original Act referred to "Rabbits Native Dogs and Marsupials" (the last meaning "Any kangaroo wallaroo wallaby or paddamelon"), but district bounty lists expanded, to include feral pigs *Sus scrofa*, hares, Red Foxes *Vulpes vulpes*, Common Wombats *Vombatus ursinus*, bandicoots, Greater Bilbies *Macrotis lagotis*, bettongs, Black-striped Wallabies *Macropus dorsalis*, Bridled Nailtail Wallabies *Onychogalea fraenata* and Brush-tailed Rock-wallabies *Petrogale penicillata* (Jarman and Johnson 1978; Short 1998).

Several of the bountied species are now extinct in NSW, but not because of bounties alone. The Act reflected the current attitudes that gave agricultural interests precedence over wildlife on pastoral lands (in debate on the Birds Protection Act Amending Bill, February 1893, De Salis was understood to say: "Whilst we had such enemies in the country as rabbits, hares and foxes interfering with the wealth of the colony, we ought to give our attention to them, and not such ridiculous trifles as the protection of a few birds." NSW Parliamentary Debates, Session 1893, 1st Series, Volume 63, p. 4569). The Act condoned the unlimited killing of individuals of any species labelled a pest. However, the validity of labelling was soon questioned, and many Boards quickly removed some species from their pest lists; but bountying remained common practice until the 1910s or 1920s in many Pastures Protection Board Districts.

19th Century knowledge

From the 1780s to about 1810, knowledge of the biota of New South Wales came from British (and French) explorers, and then from Sir Joseph Banks' collectors, including Brown, Bauer and Cayley, who sent material back to Britain (Gilbert 1981). Knowledge of Australia's biota expanded rapidly from the 1820s, thanks to Australia-based naturalists, collectors and explorers (e.g. Allan Cunningham, John Oxley, Alexander Macleay, Thomas Mitchell, Richard Cunningham) and the early stimulus of the Philosophical Society of Australasia (Hoare 1981; Gilbert 1981; Finney 1993; Hutton and Connors 1999). At the same time, the Agricultural and Horticultural Society of NSW began to develop a scientific understanding of the colony's agricultural lands.

The widespread impacts of settlement were reported quite early (Jarman 1994). Darwin (1845) in his brief visit to Australia in 1836 blamed the decline of kangaroos and emus on the southern Tablelands of NSW on settlers' greyhounds, and predicted the extinction of those species. In the 6th edition of *The Origin of Species*, among cases of introduced species displacing endemic species, Darwin reported: "In Australia the imported hive-bee is rapidly exterminating the small, stingless native bee" (Darwin 1872, Chapter 3, p. 59).

Gould (1863), from his experiences in Australia from 1838 to 1840, noted the decline of Red Kangaroos *Macropus rufus* confronted by pastoral expansion, predicting the species' "entire extirpation, unless some law be enacted for its preservation". Gould appealed for legal protection for this and other species, without which "a few years will see them expunged from the fauna of Australia" (Gould 1863).

Gerhard Krefft, collecting around the junction of the Murray and Darling Rivers in 1856 and 1857, noted the decline since stock arrived, less than 20 years earlier, of Red Kangaroo, Burrowing Bettong *Bettongia lesueur*, Greater Bilby and Pig-footed Bandicoot *Chaeropus ecaudatus*. Long-nosed Bandicoot *Perameles nasuta*, Eastern Barred Bandicoot *P. gunnii*, Southern Brown Bandicoots *Isoodon obesulus* and Eastern Hare-wallaby *Lagorchestes leporides* were still common; and Bridled Nailtail Wallaby were the commonest of the small macropods (Krefft 1866). Krefft linked the declines in abundance of marsupials unequivocally to cattle-based pastoralism.

George Bennett (1860) noted the scarcity and disappearance of some native mammals and birds. "Even in our own time, several have been exterminated; and unless the hand of man be stayed from their destruction, the Ornithorhynchus and the Echidna, the Emeu [sic] and the Megapodius....will shortly exist only in the pages of the naturalist" (Bennett 1860, Preface).

Thus, by 1865, several eminent scientists had published their view that expanding settlement was threatening the survival of some native mammals and birds. Some had publicly urged legislation to stop the killing of endangered native species; some felt that the threats comprised a suite of impacts that came with pastoral land-use.

19th Century Acclimatisation and Scientific Societies

In the 1850s and 1860s numerous plant, mammal and bird species were brought to Australia at the private expense of wealthy land-owners keen to introduce and "acclimatise" the useful plants and animals of other regions to "improve" Australia. Acclimatisation societies, established in most states from 1861 (in Victoria), encouraged further introductions, and pressed for legal protection of populations already introduced, despite the evident damage caused by some pre-Acclimatisation-Society introductions (hares and rabbits to Tasmania, and to Victoria, for example).

Acclimatisation matched the pioneering belief that the environment could be improved and that individuals should be encouraged to join in the task. This attitude under-pinned all the bird and animal protections acts up to 1903, and was reflected in the opening statement in the *Game Protection Act 1866* that: "certain persons have imported and may import from parts beyond the seas at considerable expense for the benefit of the Colony game not previously existing therein and it is expedient that encouragement should be given to such importations of game by protecting it and its progeny as also that native game should during the breeding season have a similar protection."

Many scientists strongly supported acclimatisation. Australia's most eminent 19th century botanist, Ferdinand von Mueller, first Government Botanist of Victoria (from 1853 until his death in 1896) and first Director of the Melbourne Botanic Gardens, vehemently advocated acclimatisation. Indeed, Mueller advocated the purposeful spread of blackberries in Australia until the year before his death, despite blackberries by then being declared a noxious weed in parts of Victoria (Parsons 1981; Hutton and Connors 1999).

The first scientific society in Australia was the Sydney-based Philosophical Society of Australasia, founded in 1821; this transformed into the Australian Philosophical Society in 1850, and into the Royal Society of New South Wales in 1866. The first specifically biological societies in NSW were the Linnean Society and the Zoological Society, both formed in 1879. The Natural History Association existed from 1887 to 1890, and the Field Naturalists' Society from 1890 to 1894. The Zoological Society (later renamed the Royal Zoological Society of New South Wales) was formed for "the introduction and acclimatisation of song birds and game" (as a result of dissatisfaction with the existing acclimatisation society), and lobbied for legislation to protect imported species (e.g. McLaughlin: "The Bill was introduced at the request of the Zoological Society, who had a number of imported birds which they could not let loose because they were not protected by law." Animals Protection Bill In Committee, NSW Parliamentary Debates, Session 1881, 1st Series, Volume 6, p. 2776). The Society gained its more scientific bent later (Strahan 1992). In the 19th century, none of these biological societies was dedicated to conservation, to which they turned early in the 20th century.

Changing attitudes to land uses and native vegetation

Most agricultural land in NSW east of the Darling River was taken up between the 1830s and 1860, and west of the Darling between 1850 and 1890. Many pioneers noted rapid changes in vegetation and in animals within the first decade of agricultural development. While at first tolerated or even a condition of leases, the tree destruction that accompanied pastoral expansion was publicly deplored by such as von Mueller, and the risks from ring-barking and clearing were clearly recognised by the 1890s (Hutton and Connors 1999).

Compared with agricultural produce, native plants and timber were under-valued. The Colony had no legislation for payment of royalties on timber until 1889 (and then royalties were largely ignored), and proper silviculture was not practised until 1890-3 (Walker 1991). Forestry practices became more standardised and more thoroughly administered after passage of the *Forestry Acts* of 1909 and 1916. A NSW branch of the Australian Forest League was established in 1915 to improve professional practice. Flora Reserves were established in State Forests early in the 20th Century. Since 1868, Fisheries Acts in NSW had aimed to minimise wasteful harvesting of fish and shellfish (e.g. oysters) (Hutton and Connors 1999), exemplifying attempts to protect commercial, marine biodiversity.

Some popular cut flowers, such as Christmas Bush *Ceratopetalum gummiferum* and Christmas bells *Blandfordia* spp, were protected from over-harvesting by landowners or local regulations. Bills to protect native flora, tabled in 1895 and 1897, arose from concern at the loss of native species in Central Cumberland, through commercial taking of wild flowers and their copious use in wild flower shows (Farnell, 21st August 1895, NSW Parliamentary Debates Vol 79, p 244). However, the bills were respectively declared a waste of parliamentary and public time (by Haynes and Edden, 21st August 1895, NSW Parliamentary Debates Vol 79, pp 246 & 251), and a "paltry emanation of a circumscribed intelligence" (Watson, 23rd November 1897, NSW Parliamentary Debates Volume 91, p. 4993), and lapsed. Opponents claimed that the bills would destroy the wildflower industry, that "we have millions of acres which will remain unoccupied for the next thousand years on which these flowers will grow" (Haynes, NSW Parliamentary Debates Volume 79, p. 246), that "it is impossible that the wild flowers in our bush can be exhausted for several centuries to come (Waddell, NSW Parliamentary Debates Volume 79, p. 247), and that the (Royal) National Park and Ku-ring-gai Chase should provide enough protection for native flora. The problem was thought to be confined to urbanised Sydney and inconsequential because the same flora was assumed to occur abundantly in the untouched, everlasting wilderness of the rest of NSW. Legislative protection of native flora had to wait until 1927.

In 1864, the American G.P. Marsh published *Man and Nature* (Marsh 1864), the first major work suggesting that human land-uses affected not only vegetation but also soils, water and climate. The ideas were taken up in Australia (warmly advocated by von Mueller (Powell 1976)), and led particularly to suggestions that the Australian climate would be altered by tree-clearing (e.g. Clarke 1876). South Australian Surveyor-General G.W. Goyder, in the 1860s and 1870s, encouraged preservation of tree cover and planned forestry plantations and drainage schemes, based on scientific data of the time; Goyder is best known for defining zones of climate suitable for cropping agriculture and his hurried 1865 survey to demarcate limits of drought relief (Heathcote 1981). In the 1880s, science was increasingly used to solve problems of agricultural land-use, soil, water supply and insect pests (including biological control of pests), but was not yet focused on conservation of biodiversity.

General 19th century attitudes

Walker (1991 p.18) commented that, in the 19th century, "conservation was never an important political issue"; all the NSW game, animal and native flora protection bills up to 1897 were proposed by private members, not government. The public generally supported opening up and development of the land at least to the 1880s. However, NSW experienced difficult agricultural years in the late 1880s and 1890s, including prolonged droughts and invasion by rabbits, and stock numbers fell precipitately in the 1890s. The realities of overstocking, the detrimental consequences of tree clearing, disappearance of native vegetation and ground cover, soil erosion, siltation of streams and waterbodies, plagues of vertebrate and invertebrate pests, and the disappearance of native species all became widely apparent. Testimony to the Royal Commission into the condition of farmers in the Western Division of NSW in 1901 makes stark reading still (see Lunney 1994). By the end of the 19th century, it was clear to some that land-use practices in many regions were unsustainable, and that native biota were suffering as a result. Some late-19th century authors strove to make rapidly growing scientific knowledge of native flora and fauna available to school children and the public. For example, *The Insectivorous Birds of Victoria with chapters on birds more or less useful* (Hall 1900) opens with the words: "The purpose of this handbook is to foster an interest in the insectivorous birds of our colony, and to aid in the protection of them." Such increased public understanding, education and purpose paved the way for the escalation in legislation for conservation in the 20th century.

20th Century: NSW Acts pre World War II

The schedules of the *Birds Protection Act 1901* listed mainly 'game' birds for hunting. However, the *Native Animals Protection Act 1903* at last recognised that some mammal species had declined greatly (including Red Kangaroo and Common Wallaroo *Macropus robustus*, heavily bountied under the *Pasture and Stock Protection Act*). A dozen mammal species (see Appendix 1) were given absolute protection for 14 months, and then at least periodic protection from hunting. The aim of the Act was still to limit direct killing of individuals, still the only admitted threat. Ever since the *Birds Protection Act 1881*, the Zoological Society had been granted half the fines imposed for offences; the 1903 Act was the last under which that occurred.

The *Birds and Animals Protection Act 1918* was novel in listing the birds and mammals that could be hunted (24 bird species and 21 mammal species, including several now threatened or extinct in NSW; see Appendix 1). All unlisted species were protected; however, rodents were specifically excluded from protection, as were reptiles (and, of course, frogs and invertebrates). The increasing list of protected species was still limited to birds and mammals. In the *Birds and Animals Protection (Amendment) Act 1922*, most of the previously listed native mammals were removed from the unprotected list (Walker 1991).

International pressure against trade in birds' plumes influenced that 1918 Act. Australia was not a big source of plumes, but many people had profited out of selling plumes, which often meant killing the birds in the nesting season. The Royal Australasian Ornithologists' Union (RAOU), founded at the turn of the century for "the advancement and popularisation of ornithology and ... the protection of economic and ornamental native birds" (Ryan 1907, pp. 95-6), actively lobbied for laws protecting native birds. In his President's Address, Ryan (1907, p. 96) foresaw "that each State will have to carry out its own domestic legislation regarding proper bird protection – at present, at least. It has been said that Australia is an over-governed country. So far as bird protection is concerned it is hardly governed enough." Ryan countered accusations that bird protection was a matter of sentiment with arguments for the agricultural and horticultural value of insectivorous birds and the value of birds for sport hunting. He also urged that Australia and New Zealand should not be seen to lag behind the "rest of civilised nations in matters pertaining to bird protection", arguing for the first time for Australia to match international standards in conservation laws (Ryan 1907, p. 97).

Ryan (1907) illustrated the context for the early-20th century legislation. He felt that State Game Acts should be made more effective through education, "fostering a national sentiment, and by urging the authorities to enforce the law" (Ryan 1907, p.101). Ryan recommended that schedules of protected species be made uniform between states (a new idea at that time), to prevent species being protected in one state but still vulnerable in another. He emphasised the value of creating reserves, on both Crown and private land, and the necessity of paying Wardens to manage reserves. His vision for bird conservation was in advance of Australian practice at the time, but drew upon current practice and discussion in the United States of America. The 1908 RAOU Vice-Presidential address took lessons for Australia from international agreements on the protection of birds in Europe (Campbell 1908).

The RAOU's conservation awareness was matched by the Wild Life Preservation Society of Australia (WLPSA), formed in 1909 specifically to foster conservation (Walker 1991). The WLPSA's members included senior government and university zoologists. From 1912, the Association lobbied for better protective legislation, producing their own draft of a Bill that led to the 1918 Act, and becoming the first public conservation society successfully to lobby for conservation legislation. The 1918 Act did not achieve everything advocated by the WLPSA and RAOU, but it reflected a substantial shift in public, and hence government, attitude towards conservation.

The 1918 Act for the first time allowed for the employment of rangers, and the appointment of honorary rangers; all police became rangers by virtue of their office. With government-appointed protectors, the Zoological Society no longer received half of all fines.

By 1925, after the further intensification of pastoral and cropping agriculture (e.g. in soldier-settler schemes) and the beginnings of mechanised forestry and forest clearance, biologists knew that Australia's biodiversity was being eroded by myriad, human-induced, broad-scale threats. In his introduction to the book *Save Australia. A Plea for the Right Use of Our Flora and Fauna*, Barrett (1925) listed these as settlement, the clearing of forests, introduction of exotic plants, effects of foxes and rabbits, hunting for the skin trade, 'sport' shooting, hunting for museum collections (being urgently pursued in the 1920s as Australia's wildlife was seen to be fast disappearing), and human population growth. Le Souef (1925) also identified, among the major threats to native mammals, predation by the Cat *Felis catus* and disease. He called for "preservation" through placing endangered species in reserves (preferably on islands) and a Zoological Survey, or at least rangers' reports, to counter poor information and advice to the Minister. This was one of the first suggestions for targeted research to provide a scientific basis for understanding what was happening to biodiversity, and an example of the widespread contemporary thinking that island sanctuaries should be used to preserve endangered species.

Barrett (1925, pp.14-15) gives an early example of Commonwealth control of exports being used to aid conservation: "The Minister of Trade and Customs for the Commonwealth ... has appointed Committees of Advice in the various States to assist his department in preventing export of animals and birds, dead or alive, except under license." Nicholls (1925) recognised that "the animals of Australia belong to the Commonwealth, and the Commonwealth Government should exercise its prerogative, and see that they are properly cared for" (p. 98), and called for Commonwealth or State Governments to "levy a small tax upon every skin leaving Australia, and apply all the money so collected to the establishment of a Department of Economic Biology", which would "control the fur export trade, dictate the policy of the total or partial protection of the various species, and, by scientific research, show the economic relationship that exists between animals, birds, and agriculture." Such public advocates of conservation put much faith in science.

In 1927, the *Wild Flowers and Native Plants Protection Act*, the first general law protecting native vascular plants, allowed the Minister to proclaim any native NSW plants protected in specified areas and for specified times. Again, this only protected them against picking and harvesting; it did not protect their habitat. But, notably, the *Forestry Act 1916* could not over-ride the Act, which applied on Crown land, State Forests, public land and private land unless the owner of that private land had given permission for plants to be destroyed. Protected plants could not be sold unless they had been picked from private land with the owner's permission (a wide-open loophole for flower-cutters).

The *Birds and Animals Protection Amendment Act 1930* encouraged the declaration of sanctuaries, both on private land at the owner's request and on public land, such as within a mile of every State school (to encourage children to appreciate animals; although sparrows, silvereyes, starlings, bulbuls, crows, ravens, rabbits, hares, foxes, dingoes and fruit-bats were specifically not protected in such school sanctuaries) (Walker 1991). In introducing the Bill, Captain Chaffey said: "Today there is not sufficient protection of birds against trafficking. Some of the cruellest things that could be imagined are done by hawkers of birds, who cage them under the most objectionable conditions" (NSW Parliamentary Debates, 2nd Series, Session 1929-30, Volume 121, p. 3650). Prevention of cruelty to individual animals appears to have been a major stimulus for the Act, which increased penalties for possessing protected animals, and proscribed more means of taking animals. More public servants automatically became rangers (e.g. teachers, foresters, stock inspectors, fisheries inspectors), "to establish a protective system from one end of the State to the other" (p. 3651). People could obtain licences to trade in legally taken animals.

Since 1901, the NSW laws had become more complex, reflecting a growing public regard for and awareness of the usefulness of native biota. No longer were the attitudes of sportsmen and farmers paramount; indeed, land-owners were increasingly encouraged to establish reserves or sanctuaries. Protection of wildlife was part of State education; and the decline of native mammals and birds was recognised. The first professional rangers were being paid, although they were greatly outnumbered by honorary and ex-officio rangers. In debating the 1930 Bill, several proponents hoped it would "bring the law of New South Wales regarding birds and animals into line with that of the other Australian States" (NSW Parliamentary Debates, 2nd Series, Session 1929-30, Volume 121,Tonge, p. 3651; Marks, p. 3653-4) and keep pace with conservation measures and practices in Britain, Germany and the United States of America (Tonge & Marks, pp. 3652-3). NSW legislators clearly knew of progress in conservation in other Australian states and overseas, and their awareness was affecting State legislation.

20th Century: NSW Acts post World War II

In the second half of the 20th century, land uses in NSW intensified greatly. After World War II, further soldier-settler schemes occurred and irrigation schemes (e.g. in the Riverina and Murrumbidgee Irrigation Area) pushed intensive agriculture far onto the western riverine plains. Mechanical techniques for distributing superphosphate fertilisers and seeds of exotic pasture plants, for transporting stock and crops, for clearing trees and scrub, for ploughing hitherto unploughed soils, for draining wetlands, for sinking bores and digging dams, and for extracting timber or clear-felling coupes for short-rotation forestry or plantations: all these induced rapid change in intensity and extent of modification of natural communities and set the context for the expansion of parks and reserves that characterised conservation in this period.

Up to World War II, the list of birds and mammals legally protected from being killed by humans had grown steadily (Figure 1). However, the legislation had been weakly enforced, by a handful of paid and honorary rangers, and had not recognised the full diversity of threats to, or threatened status of, biota. The *Fauna Protection Act 1948*, the first real, purpose-built conservation Act in NSW (albeit for some birds and mammals only), expressed an emerging government commitment to conservation.

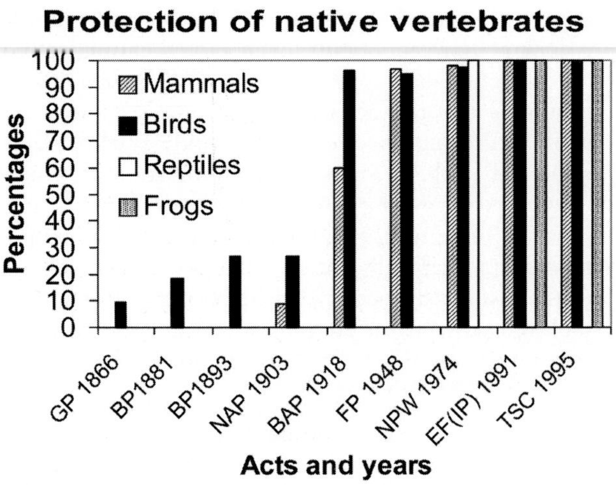

Figure 1. Changes through time in the proportions of native mammal, bird, reptile and frog species in New South Wales protected by legislation. Mammals refers to terrestrial species only. The Acts referred to by abbreviations are: GP 1866, Game Protection Act 1866; BP 1881, Birds Protection Act 1881; BP 1893, Birds Protection Act 1893; NAP 1903, Native Animals Protection Act 1903; BAP 1918, Birds and Animals Protection Act 1918; FP 1948, Fauna Protection Act 1948; NPW 1974, National Parks & Wildlife Act 1974; EF(IP) 1991, Endangered Fauna (Interim Protection) Act 1991; TSC 1995, Threatened Species Conservation Act 1995.

This Act required the appointment of a senior government administrator, the Chief Guardian of Fauna, and established a 13-member Fauna Protection Panel drawn from government departments, university, farmers, and societies constituted for "the preservation, conservation, protection or scientific investigation of fauna". The Panel was appointed to: "advise the Minister; educate ...to awaken and maintain an appreciation of the value of bird and animal life; have the care and control and management of faunal reserves; ...generally co-operate with any ... persons or bodies in the care and development of reserves for fauna; form ... local faunal societies; (and) conduct scientific and biological research in connection with the protection and care of fauna". The Act was the first to appoint a senior public servant and a committee to implement education and research for conservation.

The Act provided for the dedication, on the recommendation of the Panel, of "any Crown land as faunal reserves for ... the protection and care of fauna, the propagation of fauna and the promotion of the study of fauna." Such reserves could be revoked only by act of parliament. The Panel had to draw up a "written scheme of operations" for each reserve, which, if approved by the Minister, became the working plan for the reserve. On reserves, timber was protected, miners' rights or leases were suppressed, and any travelling stock reserve included in them could be used only with the Panel's permission. Thus the Act began to protect habitat by managing land uses, but only within Reserves on Crown land and only for fauna (often for specific species).

The Fauna Protection Act's first schedule defined *unprotected* fauna (which included both native and exotic birds and mammals) that could be killed or taken either throughout NSW or in specified localities. This schedule contained the same exotic birds as previously (and added the bulbul), the same cormorants, corvids, and mostly the same cockatoos and parrots; but added several raptors and, in some parishes, rallids were listed as unprotected (see Appendix 1). The mammals that were unprotected in the whole state were: Dingo, Ferret *Mustela putorius furo*, Red Fox, Grey-headed Flying-fox *Pteropus poliocephalus* and Little Red Flying-fox *P. scapulatus*, Brown Hare and European Rabbit; the Common Wombat was unprotected in many districts over most of its range in NSW.

The Act also specified as "protected fauna", and the property of the Crown, all mammals and birds other than those listed in schedules as unprotected. For the first time, the Governor could also declare any protected fauna to be *rare* fauna, taking of which could incur a penalty of a 50-pound fine (as for unprotected fauna) or imprisonment for up to 6 months or both. This was the first time imprisonment was set as a penalty for killing wildlife in NSW. However, the Governor could also license people to kill "any fauna found on his land which may be destroying or injuring or be likely to become injurious to his property", thus allowing for licensed control of protected species regarded locally as pests. People could also be licensed to deal in fauna or skins of unprotected species and of kangaroos, wallaroos, wallabies and possums taken under licence and on payment of royalties. The Act also ended acclimatisation, imposing penalties on: "Any person who liberates any fauna imported into New South Wales, except under .. a permit granted by the Minister." Trout continued to be liberated, fish not being subject to this Act.

The 1948 Act established the framework for late-20th century government conservation effort, centred on the reservation of land for conservation (even if expressed as the conservation of individual species). Its advisory panel represented government, scientific, land-holder and conservation-group concerns. The Act reflected growing public interest in the environment and biodiversity. For the first time it brought together under one act almost all the then-current threads of management of native birds and mammals: protection of species, control of pests, provision of reserves to protect habitat, and regulation of trade in animals and their products. Nevertheless, it still lacked a professional conservation service to implement its intentions.

A state government conservation agency: the National Parks and Wildlife Acts

When post-war (World War II) expansion and intensification of agriculture made it increasingly obvious that many species of fauna and flora, even if protected from direct killing, could not survive on land dedicated to other uses, the declaration of National Parks and other reserves became a major tool in the growing conservation effort. This reflected four things: an acceptance that production land-uses could be detrimental to at least some fauna and flora; recognition by government that dedicating Crown land for conservation was politically easier than curbing land-holders' management of private land; the urban public's growing interest in recreational uses of natural bush; and the first ecological insights (ecology having emerged as a quantitative scientific discipline in the 1930s) into a population's requirements for habitat and resources.

The 1967 and 1974 *National Parks and Wildlife Acts* both centred on establishing the NPWS as the State's conservation authority. The 1967 Act consolidated, without immediately expanding, the taxonomic coverage of the *Fauna Protection Act 1948* and the *Wild Flowers and Native Plants Protection Act 1927,* listed the pre-existing12 National and seven State Parks, and re-named pre-existing faunal reserves as Nature Reserves. Most importantly, the *National Parks and Wildlife Act 1967* established a framework for a government agency, the NSW National Parks and Wildlife Service, charged with conserving both fauna and flora. The Fauna Protection Panel and office of Chief Guardian of Fauna, established under the *Fauna Protection Act 1948*, were abolished and their functions transferred to the new National Parks and Wildlife Service. Similar Acts were passed in the early 1970s in most other Australian States and Territories, and a Commonwealth Act brought about an Australian National Parks and Wildlife Service.

The *National Parks and Wildlife Act 1974* listed *unprotected* fauna including, for the first time, deer and antelope, which had been specifically protected a century earlier by the *Game Protection Act 1866*. Rusa Deer *Cervus timorensis* was specifically excluded from the unprotected list, showing that sentiment could still selectively override the principle of listing as unprotected all exotic, free-ranging vertebrates. However, their continued protection may have helped prevent poaching in Royal National Park (the only home of free-ranging Rusa Deer in NSW). The unprotected species included two native mammal species (Grey-headed and Little Red Flying-foxes) and 13 native bird species (four cormorants, three corvids, Pied Currawong *Strepera graculina*, Galah *Cacatua roseicapilla*, Sulphur-crested Cockatoo *C. galerita*, Crimson Rosella *Platycercus elegans*, Eastern Rosella *P. eximius* and Silvereye *Zosterops lateralis*) (see Appendix 1 for the full list). Thus the unprotected schedule was still used to allow the killing of native vertebrates seen as pests.

The Act also listed (in its Schedule 12) eight native mammal and 28 bird species as *endangered.* Some of these had previously been subjected to legally sanctioned sport-hunting or pest control (Bridled Nailtail Wallaby, Grey Falcon *Falco hypoleucos*), although others had intermittently received protection since the *Game Protection Act 1866* (e.g. Brolga *Grus rubicundus*, bustard, Malleefowl). Now they were protected because of concerns about their status, although the criteria by which they were judged to be endangered was not spelled out in the Act. The *National Parks and Wildlife Act 1974* also for the first time in NSW protected reptiles and included a reptile (Broad-headed Snake *Hoplocephalus bungaroides*) among the listed endangered species. All other snakes received protection unless they were thought to be endangering people, such as when a venomous snake was seen under a back step or in a school yard.

The schedule of protected native plant species was still intended to prevent over-harvesting of commercially valuable species. Thus it specified sphagnum moss, many ferns, several palms, all orchids, all *Telopea*, *Pandanus*, *Doryanthes*, *Eriostemon*, *Crowea*, *Blandfordia* and *Boronia* species, and a number of species commonly sold as cut flowers (e.g. Flannel Flower *Actinotus helianthi*, Christmas Bush *Ceratopetalum gummiferum*, and Sturt's Desert Pea *Clianthus formosus*). However, while a few species (e.g. *Grevillea aspleniifolia*, *G. caleyi*, *G. longifolia*) may have been listed because of their rarity, rarity was not a common reason for listing a plant species for protection.

The conservation tools available under the 1974 Act remained the relatively crude ones of protecting a species from being directly killed or harvested (including regulation of trade, display and captive keeping), and of declaring and managing parks and reserves. Many native bird and mammal species were still unprotected because of their assumed damage to agriculture, horticulture or fisheries. Other species could be declared pests in specific circumstances, although their consequent treatment often ignored the context of damage (e.g. the kangaroo management program did little to ensure that kangaroo culling targeted site-specific problems). To a great extent, the 1974 Act still reflected the premise that threatened species could be conserved if only the law could stop people killing them. However, the numbers of parks and reserves rapidly increased under the Act, reflecting, in part, a feeling that habitat modification was a threat most easily countered by reserving government-managed land for conservation. Private land and its management remained largely beyond the reach of the Act. Although NSW governments through the 1970s and early 1980s (and again in the mid- to late-1990s) seized opportunities to convert other Crown lands into National Parks and Nature Reserves, the emerging conserved estate was not necessarily "comprehensive, adequate and representative" (to quote the catch-cry of the times) (Pressey 1990). Despite effective laws, this opportunistic creation of parks, ignoring the best science, came at a longer-term cost to conservation (Whitehouse 1990; Pressey *et al.* 1993). Whitehouse (1990) attributed the "success of the reserve expansion programme in New South Wales" to "the interaction and co-operation between the National Parks and Wildlife Service and the non-government conservation movement". All later conservation legislation has included mechanisms to facilitate such co-operation.

The growing belief that destruction of habitat and resources threatened endangered species was tested in the Land and Environment Court (Corkill v Forestry Commission of New South Wales (1991) 73 LGR 126). The judgment, that logging would have put at risk a number of endangered and protected fauna species, meant that the Act could be invoked to prevent, not just direct killing or harvesting, but also indirect threat through habitat and resource destruction. This triumph for the North East Forest Alliance exemplifies the ability of community groups to affect the interpretation and implementation of conservation law. Extension of provisions under the existing National Parks and Wildlife Act to protect habitat of endangered species outside parks and reserves matched growing public sentiment and scientific understanding of what was needed for conservation. But it required the State's conservation agency to impose its will in areas such as planning, and upon other land-uses, such as forestry. The Act's inappropriateness for that, and the developing views about conservation, led to new legislation.

Modern Australian threatened-species legislation

Starting with the Victorian *Flora and Fauna Guarantee Act 1988* and continuing through the 1990s, most States, Territories and the Commonwealth introduced modern Acts to conserve threatened biodiversity. In NSW, the *Endangered Fauna (Interim Protection) Act* (EF(IP) Act) *1991* was rushed through a hostile parliament by a combination of the Labor opposition and independents, partly in response to the outcome of Corkhill v. Forestry Commission of New South Wales (1991) and the consequent need to license logging in State Forests. The objects of the Act included: "to provide urgently an objective scientific evaluation of the conservation status of fauna in New South Wales" (mammals, birds, reptiles and, for the first time, frogs; but still not invertebrates and flora), "to divide species of fauna into endangered, protected and unprotected species", and "to ensure endangered species of fauna are only harmed with the informed consent of the Director of National Parks and Wildlife", but to allow 'harming' of protected fauna if approved under the Environmental Planning and Assessment Act 1979 and if the impact upon the protected fauna had been examined. (The concept of minimising impacts of developments upon wildlife in general had been in place since the government's 1972 Environmental Impact Policy, and the NSW *Environmental Planning and Assessment Act 1979*.) The three-person Scientific Committee under the EF(IP) Act was the first in NSW to apply criteria, as set out in the EF(IP) Act, by which to judge the endangered status of fauna, and their lists became entrenched in the schedules of the subsequent *Threatened Species Conservation Act 1995* (Lunney *et al.* 2004).

The EF(IP) Act was replaced by the *Threatened Species Conservation Act 1995* (TSC Act) for 'non marine' organisms. The *Fisheries Management Act 1994* (FM Act) similarly protects 'marine' organisms and together these two Acts cover NSW biota. At a national scale, the *Endangered Species Protection Act 1992* (ESP Act) first covered all biota and was superseded by the *Environmental Protection and Biodiversity Conservation Act 1999* (EPBC Act).

The TSC Act and its cohort of contemporary Acts have common objectives, namely to:

- Conserve biological diversity in all its forms;
- Prevent extinction and promote recovery of endangered species, populations and ecological communities;
- Eliminate or manage threatening processes;
- Ensure proper assessment of the impact of actions affecting threatened taxa and communities; and
- Encourage conservation of threatened species, populations and ecological communities.

In NSW, the TSC Act and FM Act covered a much greater range of taxa than the 1974 NPW Act, including, not just all vertebrates, but also invertebrate animals, and vascular and non-vascular plants. Any species could, if a review of its status warranted it, be listed as "threatened" at one of several levels. The intent of these Acts was to include all biota. The long initial (January 1996) lists of threatened species under the TSC Act included all the vertebrates placed on the schedules in 1992 under the *Endangered Fauna (Interim Protection) Act 1991*, and a few more species added when it passed through parliament in December 1995. In 1992, there were 40 vertebrate species listed as extinct, 37 as endangered, and 156 as vulnerable; by December 2000, these numbers had expanded slightly to 40 extinct, 52 endangered and 155 vulnerable (Lunney *et al.* 2000). In January 1996, there were 41 plant species presumed extinct; 213 plant species listed as endangered and 196 as vulnerable; the lists have been extended greatly since then.

Moreover, the TSC Act considered for listing as threatened, not just whole species, but also separately identified *populations* of species, and whole *ecological communities*. It recognised the need to conserve *critical habitat* and, for the first time, it recognised *key threatening processes*. Furthermore, anyone could propose (to the Scientific Committee; see below) any species, population or ecological community for listing as threatened, or any process as a threatening process, and the public was invited to comment on the preliminary determinations that arose from such proposals. The initial lists under the *Endangered Fauna (Interim Protection) Act 1991* were opened for public comment after the public promulgation in March 1992. Similarly, the public could nominate for listing under the *Fisheries Management Act* threatened 'marine' species, and the nomination would be considered by the Fisheries Scientific Committee. (That terrestrial and aquatic biota should be conserved under separate legislation is an anomaly that reflects the evolution of State laws, rather than rational choice by conservation legislators.) This accessibility has facilitated individuals and community groups contributing to listing of threatened 'taxa' and threatening processes.

The process of proposal, consideration and listing was made transparent, and, for the listing of species not in the schedules when the TSC Act passed through parliament in December 1995, used internationally accepted criteria (e.g. the IUCN Red List Categories. IUCN 1994). However, the listing of threatened ecological communities and key threatening processes relied largely upon criteria drawn up first for the Victorian *Flora and Fauna Guarantee Act* and Commonwealth ESP Act and modified for the NSW Acts. Australia, which at the start of the 20th century had been urged not to fall behind the developed world in its protective legislation (Ryan 1907), broke new ground in formulating legally workable definitions of threatened ecosystems and threatening processes. Achieving these workable definitions was difficult because laws more easily handle concrete objects (like individual animals) or classes of objects (species) than they do dynamic ecosystems that are, to an important extent, characterised by the processes of interaction between objects or classes of objects.

Decisions (not just recommendations) to list under the TSC Act were made, on the best available evidence, by the Scientific Committee, whose membership was drawn largely from outside the NPWS conservation agency. This power of decision gave a greater independence (from the Minister responsible for conservation) in the process of determining threats and conservation status for non-marine biota in NSW than in other States or the Commonwealth or for marine species in NSW under the FM Act. The effect of listing was to require the government to plan, and implement the plan, to manage the threatened taxon, ecological community or threatening process. Thus, listing (in theory) compelled the government to take conservation action.

Recognition of *the need to manage ecological processes* was perhaps the most novel legal feature of the TSC Act and its counterparts in other jurisdictions. It marked a maturing of scientific and public understanding, moving beyond belief that direct killing is the only threat to endangered species, and that conservation reserves alone would ensure the long-term security of widely distributed species. A system of disjunct reserves, separated by inhospitable matrix, is unlikely to preserve the ecological and evolutionary integrity of metapopulations of endangered species whose survival and continuing evolution depend upon landscape-scale dispersal. Maintaining opportunities for evolution is a key feature of the Commonwealth's National Strategy for the Conservation of Australia's Biological Diversity (Department of Environment, Sport and Territories 1996) and is reflected in the various current State and Commonwealth Acts. Achieving that objective for an artificially fragmented population may require active management of its gene flow, for example.

The TSC Act, recognising that government-owned conserved areas are not enough, sought "to encourage the conservation of threatened species, populations and ecological communities by the adoption of measures involving co-operative management" (TSC Act 1995, page 2, section 3, Objects of Act (f)) presumably with private land-owners. Specific means were not spelled out and this remains a pious hope rather than positive action. However, provision was made for NPWS to enter into joint management agreements with other public authorities to manage actions that might jeopardise "the survival of a threatened species, population or ecological community" (TSC Act 1995, page 64, section 121).

The TSC Act also reflected the increased importance of conservation in the life of the community. Sections of the Act attempted to deal pre-emptively with possible impacts that developments and other land-use activities might have upon protected wildlife or communities, and the possibility that such activities might constitute threatening processes. In this regard, it moved far beyond reserves being the only places where conservation of wildlife took priority, and established the principle that conservation of wildlife must be considered in every proposed environment-altering activity, almost everywhere.

The NSW *Threatened Species Conservation Act 1995* was amended by the *Threatened Species Conservation Amendment Act 2002*, partly in response to the Commonwealth's *Environment Protection and Biodiversity Conservation Act 1999* (see below). At the time of writing, the Threatened Species Conservation Amendment Act is itself undergoing review and amendment.

Conservation of threatened species and communities affects, and is affected by, actions under other laws relating to land-use, especially development planning, forestry, agriculture, fisheries, mining, provision of water, and other extensive industries. Thus, threatened-species legislation must be integrated with those Acts. For example, if a species is listed under the TSC Act, its wellbeing must be considered in planned actions under the *Environmental Planning and Assessment Act 1979*. The ability of land-uses to yield commercial products as well as conserve biodiversity is increasingly safeguarded by encompassing acts such as the *Native Vegetation Conservation Act 1997*. That Act also extends responsibility for conservation beyond the traditional government conservation agency, the former National Parks and Wildlife Service (incorporated into a Department of Environment and Conservation in September 2003), and into the Department of Infrastructure Planning and Natural Resources. Similarly, NSW State Forests and NSW Fisheries (from April 2004 both within a Department of Primary Industries) carry considerable responsibilities for conservation in areas under their control. This greater spread of responsibility for conservation has required integration between departments in their conservation approaches and actions. The increasing scope of threatened species legislation has also meant that threatened species must be considered by all levels of government, from Federal to Local.

Co-operation, integration and internationalisation: 1970s-2003

Commonwealth conservation legislation (other than control over the export of biota) only became effective in the 1970s, with passage of the *Environment Protection (Impact of Proposals) Act 1974*, the *National Parks and Wildlife Conservation Act 1975*, the *Great Barrier Reef Marine Park Act 1975* and the *Australian Heritage Commission Act 1975*. Despite some early conflict between Commonwealth and some States over use of these Acts, since the 1970s approaches to conservation have developed in parallel in all Australian States and Territories, and the Commonwealth has led the way in developing some aspects of conservation. In particular, the Commonwealth has fostered inter-state co-operation in the conservation of threatened species (while holding the purse-strings), and has signed a variety of international agreements and conventions pertaining to "off-shore" and international conservation: e.g. JAMBA (the *Agreement between the Government of Australia and the Government of Japan for the Protection of Migratory Birds and Birds in Danger of Extinction and Their Environment*; signed in 1974, in effect from 1981); the Ramsar Convention (*Convention on Wetlands of International Importance especially as Waterfowl Habitat*; signed in 1971); CITES (the

Convention on International Trade in Endangered Species of Wild Fauna and Flora; signed in 1973); the *Convention for the Protection of the World Cultural and Natural Heritage* (in effect from 1972); the *Convention on Biological Diversity* (in force from 1993). The Commonwealth has also put together its own *National Strategy for the Conservation of Australia's Biological Diversity* (Department of Environment, Sport and Territories 1996), reflected in its *Environment Protection and Biodiversity Conservation Act 1999* (EPBC Act).

A major effect of these agreements has been to bring Australia's conservation into line with international procedures (as the RAOU President had hoped in 1907) and to allow the Commonwealth some influence on and between the States. For example, whereas in the 1970s each legislature in Australia had to define for itself what constituted a rare or endangered species, now all States and Territories and the Commonwealth accept (with minor variations) the internationally used definitions promoted by IUCN. This standardisation means that Australia's, and each State's, conservation efforts and the status of its biota can be monitored, audited and compared with those of any other State or country (although in practice the breadth and efficacy of taxonomic coverage still varies between States).

The Commonwealth-inspired co-operation between States and Territories has ensured that all states have a chance to contribute to the conservation of those threatened species whose ranges cross state borders. Changes in status of species in one State will now usually trigger a review of their listing by other range States and the Commonwealth. Conservation is better integrated and its legislation is more effectively complementary across Australia than at any time since European settlement.

Summary of changes

Changes in objectives of protection

The objectives of laws to protect biodiversity have changed through time (Tables 2 and 4). The objectives of the hunting laws were to ensure continued sport, by limiting the risk that hunters would overkill a population of a game animal. Laws limiting harvesting of plants had similar objectives: maintenance of the opportunity to harvest. Conservation of the species was a by-product. The objectives of the pest control acts were simply to facilitate suppression or eradication of pests, and many native vertebrate species were legally suppressed as pests. However, one objective of the *Rabbit Nuisance Act 1883* was to protect species that were natural enemies of rabbits, because they were useful rather than for their intrinsic value. Thus protection arose from people wanting to use the species, rather than as an end in its self.

The objectives of protective acts, like the *Fauna Protection Act 1948* and the *National Parks and Wildlife Act 1967*, were to prevent needless killing of species (of mammals or birds), and also to provide reserves where the habitat of native vertebrates and plants would be safe from direct human interference. The impact of indirect human effects was either not understood or was ignored. The objective of modern threatened species legislation is the mandatory and active conservation of species (or populations or ecological communities) identified as threatened. Such conservation may include intensive management of habitat and resources and ex-situ conservation. The objectives of the NSW *Threatened Species Conservation Act 1995* and its successors are far-reaching and far-sighted, seeking to prevent extinction and allow evolution of all native biodiversity to continue to occur, by ameliorating multiple, identified, general, threatening processes.

Table 2. The objectives of some NSW Acts relating to the protection or conservation of biodiversity, listed chronologically.

Year	Short Title	Objectives
1866	Game Protection Act	To encourage "importations of game by protecting it and its progeny" and also "native game" during the breeding season.
1881, 1893 and1901	Birds Protection Act	To "encourage the importation and breeding of certain Birds not indigenous to the Colony of New South Wales and also to prevent the destruction of certain Native Birds during the breeding season" (1881 Act)
1903	*Native Animals Protection Act*	To "protect native animals" by declaration of close seasons and absolute protection for scheduled marsupials and monotremes
1918	*Birds and Animals Protection Act*	To protect mammals and birds, except those on schedules of species seen as pests.
1927	*Wild Flower and Native Plants Protection Act*	To protect native plants from picking and selling; aimed at destructive commercial harvesting on Crown or private land.
1948	*Fauna Protection Act*	To "make provisions for the protection and preservation of fauna" (meaning birds and mammals, mostly but not exclusively native).
1967 and 1974	*National Parks and Wildlife Act*	To reserve parks, etc., to establish and run the National Parks and Wildlife Service, and to protect "certain fauna (and) native plants" (1974 Act).
1991	*Endangered Fauna (Interim Protection) Act*	To evaluate conservation status of fauna (mammals, birds, reptiles and amphibians) in NSW and divide fauna into endangered, protected and unprotected species; to protect endangered fauna; to include cane toad on unprotected list.
1995	*Threatened Species Conservation Act*	To "conserve threatened species, populations and ecological communities of animals and plants".
1997	*Native Vegetation Conservation Act*	To conserve and manage native vegetation on a regional basis; protect native vegetation of high conservation value; improve the condition of existing native vegetation and encourage revegetation and rehabilitation of land; and to prevent clearing of native vegetation.

Changes in the objects being protected

The objects (components) of biodiversity protected by law have also changed through time (Table 3; Figure 1). Game laws protected, for hunting, very short lists of birds and mammals, some native and some introduced. Those lists were then expanded to protect some species that were useful or valued for intrinsic qualities. After World War II, all native mammals and birds received protection unless they were listed as pests, and some species were singled out for special treatment because they were 'rare'. From 1974 onwards, (almost) no exotic vertebrates were any longer protected, but (almost) all native vertebrates were. All native vertebrates and vascular plants could also be listed as threatened, at one of a number of levels, and hence receive additional protection. Finally, from 1995 onwards, almost all native species of animals, vascular and non-vascular plants, and fungi can potentially be scheduled as threatened (although no unicellular organism has yet been listed), and hence receive conservation action. Moreover, for any taxon, a sub-population can be deemed in need of protection regardless of the status of the whole species.

In the first half of the 20th century, the need to protect habitat was recognised scientifically, but not legally except as part of a reserve. From 1974 'critical habitat' was legally recognised and protected and, under the 1995 Act, whole ecological communities could be listed as threatened and thus be protected.

Thus the taxonomic coverage of threatened species legislation has grown in extent (numbers of taxa potentially listed as threatened) and depth (levels from genetically defined sub-population conservation units up to ecological communities).

Changes in kinds and integration of legislative protection

Hunting, pest-control and early protective laws had to be authoritarian and punitive to stop (or to ensure, in the case of pest species) adverse things being done directly to individual animals. Once whole species were recognised as threatened by indirect and often general effects of broad-scale land-uses, legislation changed to become facilitative, to reserve pristine areas where government could dictate land-uses and hence protect native biota from harmful land-use effects.

With growth in understanding that isolated reserves were not enough to conserve metapopulations, it became clear that landholders across whole landscapes had to be engaged in conservation of threatened taxa and communities. Thus modern legislation is charged with encouraging conservation on and off reserves, through cooperative, community-based conservation in partnership with government. The laws supporting this approach are far less authoritarian and paternalistic, and better organised to attract community input. This modern generation of laws also makes it mandatory that the government plan and execute appropriate management for identified threatened taxa and threatening processes.

Table 3. The biota receiving legal protection under some NSW Acts relating to the protection or conservation of biodiversity, listed chronologically, and those biota listed as having special status (game, pest, protected, unprotected, or categories of threatened).

Year(s)	Protected Biota	Biota given special status
1866	None, except some native and many exotic bird and mammal 'game' species.	Game species
1881, 1883, 1893	As above, but also some useful or aesthetic native birds; plus 'natural enemies' of rabbits. (Some exotic and native mammals and birds bountied as pests.)	Game species. Pest species, at State or district level.
1901, 1903	Many exotic and some native bird species protected as game or as useful or aesthetic species; a few native mammal species (1903).	Schedules of protected bird and mammal species.
1918	Native and exotic birds and mammals (but not rodents or reptiles), unless scheduled as pests or under declared open season (game).	Schedule of unprotected mammal and bird pests.
1927	Proclaimed native plants (mainly commercial species).	Proclaimed native plants.
1948	Any mammal or bird species except specified unprotected (pest, native and exotic) species. Some mammal and bird species declared rare. Some exotics still protected.	Schedule of unprotected species. Declared Rare species.
1974	All native mammal, bird and reptile species, except listed unprotected species (native and exotic, plus snakes when threatening life); and a schedule of "protected native plants".	Schedules of: Unprotected fauna; Endangered fauna; and Protected native plants.
1991	All native mammal, bird, reptile and amphibian species.	Endangered fauna on schedule as threatened, vulnerable or rare.
1994-5	All biodiversity, but still bias to vertebrates and flowering plants; emphasis on threatened taxa (species, populations or ecological communities).	Threatened taxa as endangered, presumed extinct, or vulnerable.
1997	Native vegetation: indigenous "trees, understorey plants, groundcover (and) plants occurring in wetlands", but not mangroves, sea grasses, etc., which were covered by the *Fisheries Management Act* 1994.	

Table 4. Time-lines for actions made possible under NSW legislation. A tick indicates that an action was facilitated by legislation introduced or in force in the decade ending in the year at the head of the column. Relevant Acts are listed in Table 1. The subscripts to the ticks refer to the explanatory footnotes, where the abbreviations for the Acts follow those given in Figure 1, with the addition of WFNP 1927 for the *Wild Flowers and Native Plants Protection Act* 1927 and NV 1997 for the *Native Vegetation Act* 1997. "List" indicates that the protected or unprotected species or species-groups are named in the legislation.

	1860	1870	1880	1890	1900	1910	1920	1930	1940	1950	1960	1970	1980	1990	2000
Encourage import & release of exotic fauna & flora		✓	✓												
Ban import & release of exotic fauna & flora										✓$_{10}$	✓$_{10}$	✓$_{10}$	✓$_{10}$	✓$_{10}$	✓$_{10}$
Control hunting; e.g. close or open seasons		✓	✓	✓	✓	✓	✓	✓	✓	✓	✓	✓	✓	✓	✓
Suppress animal pest species	✓	✓	✓	✓	✓	✓	✓	✓	✓	✓	✓	✓	✓	✓	✓
List protected exotic fauna species		✓$_{1}$	✓$_{1}$	✓$_{2}$	✓$_{2}$	✓$_{2}$	✓$_{3}$	✓$_{3}$	✓$_{3}$	✓$_{3}$	✓$_{3}$	✓$_{3}$	✓$_{3}$	✓$_{3}$	✓$_{3}$
List unprotected exotic fauna species							✓	✓	✓	✓	✓	✓	✓	✓	✓
List protected native fauna species		✓$_{1}$	✓$_{1}$	✓$_{2}$	✓$_{2}$	✓$_{2,4}$	✓$_{5}$	✓$_{5}$	✓$_{5}$	✓$_{6}$		✓$_{13}$	✓		✓$_{14}$
List unprotected native fauna species							✓	✓	✓	✓	✓	✓	✓$_{7}$	✓	
Control commercial taking of native fauna													✓	✓	✓
Control commercial taking of native flora													✓	✓	✓
List protected native flora species								✓$_{8}$	✓$_{8}$	✓$_{8}$	✓$_{8}$	✓$_{12}$	✓	✓	✓
Control vegetation clearance & habitat alteration															✓$_{15}$
Declare species threatened										✓$_{9}$	✓$_{9}$	✓$_{11}$	✓	✓	✓
Declare populations threatened															✓
Declare ecological communities threatened															✓
Declare threatening processes															✓
Consider threatened species, etc. in planning consent													✓$_{17}$	✓	✓
Facilitate government conservation reserves				✓$_{16}$	✓$_{16}$	✓$_{16}$		✓	✓	✓	✓	✓	✓	✓	✓
Encourage private/community conservation reserves				✓$_{16}$	✓$_{16}$	✓$_{16}$		✓	✓	✓	✓	✓	✓	✓	✓
Appoint government Rangers							✓	✓	✓	✓	✓	✓	✓	✓	✓
Appoint Chief Guardian of Fauna										✓	✓				
Appoint Director and form NPWS											✓	✓	✓	✓	✓
Establish Scientific Committee or equivalent										✓	✓	✓	✓	✓	✓

1 Lists of 'game' birds; protection from hunting in close season only GP 1866, 1886
2 Game birds, but also some aesthetic bird species, protected in close season BP 1881
3 Only Rusa deer (Royal National Park) protected BAP 1918 onwards
4 Some marsupial species protected under NAP 1903
5 All mammal & bird species protected except rodents and listed unprotected species BAP 1918
6 All mammal & bird species protected except listed unprotected species FP 1948
7 Only two flying-fox species NPW 1974
8 Protected named plant species against harvesting on Crown land WFNP 1927
9 Native mammal & bird species listed as threatened FP 1948 onwards
10 Except trout species
11 Reptile species listed as threatened NPW 1967 onwards
12 Some native plant species listed because of rarity NPW 1967 onwards
13 Included reptiles NPW 1967 onwards
14 Included frogs EF(IP) 1991, then all native biota TSC 1995
15 Habitat of threatened spp protected EF(IP) 1991; more general protection NV 1997
16 Reserves for hunting, rather than conservation in a modern sense BP 1881, 1893, 1901
17 Only through Environmental Planning and Assessment Act 1979

Early hunting, pest-control and protection laws were only lightly integrated with other NSW laws. They often imitated similar laws in other States, but without overt integration across state boundaries. Integration with Commonwealth legislation (and hence with some international conventions and agreements) has become increasingly necessary since the 1970s. NSW conservation legislation is now integrated with and embedded in a nation-wide set of laws and supporting practices that facilitate effective national management of biodiversity.

Changes in public perceptions

People's general awareness of our biodiversity has changed greatly through time, encouraged by enlightened educators and those who saw education and publicised information as keys to progress in conservation (e.g. Ryan 1907; Barrett 1925). Since World War II, public interest in and education about biodiversity has grown rapidly. Since 1960, biodiversity has gone from being politically ignorable to being an occasionally potent vote-catcher. Moreover, the public, as it has become better informed, has become more discriminating and has engaged in conservation at all levels, from local to international organizations. Modern legislation has allowed for more direct community in-put to conservation decision-making, and the decline (at least temporarily) of paternalism in government-directed conservation.

The Future

Given the history of threatened species legislation, we can expect such laws to keep changing, as our understanding changes, as species' populations change, as ecological communities change, and as ecological processes change under the effects of an increasing human population.

The increased taxonomic scope of threatened-species legislation arises from our modern perception that conservation objectives should apply to all native taxa. In practice, however, the application of the laws is still taxonomically uneven, with the large and charismatic biota getting most attention. A re-emerging utilitarian view ("ecosystem services") may see the balance redressed, with invertebrates, lower plants and microbes gaining more attention. However, that attention is likely to focus on the ecosystem processes in which these biota are involved, rather than on the biota themselves as separable components.

Modern predictive ecology emerged only in the 1930s, although environmental planners and decision-makers, or those who wanted to influence them, used ecological explanations of systems processes at least by the 1850s. Modern threatened-species legislation partly relies upon ecological understanding, particularly of population processes. Our grasp of processes in community ecology is under-used in conservation, but the identification of 'key threatening processes' is an excellent example of its potential. Our understanding of community processes might more often be applied in future, for example, to devising process-related (rather than component-related) criteria for initiating management actions or evaluating their outcomes.

In current NSW legislation, an ecological community is defined as an assemblage of species; but does this serve conservation well? The ecological processes, not just the component species, might better identify an ecologically meaningful community. For example, the effective presence of top carnivores limiting the population densities of preferred prey species, or the occurrence of fires at appropriate intervals, may be critical determinants of the persistence of particular communities. Good monitoring of the status of threatened ecological communities also requires better ways to evaluate ecosystem processes, not just their outcomes. We also envisage the need for legislation that allows for the identification, not just of first-order threatening processes, but also of the synergies between threats (for example, the synergy between clearing vegetation and exposure of ground-nesting birds and critical-weight-range mammals to exotic predators).

Evolutionary understanding lags even further behind in its application to conservation, although genetics was quick off the mark in setting guidelines for conservation action. Increasing concern for preserving the opportunity for evolution and better understanding of population genetics may require that genetically defined conservation units be incorporated more firmly in legislation (as Evolutionarily Significant Units), a further increase in the depth of the legislation. Ecological and evolutionary genetics are sure to play increasing parts in future implementation of threatened species legislation.

Finally, we expect that current legislation will need to be modified as we find better ways (including financial rewards) to engage all landholders and other interested groups in broad-scale conservation. The emphasis of current laws upon conserving threatened species has been criticised for diverting attention and funding from the conservation of whole communities (for example, Lunney *et al.* 2003); and the laws may need modification to redress this imbalance. Any legislation is only as good as its application. Since the earliest NSW laws to preserve biota (the game laws), people have complained that their application fell short of their intent. For example, the Hon. G. Thornton in debate on the Birds Protection Act Amending Bill, 23rd February 1893: "The act of 1881 is...almost a dead letter. It is a notorious fact that in many parts of the country, although wild duck are prohibited from being shot during the close season, they are nevertheless shot indiscriminately in hundreds and thousands" (NSW Parliamentary Debates Session 1893, 1st Series Volume 63, p. 4566). Failure to apply the law fully is still an impediment to effective conservation; see, for example, Tsamenyi, Rose and Castle (2003) on the defects of implementation in Australia of the full intent of international Conventions affecting conservation. Laws may need modification to make them more readily applicable.

Threatened-species legislation in NSW and Australia has changed in purpose, scope and function as society's knowledge about and attitudes towards biodiversity have developed and changed. We foresee that such legislation will continue to evolve, reflecting changing societal needs and perceptions. Scientists, managers and informed lay-people, as well as legislators, will be involved in bringing about those changes. They can contribute most effectively if they understand how legislation has been crafted to complement scientific understanding of biodiversity.

Acknowledgments

We thank the editors of this volume, Dan Lunney, Chris Dickman and Pat Hutchings, and two anonymous referees for their constructive criticism of an earlier version of this paper.

References

Barrett, J. 1925. *Save Australia. A Plea for the Right Use of our Flora and Fauna*. MacMillan and Co., Melbourne.

Bennett, G. 1860. *Gatherings of a Naturalist in Australasia: being observations principally on the animal and vegetable productions of New South Wales, New Zealand and some of the Austral Islands*. John van Voorst, London. (Facsimile edition 1982 The Currawong Press, Milson's Point, NSW)

Campbell, A.J. 1908. Vice-Presidential Address: Bird Protection in the Old World. *The Emu* **7**: 126-136

Clarke, W.B. 1876. Effects of forest vegetation on climate. *Journal and Proceedings of the Royal Society of New South Wales* **10**: 179-232

Darwin, C. 1845. *The Voyage of the Beagle*. Collier and Sons, New York.

Darwin, C. 1872. *The Origin of Species by Means of Natural Selection, or the Preservation of Favoured Races in the Struggle for Life*. 6th edition. John Murray, London.

Department of Environment, Sport and Territories 1996. *National Strategy for the Conservation of Australia's Biological Diversity*. Australian Government Publication, Canberra

Finney, C. M. 1993. *Paradise Revealed. Natural History in nineteenth-century Australia*. Museum of Victoria, Melbourne.

Gilbert, L.A. 1981. Plants, politics and personalities in Colonial New South Wales. Pp 220-258 in *People and Plants in Australia*, edited by D.J. Carr and S.G.M. Carr. Academic Press, Sydney.

Gould, J. 1863. *The Fauna of Australia*, Volume II, *Macropodidae*. Published in facsimile as *Kangaroos: John Gould*, with commentary by J.M.Dixon, Macmillan, Melbourne and Sydney, 1973

Hall, R. 1900 *The Insectivorous Birds of Victoria with chapters on birds more or less useful*. Published by the author, Melbourne.

Heathcote, R.A. 1981. Goyder's Line – a Line for all Seasons? Pp 295-321 in *People and Plants in Australia*, edited by D.J. Carr and S.G.M. Carr. Academic Press, Sydney.

Hoare, M. 1981. Botany and Society in Eastern Australia. Pp 183-219 in *People and Plants in Australia*, edited by D.J. Carr and S.G.M. Carr. Academic Press, Sydney.

Hutton, D. and Connors, L. 1999. *A History of the Australian Environment Movement*. Cambridge University Press, Cambridge and Melbourne

IUCN. 1994. *IUCN Red List Categories*. Prepared by the IUCN Species Survival Commission. IUCN, Gland, Switzerland.

Jarman, P.J. 1994. *Kangaroos: two centuries of ecosystem management*. Inaugural lecture, University of New England, Armidale, NSW.

Jarman, P.J. and Johnson, K.A. 1978. Exotic mammals, indigenous mammals, and land use. *Proceedings of the Ecological Society of Australia* **10**: 146-166

Krefft, G. 1866. On the vertebrated animals of the lower Murray and Darling: their habits, economy and geographical distribution. *Transactions of the Philosophical Society of New South Wales* 1862-65, 1-33.

Le Souef, A.S. 1925. The Australian native animals. How they stand today, and the cause of the scarcity of certain species. Pp175-184 in *Save Australia. A plea for the right use of our flora and fauna*, edited by J. Barrett. MacMillan and Co., Melbourne.

Long, J.L. 1981. *Introduced Birds of the World*. Reed, Sydney, NSW.

Lunney, D. 1994. Royal Commission of 1901 on the Western Lands of New South Wales – an ecologist's summary. Pp 221-40 in *Future of the Fauna of Western New South Wales*, edited by D. Lunney, S. Hand, P. Reed and D. Butcher. Royal Zoological Society of New South Wales, Mosman, NSW.

Lunney, D., Curtin, A. L., Ayers, D., Cogger, H. G., Dickman, C. R., Maitz, W., Law, B. and Fisher, D. 2000. The threatened and non-threatened native vertebrate fauna of New South Wales: status and ecological attributes. *Environmental and Heritage Monograph Series No. 4*. Pp 1-132. National Parks and Wildlife Service, Hurstville.

Lunney, D., Matthews, A., Stein, J.A. and Lunney, H.W.M. 2003. Australian bat research: the limitations of The Action Plan for Australian Bats in determining the direction of research. *Pacific Conservation Biology* **8**: 255-60.

Lunney, D., Matthews, A., Cogger, H. and Dickman, C.R. 2004. The neglected 74% - the non-threatened vertebrates - and a reflection on the limitations of the process that fashioned the current schedules of threatened species in New South Wales. Pp 145-157 in *Threatened Fauna Legislation: is it just an Act?* edited by P. Hutchings, D. Lunney and C. Dickman. Royal Zoological Society of NSW, Mosman, NSW.

Marsh, G.P. 1864. *Man and Nature, or, Physical Geography as Modified by Human Action*. New York. (Re-printed edition, edited by D. Lowenthal,1964, Cambridge, Massachusetts)

Nicholls, B. 1925. The Mammals of Australia. Pp 85-99 in *Save Australia. A plea for the right use of our flora and fauna*, edited by J. Barrett. MacMillan and Co., Melbourne.

Parsons, W.J. 1981. The history of introduced weeds. Pp 179-193 in *Plants and Man in Australia*, edited by D.J. Carr and S.G.M. Carr. Academic Press, Sydney.

Powell, J.M. 1976. *Environmental Management in Australia, 1788-1914. Guardians, Improvers and Profit: an Introductory Survey*. Oxford University Press, Melbourne.

Pressey, R.L. 1990. Reserve selection in New South Wales: Where to from here? *Australian Zoologist* **26**: 70-75.

Pressey, R.L., Humphries, C.J., Margules, C.R., Vane-Wright, R.I. and Williams, P.H. 1993. Beyond opportunism: key principles for systematic reserve selection. *Trends in Ecology and Evolution* **8**: 124-128.

Ryan, C.S. 1907. The President's Address: The Protection of Native Birds. *The Emu* **6**: 95-103.

Short. J. 1998. The extinction of rat-kangaroos (Marsupialia: Potoroidae) in New South Wales, Australia. *Biological Conservation* **86**: 365-377.

Strahan, R. 1992. The origin and early history of the Royal Zoological Society of New South Wales. *Australian Zoologist* **28**: 6-10.

Tsamenyi, M., Rose, G. and Castle, R. 2003. International marine conservation law and its implementation in Australia. Pp 1-17 in *Conserving marine environments. Out of sight, out of mind*, edited by Pat Hutchings and Daniel Lunney. Royal Zoological Society of New South Wales, Mosman, NSW.

Walker, R.B. 1991. Fauna and Flora Protection in New South Wales, 1866-1948. *Journal of Australian Studies* **28**: 17-28.

Whitehouse, J.F. 1990. Conserving what? The basis for nature conservation reserves in New South Wales 1967-1989. *Australian Zoologist* **26**: 11-21.

APPENDIX 1

Appendix 1. Species listed as protected or unprotected under the original NSW Acts pertaining to biodiversity, 1866 to 1974

a. Species listed for protection under the original *Game Protection Act 1866*. The First Schedule is of imported species, and the Second is of native bird species. The names are given as published.

The First Schedule

"Pheasants, Partridges, Grouse, Hares, Deer, Swans, Antelope, All birds not indigenous to Australia and their produce".

The Second Schedule

"Wild ducks of any species, Teal, Emu, Native Companions, Wild Turkey or Bustard, Black Swans, Wild Geese, Bronze-winged and other wild Pigeons, Mallee Hen, Quail, Land Rail, Curlew, Tallegalla or Brush Turkey, Plover of any species, Great Kingfisher commonly known as the Laughing Jackass."

b. Species listed for protection under the original *Birds Protection Act 1881*. The First Schedule is of "Imported Game" species, the Second is of "Native Game" species, and the Third is of "Song Birds – Native and Imported". The names are given as published.

The First Schedule

"Pheasant, Partridge, Grouse, California Quail, Chinese Quail, White Swan".

The Second Schedule

"Wild Duck of any species, Teal, Emu, Native Companion, Wild Turkey or Bustard, Black Swan, Wild Geese, Mallee Hen, Lyre Bird, Quail, Land Rail, Tallegalla or Brush Turkey, Plover Spur-wing, Bronze-winged and other Wild Pigeons and Doves."

The Third Schedule

"Robin, Nightingale, Skylark, Linnet, Thrush, Starling, Blackbird, Goldfinch, Chaffinch, Blue Wren, Wagtail, Honeysucker, Diamond Sparrow, Flycatcher, Great Kingfisher – commonly known as the Laughing Jackass, Swift, Peewit, Reed Sparrow, Colonial Lark, Indian Minah, American Mocking Bird, Magpie other than the black Magpie, Curlew, Regent Bird, Rifle Bird".

c. Species listed for protection under the original *Birds Protection Act* 1893. The First Schedule is of species "absolutely protected for five years and after that during the Close Season", and lists "Foreign" and "Australian" birds separately; the Second is of "Australian Game Birds"; and the Third Schedule is of "Australian Birds other than game protected during the close season". The names (both colloquial and binomial) are given as published.

First Schedule

Foreign Birds. "Skylark Alauda arvensis, Chaffinch Fringilla coelebs, Goldfinch Carduelis elegans, Linnet Linaria cannabina, Starling Sturnus vulgaris, Thrush Turdus musicus, Blackbird Turdus merula, Nightingale Sylvia luscinea, Pheasant Phasianus colchicus, Grouse of every species Tetrao and Lagopus, Partridges of every species Perdix, White swans Cygnus, California Quail Callipepla Californica".

Australian Birds. Fish Hawks or Sea Eagles Haliaetus Haliastur and Pandion, Lyre Birds Menura, Regent Bird Sericulus melinus, Rifle Bird Ptilorus paradisea, Bell Bird Manorhina melanophrys, Black Cockatoos Calyptorhynchus, Native Companion Grus Australiensis, Nankin Crane Nycticorax Caledonicus, Laughing Jackass Dacelo gigas, Kingfishers Halcyon and alcyone, Emu Dromaius Novae Hollandiae, Seagulls of every description".

Second Schedule

Brush Turkey Talegallus Lathami, Mallee Hen Leipoa ocellata, Land Curlew Oedicnemus grallarius, Bittern Botaurus poicilopterus, Land Rail Hypotaenidia Phillipeusis, Black Swan Cygnus atratus, Bustard or Plains Turkey Eupodotis Australis, Wild ducks of every species, including Teal, Plovers of every species, Pigeons and doves of every species, Wild geese of every description, Quail of every species, including Turnix, Pedionomus, Coturnix, Synoïcus, and Excalfatoria".

Third Schedule

Dollar Bird Eurystomus pacificus, Wood Swallows Artamus, Magpies (other than the Black Magpie) Gymnorhina, Peewit or Magpie Lark, Grallina Australis, Thrush Collyriocincla harmonica, Wagtails Sauloprocta and Seisura, Jackey Winter Micraeca fascinans, Robins Petroeca, Yellow Robins Eopsaltria Australis, Blue Wrens Malurus cyaneus and Malurus Lamberti, Emu Wren Stipiturus malachurus, Lark Anthus Australis, Skylark or Nightingale Cinchloramphus, Diamond Sparrow Estrilda guttata, Gang Gang Cockatoo Callocephalon galeatum, Indian Minah Acridotheres tristis, Butcher Bird Cracticus torquatus, Coachwhip Bird Psophodes crepitans, Honeysucker Meliornis, Mocking Bird Anellobia mellivora, Swamp Pheasant, King Parrots Aprosmictus and Ptistes, Ibis Geronticus and Threskiornis, The bird commonly known as "The Happy Family," or "The Twelve Apostles"."

d. Species listed for protection under the original *Birds Protection Act 1901*. This Act contained only one Schedule, which combined, without alteration, all species listed in all three Schedules in the *Birds Protection Act* 1893.

e. Species listed for protection under the *Native Animals Protection Act 1903*. This Act contained one Schedule, as follows. The names (both colloquial and binomial) are given as published.

Macropus rufus Red Kangaroo; *Macropus robustus* Wallaroo; *Phascolarctus* [sic] *cinereus* Native Bear; *Phascolomys mitchelli*, *Phascolomys ursinus*, *Phascolomys latifrons* Wombats; *Ornithorhynchus anatinus* Platypus; *Tachyglossus aculeatus* Echidna or Native Porcupine; *Petaurus sciureus*, *Petaurus breviceps*, *Petaurus australis* Sugar Squirrels; *Petauroides volans* Flying Opossum".

f. Species listed as unprotected under the original *Birds and Animals Protection Act 1918*. All other bird and mammal species, with the specific exceptions of "domestic birds or animals" or reptiles or "rats or mice of any species", were protected. The First Schedule is of unprotected bird species, and the Second of unprotected mammal species.

First Schedule

"Sparrow *Passer domesticus*, Silver Eye *Zosterops coerulescens*, Black Cormorant or Shag *Phalacrocorax carbo*, White-breasted Cormorant *Phalacrocorax gouldi*, Pied Cormorant *Phalacrocorax hypoleucus*, Little Black Cormorant *Phalacrocorax sulcirostris*, Little Cormorant *Phalacrocorax melanoleucus*, Crow *Corvus coronoides*, Raven *Corone australis*, Friar-bird or Leatherhead *Tropidorhynchus corniculatus*, Yellow-throated Friar-bird *Philemon citreigularis*, Garrulous Honey Eater or Miner or Soldier Bird *Myzantha garrula*, Sulphur-crested or White Cockatoo *Cacatua galerita*, Rose-crested Cockatoo or Galah *Cacatua roseicapilla*, Lory *Platycercus elegans*, Rose Hill or Rosella Parrot *Platycercus eximius*, Blue-bellied Lorikeet or Blue Mountain Parrot *Trichoglossus novae-hollandiae*, Red-rumped Grass Parakeet *Psephotus haematonotus*, Wedge-tailed Eagle or Eagle Hawk *Uroaetus audax*, Every species of Snipe, Gill-bird, Starling, White-throated Falcon."

Second Schedule

"Rabbit *Lepus cuniculus*, Hare *Lepus europaeus*, Dingo *Canis dingo*, Fox *Vulpes alopex*, Fruit Bat or Flying Fox *Pteropus poliocephalus*, Tiger Cat *Dasyurus maculatus*, Native Cat *Dasyurus viverrinus*, Common Kangaroo Rat *Potorous tridactylus*, Brush-tailed Kangaroo Rat *Bettongia penicillata*, Gaimard's Kangaroo Rat *Bettongia gaimardi*, Rufous Kangaroo Rat *Aeprymnus rufescens*, Hare Wallaby *Lagorchestes leporoides*, Nail-tailed Wallaby *Onychogale frenata*, White-throated Wallaby *Macropus parma*, Paddymelon *Macropus thetidis*, Red-legged Wallaby *Macropus wilcoxi*, Black-striped Wallaby *Macropus dorsalis*, Wombat *Phascolomys mitchelli*, Long-nosed Bandicoot *Perameles nasuta*, Common Bandicoot *Perameles obesula*, Rabbit Bandicoot *Thylacomys lagotis*."

g. Species listed as unprotected under the original *Fauna Protection Act* 1948. The Schedule **is of unprotected bird and mammal species, those marked with an asterisk being unprotected in named parishes only. All other bird and mammal species were protected.**

The names (both colloquial and binomial) are given as published.

Birds

"Goldfinch Carduelis carduelis, Greenfinch Ligurinus chloris, Grey Butcher Bird Cracticus torquatus, Sparrow Passer domesticus, Silver Eye Zosterops lateralis, Black Cormorant or Shag Phalacrocorax carbo, White-breasted Cormorant Phalacrocorax fuscesens, Pied Cormorant Phalacrocorax varius, Little Black Cormorant Phalacrocorax ater, Little Cormorant Microcarbo melanoleucus, Crow Corvus Cecilae, Raven Corvus coronoides, Black Magpie or Currawong Strepera graculina, Friar-bird or Leatherhead Philemon corniculatus, Yellow-throated Friar-bird Philemon citreigularis, Sulphur-crested or White Cockatoo Kakatoe galerita, Rose-breasted Cockatoo or Galah Kakatoe roseicapilla, Lory or Crimson Rosella Platycercus elegans, Rose Hill or Rosella Parrot Platycercus eximius, Wedge-tailed Eagle or Eagle Hawk Uroaetus audax, Common Starling Sturnis vulgaris, Red-whiskered Bul Bul Otocompsa emeria, Little Falcon Falco longipennis, Black Falcon Falco subniger, Grey Falcon Falco hypoleucus, Peregrine or Black-cheeked Falcon Falco peregrinus, Grey Goshawk White Goshawk – Albino Form of Grey Goshawk Astur novae-hollandiae, Australian Goshawk Astur fasciatus, Bald Coot Porphyrio melanotus*, Coot Fulica atra*, Dusky Moorhen Gallinula tenebrosa*, Black-tailed Water Hen Tribonyx ventralis*.

APPENDIX 1

Mammals

"Dingo Canis dingo, Ferret Mustela putorius, Fox Vulpes alopex, Fruit Bat or Flying Fox Pteropus poliocephalus and Pteropus scapulatus, Hare Lepus europaeus, Rabbit Oryctolagus cuniculus, Wombat Vombatus hirsutus*."

h. Species originally listed as unprotected under the *National Parks and Wildlife Act 1974*. These species appeared in Schedule 11 Unprotected Fauna. Names are as listed.

Mammals

"Carnivora other than Pinnipedia (Bears, lions, dogs, etc.); Insectivora (Moles, hedgehogs); Artiodactyla, other than *Cervus timorensis* (Javan Rusa deer) (Cloven hoofed animals); Perissodactyla (Horses, donkeys, etc.); Primates (Apes, monkeys); Subungulates (Elephants); *Pteropus policephalus* and *Pteropus scapulatus* (Fruit bat or flying fox); *Lepus europaeus* (Hare); *Oryctolagus cuniculus* (Rabbit); *Sciurus palmarum* (Indian Palm Squirrel)".

Birds

As for the 1948 list, with some up-dating of nomenclature, and the following amendments:

Grey Butcher Bird, White-breasted Cormorant, both Friarbird species, all raptors and all rallids omitted; and "Blackbird *Turdus merula*, Tree Sparrow *Passer montanus*, Indian Myna *Acridotheres tristis*, Indian Turtle Dove *Streptopelia chinensis*, Little Crow *Corvus bennetti*, Spice (nutmeg) Finch *Lonchura punctulata*, Black-headed Mannikin *Lonchura ferruginosa*, Mallard Duck *Anas platyrhynchos*" added.

Expert frailties in conservation risk assessment and listing decisions

Mark Burgman

School of Botany, University of Melbourne, Parkville, 3010, Australia

Email: markab@unimelb.edu.au

ABSTRACT

Listing decisions are a form of risk assessment supported largely by expert judgement. Expert judgements of rare events in novel circumstances are error prone. Experts are susceptible to social influences and their views are shaped by context, framing, and personal values. Expert judgment is a legitimate source of knowledge, but only if it is fallible. A range of behavioural and numerical methods may be employed to elicit, combine and communicate opinions in deliberation processes by expert groups. Expert groups provide an efficient and convenient way of synthesising a broad range of knowledge but their advice may be improved by training, feedback of performance, consistent communication formats, and awareness of motivational and other sources of bias. A suite of behavioural and numerical methods to aggregate opinions and communicate uncertainty provides new opportunities.

Key words: Conservation, listing, expert judgment, uncertainty, aggregation, expert groups

Introduction

Risk is the chance, within a time frame, of an adverse event with specific consequences. In our personal lives, we perform risk assessments every day and carry the burden of the consequences. Conservation biologists evaluate risks to species, communities, and ecosystem processes. Epidemiologists, toxicologists, engineers, ecologists, geologists, chemists, sociologists, economists, foresters and others also conduct environmental risk assessments routinely.

Under most legislation, listing decisions are recommendations, usually by panels of scientific experts, regarding proposals for species or ecological communities to be afforded protection under threatened species legislation. For example, Table 1 gives the numbers of species of plants and animals thought to be at risk in several countries. Such lists are published and used routinely by governments for a number of purposes (Possingham *et al.* 2002a). They are created largely on the basis of expert opinion.

Table 1. The number of species listed nationally as endangered in 2001 in Australia, USA and China (each of these countries has relatively large numbers of endemic species) compared to the number of species listed as endangered in the UK and the number thought to exist globally (data from Groombridge 1994, Anonymous 1995, May *et al.* 1995, Burgman 2002).

Taxon	Estimated total number of species in the world[1]	Number listed as 'Threatened'[2]			
		Australia	*USA*	*China*	*UK*[2]
Fungi	1,000,000	0	0	0	46
Nonvascular plants	100,000	0	0	0	70
Vascular plants	250,000	1597	1845	343	61
Invertebrates	5,000,000	372	860	13	171
Fish	40,000	54	174	16	7
Amphibians	4,000	20	16	0	3
Reptiles	6,000	42	23	8	6
Birds	9,500	51	46	86	25
Mammals	4,500	43	22	42	18

1. Estimates of the global numbers of species in each taxon were estimated crudely from numbers of currently described species and expert judgments of the proportion remaining to be described (see May *et al.* 1995, Burgman 2002).

2. The numbers provide a good approximation of the totals listed nationally as threatened in each country in 2001.

3. Numbers from the UK represent all those species in the UK's Biodiversity Action Plan for which conservation action plans were written (Anonymous 1995).

Pp 20 - 29 in *Threatened species legislation: is it just an Act?* 2004, edited by Pat Hutchings, Daniel Lunney and Chris Dickman. Royal Zoological Society of New South Wales, Mosman, NSW, Australia.

Decisions to include species on a threatened species list are sometimes mistaken. The uncertainty is reflected in turnover of species on the lists through time. For instance, Briggs and Leigh (1996) created a list of 'threatened' Australian plants, estimating 4955 species of vascular plants considered to be '*at risk (i.e., endangered, vulnerable, rare or poorly known and thought to be threatened)*'. Briggs and Leigh's (1996) list was modified as new information came to hand, as people and agencies responsible for the list changed, and as taxonomists revised species descriptions. In 2001, only 65% of the species earlier listed as extinct, endangered and vulnerable remained on the official national list (Burgman 2002).

The lists are also biased. May *et al.* (1995) called them popularity polls. For instance, the list from the UK (Table 1) has relatively large numbers of fungi, nonvascular plants and invertebrates. In contrast, the proportions of threatened species in these groups in other countries are much smaller. The total number of plants and animals in the UK is relatively modest and survey and taxonomic effort per species has been high. Scientists and others may be guided by funding opportunities and personal interests. Numbers of threatened insects, non-vascular plants and fungi are low in other countries because of scientific and popular lack of interest.

Some of the reasons for changes are illustrated in Table 2. Changes in attitude to extinction and taxonomic uncertainty drove many changes in lists of plants thought to be extinct. In some instances, the listing authority became more convinced of a taxon's extinct status after reconsidering existing literature or specimens, and broader expert consultation. Changes in attitude towards taxonomic status were usually due to advice from taxonomic specialists. Expert judgment weighed heavily on these outcomes.

Table 2. Reasons for changes to federal lists of extinct Australian vascular plants since 1981 (after Keith and Burgman 2004).

Reason for change in listing status	Additions	Deletions
Presumed extinction during 1981-2001	8	
Not previously evaluated	48	
Change in extinction uncertainty[1]	46	1
Taxonomic revision[2]	6	25
Change in taxonomic uncertainty	7	15
Rediscovered		89
Correction		34
Change in opinion on native status		3
Total changes in status	115	167
Originally listed in 1981	113	
Remaining listed in 2001		61[3]
Total listings	**228**	**228**

1. Triggered by new information (unsuccessful searches) or reconsideration of literature and expert opinion.
2. Taxonomic revision refers to newly described taxa based on old collections. Change in taxonomic uncertainty means that the listing authority became more convinced of a species status as a distinct taxon due to expert advice. Species were moved from poorly known or data deficient to presumed extinct on the basis of this advice.
3. Includes 12 species remaining listed despite rediscovery.

The considerable uncertainty in the status of listed species is not communicated. The consequence of failing to acknowledge and communicate uncertainty is that people are misled into believing the lists are reliable and unbiased.

Often, different kinds of information are treated as though they are equivalent. In most listing decisions, direct measurements are almost always missing. Expert opinion is usually available. To work with experts, one must,

1. define necessary knowledge and skills for the problem at hand (decide who might be considered an expert),
2. select the experts,
3. elicit information,
4. evaluate the reliability of the information,
5. combine information from different experts, and
6. use the information to reach decisions.

Scientific experts and groups are expected to make reliable, transparent and consistent judgments on behalf of the broader community. Listing determinations are published but lists are not accompanied by measures of the reliability of the decisions. So, how reliable is expert judgment? Has it been assessed? Is it used in the same way in other disciplines? This paper explores the frailties of expert groups, particularly those involved in conservation risk assessments, and how their errors may propagate in decision-making. It outlines some of the options for using expert judgment that may make it more reliable and transparent.

Kinds of uncertainty

Many taxonomies of uncertainty distinguish between natural variability and 'incertitude'. Incertitude reflects lack of knowledge about nature and can be reduced with additional samples. Natural variation can be better estimated by additional samples, but cannot be reduced (Burgman 2004). Regan *et al.* (2002) created a taxonomy of uncertainty that, at the highest level, distinguished between epistemic and linguistic uncertainty. Epistemic uncertainty exists because of the limitations of measurement devices, insufficient data, extrapolations and interpolations, and variability over time or space. There is a fact, but we do not know it exactly. For instance, a population's size may in fact be 281, but be estimated to be 300, with 95% confidence intervals of 250 to 350. This is the domain of statistics and conventional scientific training.

Linguistic uncertainty, on the other hand, arises because natural language, including scientific vocabulary, is not exact. Linguistic uncertainty can be classified into several distinct types.

Vagueness arises because language permits borderline cases. For example, the term endangered in general usage refers to a continuum of threat but is defined sharply by the IUCN (1994, 2001), for example, as a species that is estimated to number fewer than 250 mature individuals (Rule D). Context dependence arises from a failure to specify the context in which a proposition is to be understood. For example, if a range reduction is said to be 'small', without specifying the context, it is not possible to say whether the reduction is small on the scale of a local region or a continent. Ambiguity arises when a word has more than one meaning and it is not clear which meaning is intended. Knowledge is under-specific when there is unwanted generality in a statement or where data could have been obtained but are no longer available. For example, fauna and flora collections often are associated with very imprecise locations such as 'north of Esperance'.

All risk assessments involve at least some elements of linguistic uncertainty. Imprecise language clouds the perceptions of experts and limits their ability to communicate. Expert assessments may be made substantially more reliable if care is taken to eliminate as many sources of linguistic uncertainty as possible, prior to the commencement of an elicitation process. The preamble developed by the IUCN (2001) is a good example of a careful attempt to minimise linguistic uncertainty.

Who's an expert? ... philosophical and legal definitions

Hart (1986; see also Meyer and Booker 1990) identified three attributes of an expert:

1. effectiveness; they use knowledge to solve problems with an acceptable rate of success,
2. efficiency; they solve problems quickly, and
3. awareness of limitations; they are willing to say when they cannot solve a problem.

This definition is superficially appealing but it is difficult to put into operation. It can be very difficult to distinguish between a competent expert and an incompetent one, and to have a clear picture of what an expert knows. Experts usually are assumed to have both substantive expertise resulting from technical experience and training, and normative expertise, which is the ability to communicate and knowledge of the statistical principles and jargon of a field.

Lawson (1900) wrote the foundation rules for expert and opinion evidence for the US legal system. In these rules, opinion is not admissible in evidence except '*...on questions of science ...persons instructed therein by study or experience may give their opinions. Such persons are called experts.*' (p. 1-2). Expert opinion is seen as necessary to inform a court completely as possible about facts that might be otherwise unattainable because they are future probabilities, contingencies or facts '*not within positive knowledge*' (Lawson 1900, p. 236). In Australian Federal courts, opinion evidence is admissible if it assists '*...the trier of fact in understanding the testimony, or determining a fact in issue*' (ALRC 1985, p. 739-740). Science is presumed to act as a check on bias or prejudice.

Courts have the luxury of being able to question the reliability of experts and to reject them. The qualifications of an expert may be tested by the opinions of other experts. In adversarial systems, the accuracy of testimony may be tested under cross-examination. The substantive knowledge of an expert may be tested by hypothetical questions. Opinions may be 'impeached' by proof that on a former occasion, an expert expressed a different opinion. The ability to question experts is an essential part of the provision and use of expert judgment.

In the adversarial legal systems, potential expert witnesses are selected overwhelmingly for their credentials and for the strength of their support for the lawyer's viewpoint (Shuman *et al.* 1993, in Freckelton 1995). Lawyers search for appropriate attributes in an expert and develop strategies to maximize their chance of winning a case. This view suggests that scientists are advocates of a scientific position (Walton 1997), whereas scientists usually see themselves as objective interpreters of fact, protected from prejudice by scientific method.

Elicitation

The reliability of expert judgments depends, to some extent, on how the experts are interrogated. Methods for elicitation may include correspondence, personal interviews, traditional meetings, structured group meetings aimed at achieving consensus, and meetings that combine consensus with numerical aggregation. Meyer and Booker (1990), Morgan and Henrion (1990), Cooke (1991), Vose (1996), Ayyub (2001) and Walley and DeCooman (2001) provide practical advice on how to conduct scientific elicitation and analyse the information they produce.

Context may affect the quality of expert judgment. Experts sometimes make judgments just to be helpful, or to retain the semblance of expert respectability (because the experts have been brow-beaten into providing an answer, or won't admit they can't). Experts tire during prolonged elicitation exercises and fatigue may make their knowledge unavailable.

Conservation listing decisions often involve novel circumstances in which data and experience are limited, conditions under which expert judgments are particularly error-prone. Errors are exacerbated when the topic involves very low probability events such as catastrophic failure of a system, extreme weather, coincidences of independent events, and so on. Sometimes, it may be possible to compare the situation with other events that are better defined, or it may be possible to disaggregate the rare event into a sequence of more likely events that are easier to estimate, the combination of which generates the outcome in question (Bier *et al.* 1999).

Expert frailties

In the 1970s, Kahneman and Tversky began experiments on the ways in which people perceive and react to risks. One of their discoveries was that experts are susceptible to context and how questions are framed. A framing effect occurs when a change in the presentation of a choice influences choice behaviour, even when the objective characteristics (outcomes or probabilities) are not changed (Kahneman and Tversky 1979, 1984). Scientific training provides only limited protection against these weaknesses (Fischhoff *et al.* 1982, Fischhoff 1995, Morgan *et al.* 1996, Freudenburg 1996, Slovic *et al.* 2000, Gigerenzer 2002, Trumbo and McComas 2003).

Conservation risk assessments are subject to distorting influences, perhaps more so than other types of scientific analysis, because of the politically charged and value-laden nature of many of the problems. Typically there is considerable pressure to produce results to diffuse social tension. Unfortunately, experts are themselves susceptible to the same set of pressures, even in circumstances in which they believe themselves to be considering only the scientific dimension of a decision. They cannot occupy the independent, objective ground that politicians and policy makers wish them to.

Biases may arise because of the relative ease with which different kinds of information is remembered by experts. The reliability of an estimate is determined by the expert's experience, their ability to recollect accurately and the similarity of past experiences to the problem at hand (Meyer and Booker 1990, Vose 1996).

Optimism is a pervasive feature of expert assessments (Plous 1993, Tufte 1997). Typically there is little relationship between confidence and accuracy, including eyewitness testimony in law, clinical diagnoses in medicine, and answers to general knowledge questions by people without special training. Most experts have a region of overconfidence, a domain between the subset of facts they have learned (Figure 1) and the subset they think they know. It varies between experts and depends on the elicitation process.

Baran (2000) described a 40 ha patch of sclerophyll forest and asked a group of ecologists at a professional meeting to estimate the number of 0.1 hectare quadrats that would be necessary to sample 95% of the perennial, vascular plant

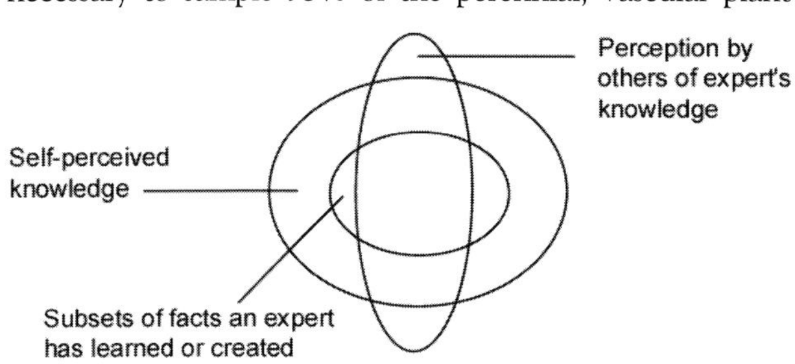

Figure 1. Expert knowledge and the region of overconfidence, which lies between the subset of facts known by the expert and the subset they believe they know (modified from Ayyab 2001; Burgman 2004).

species. She also asked them to provide 90% credible bounds on their estimates. The context was one in which the experts may have been expected to provide reliable answers. Baran (2000) sampled the area intensively to validate their answers. Unfortunately, only 2 of 22 intervals captured the correct answer. The median response substantially underestimated the number of samples required. Experts are often insensitive to sample size and there is no simple feedback between estimation and outcome in plant surveys. As a result, experts may be poor judges of the effort required to achieve a given level of reliability.

The surprise index is the frequency with which a true value falls outside the limits of a judgment. If experts are asked to provide 1st and 99th quantiles, for instance, surprises should occur roughly 2% of the time. Cooke (1991) and Vose (1996, p. 156-158) found in a range of settings that experts provide intervals that contain the truth about 40% of the time, when asked for intervals that will contain the truth about 90% or more of the time. Morgan and Henrion (1990) concluded that all methods for elicitation show a strong tendency towards overconfidence, generating far more surprises than expert judgments lead us to expect. Freudenburg (1999, p. 108) called it 'insufficient humility about what we do not know'.

Motivational biases are tied to ethical positions and experts often are unaware of the value-laden nature of their opinions. Such biases may arise during listing because the scientists involved may have a view of what is to be gained or lost by a decision that influences their interpretation of uncertain information. In a different context, but one that illustrates the issue, Campbell (2002) interviewed experts interested in the conservation and management of marine turtles. They were asked about the sustainable, consumptive use of turtles and their eggs. Campbell's results revealed four 'positions' on use, all of which were defensible on scientific grounds (Table 3).

A total of 19 experts thought that uncertainty dictates caution, so that managers should err on the side of non-use. Three were implacably opposed to commercial use, irrespective of uncertainty. Most experts saw opposing views as influenced by 'emotions', at the same time claiming dispassionate scientific objectivity for their own views, irrespective of their positions. Participants attributed emotional involvement to conservationists without biological training, as though scientific training protects people against emotion. Few propositions are so plainly self-deluded. Almost all experts have personal, value-laden opinions about the outcomes of environmental decisions. It is naïve to think that scientists are anything but advocates of scientific positions and personal value systems.

Table 3. Positions on use and uncertainty, and institutional affiliation of 36 marine turtle experts (after Campbell 2002).

Positions	NGO	University	Government	Total
Consumption is supported, learn through use	3	5	2	10
Support limited consumptive use	2	2	0	4
Uncertainty dictates caution, consumptive use is not supported	5	7	7	19
Consumptive use is unacceptable	0	2	1	3

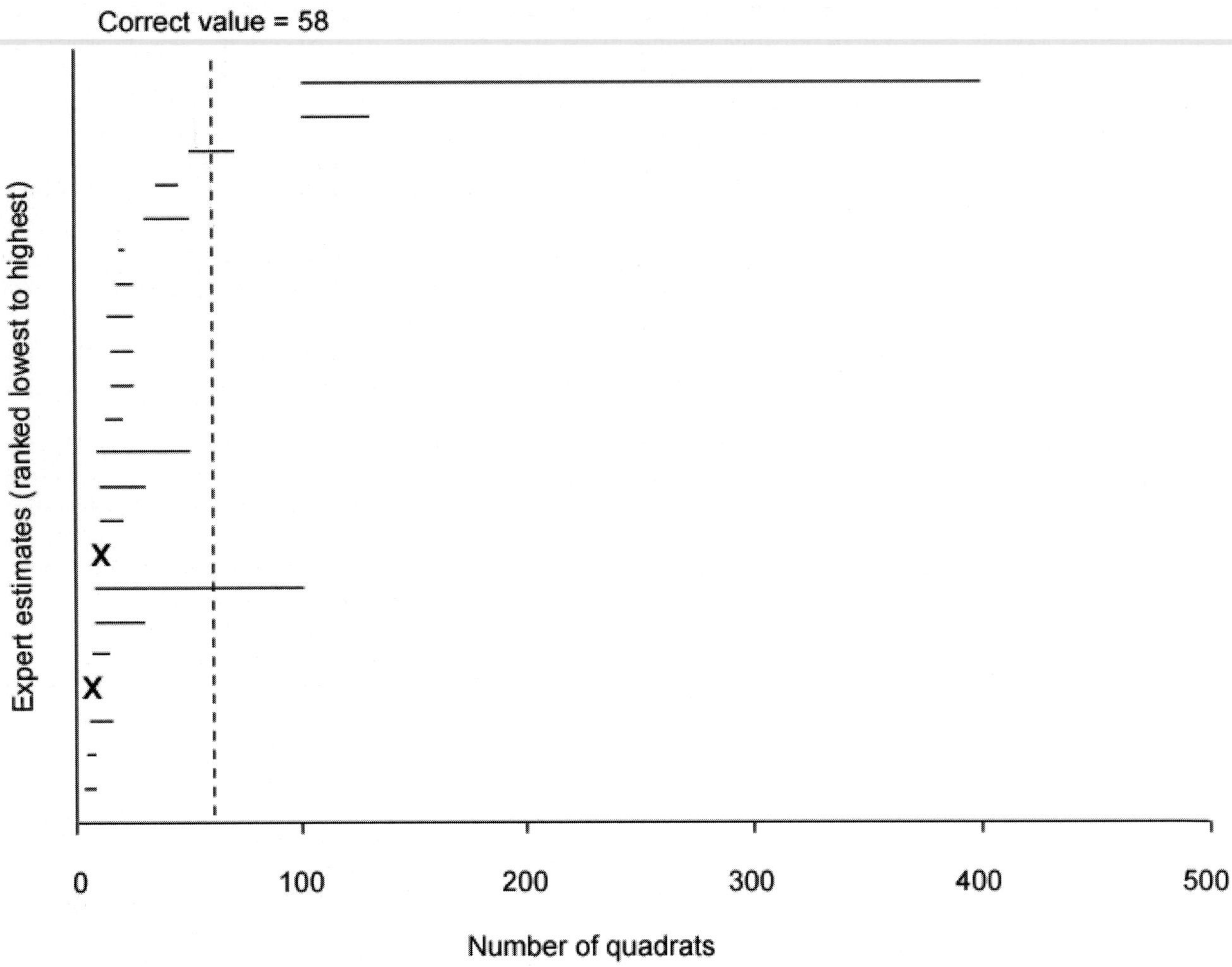

Figure 2. Expert estimates of the number of 0.1 ha quadrats necessary to sample 95% of the vascular plant species in a 40 ha patch of dry sclerophyll forest containing about 120 species, and their 90% credible bounds (from Baran 2000). The analyses assume plants were randomly distributed in the 40 ha patch, with respect to the sample quadrats. Two people failed to provide bounds, marked by a 'x'.

Are expert judgments any good?

Expert judgments are used because estimates are needed in circumstances in which it is difficult or impossible to acquire direct data. There are a few direct comparisons of expert estimates with actual probabilistic outcomes. Most are from engineering (see Cooke 1991). For example, Hynes and Vanmarcke (1975, in Krinitzsky 1993) found experts' risk assessments to be optimistic (their confidence intervals rarely captured the true value) and sometimes substantially biased (Figure 3).

Training can alleviate some of these problems. Routinely reliable estimates have only ever been demonstrated in people who make frequent, repeated, easily verified, unambiguous predictions so that they learn from feedback, such as weather forecasters (Figure 4). Expert engineers underestimated the lifetimes of long-lived components and over-estimated the lifetime of short-lived components. Subjective probabilities made by army doctors that patients have pneumonia were poor predictors of the outcomes of more reliable diagnoses based on radiography. Performances of experts in business and military intelligence are likewise mixed and unimpressive (Morgan and Henrion 1990).

Most technical disciplines have conventional (shared) wisdom. If the technical base of expertise is narrow, or if people are selected by inter-personal relationships, they may represent a single school of thought or be influenced by the same data (Bier 2002).

Experts do somewhat better than untrained people, within their domain of expertise. Wright *et al.* (2002) concluded that if objective models and reliable feedback are available, it may be possible for experts to improve their performance above the performance of an untrained person. However, judgments by some professionals may be unlearnable, especially if they involve novel or infrequent events. Demands for unlearnable (and untested) judgments, together with unquenchable expert optimism, create a spectre of widespread, unreliable expert opinion that wears a mantle of scientific reliability. The documented unreliability of expert opinion in other disciplines signals the possibility that unreliable expert opinion may be prevalent in environmental domains where novel events are common and objective models, monitoring and feedback are rare.

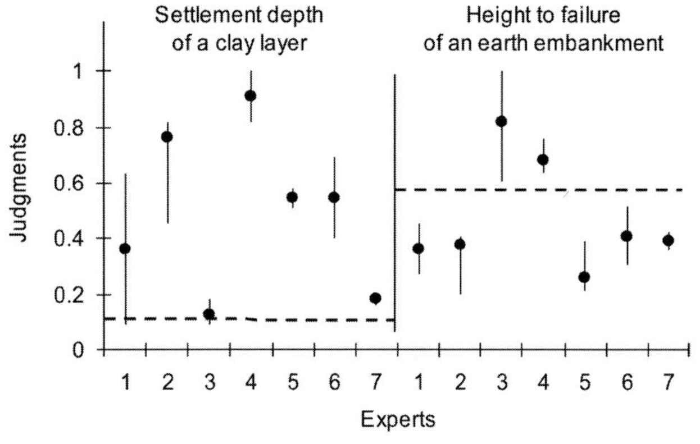

Figure 3. Opinions of geotechnical experts on two standard problems. The correct (measured) values for settlement depth was 1.5 cm and for added height to failure was 4.9 m. The x-axis for both was rescaled so the maximum value was 1. Correct values are shown as dashed horizontal lines. The intervals show 'minimum' and 'maximum' values reported by the experts (after Hynes and Vanmarcke 1975 in Krinitzsky 1993).

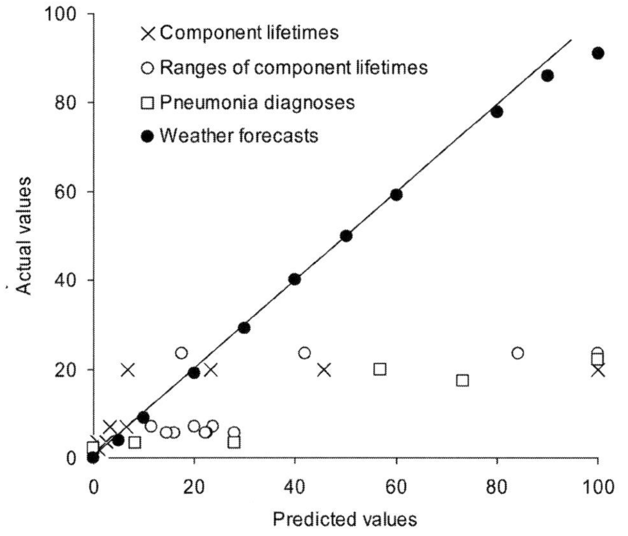

Figure 4. Calibration curves: expert predictions plotted against actual outcomes. Crosses are estimates by engineers of the mean lifetime of components in nuclear power systems, versus measured lifetimes. Open circles are estimates of the ranges for the mean lifetimes of the same components, versus measured ranges. Ranges are expressed as the maximum divided by the minimum. The components included pumps, valves and heat exchange units (after Mosleh 1987, in Cooke 1991). The squares are army doctors' subjective probabilities that patients have pneumonia, versus more reliable diagnoses based on radiography (after Christensen-Szalanski and Bushyhead 1981). Solid circles are meteorologists predictions for the probability of rain on the following day, against the observed relative frequencies (after Murphy and Winkler 1984, in Plous 1993). The diagonal line provides the line of correct estimation for all sets of observations. Values are scaled so that the maximum value in each set is 100.

Salvation in expert groups?

Many disciplines use expert groups to provide independent, technical advice. In conservation biology, disagreements among experts involved in listing decisions are rarely communicated. Regan *et al.* (2004) documented uncertainty among experts in listing assessments. They asked 18 conservation biologists to assess the status of 13 species from a broad range of taxonomic groups. The assessors were provided with standard information sets and interpreted them using the IUCN (2001) logical rules (Figure 5).

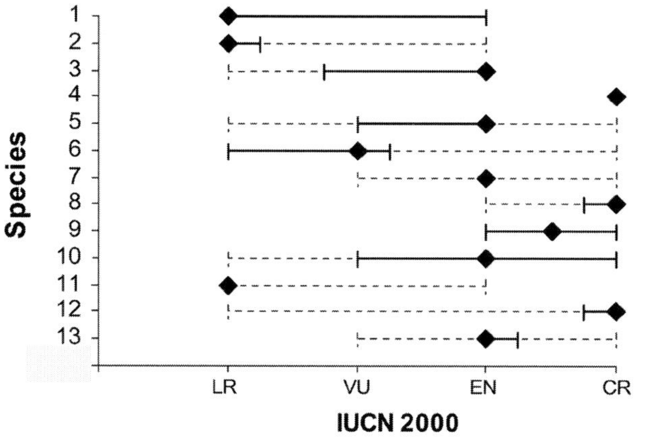

Figure 5. Assessments of the conservation status of 13 species made by 18 assessors (Regan et al. 2004). The diamond represents the median assessment. 50% of assessments are contained within the solid bars. The dashed lines show the full range of assessments. LR is low risk, VU is vulnerable, EN is endangered, and CR is critically endangered.

Some participants were unwilling to make inferences, due to different attitudes and interpretations of language in the guidelines and the information sets. Some parameter estimates spanned thresholds in the rule set. Variation in qualitative judgments, inconsistent logic and mistakes entering data all played a role. The lesson from Regan *et al.*'s (2004) study was that expert judgments can differ, even when the only source of uncertainty is the subjective interpretation of language and data, a largely unavoidable element in almost all conservation risk assessments.

Careless use of expert groups can lead to unacknowledged uncertainty and bias. Uncertainty will be underestimated if the experts involved share common values, experiences, professional norms, context, and cultural background, so that they stand to gain or lose in similar ways from the outcomes of decisions and hold the same motivational biases (Bier 2002). Anonymous opinions may lead to bias because experts cannot be held personally accountable (Krinitzsky 1993). Strong personalities influence outcomes, participants promote decisions before examining the problem, views are anchored, change is resisted, people hold covert opinions, and there is substantial pressure for conformity (e.g., Ruckelshaus *et al.* 2002). Group judgment may lead to overconfidence and polarization of opinions in which groups adopt a position that is more extreme than any individual member (Clemen and Winkler 1999). Despite these frailties, some experts persist in the belief that expert-group elicitation of risks is largely detached from subjective, social and emotional influences (e.g., Kerr 1996).

If these issues are managed explicitly, if participants act rationally (from the perspective of classical probabilities) and are provided with information that reflects utilities and expectations in an unbiased way, then in general, group (consensus) judgments can be expected to be better than individual ones. Multiple experts with different backgrounds increase the extent of knowledge and experience contributing to an assessment (Clemen and Winkler 1999). Groups may improve the interpretation of evidence: experts revise opinions when they are alerted to new and relevant issues or errors in their logic or interpretation of language. Stratifying experts to include a range of demographic and social attributes increases the chances of disagreement, but reduces the chances of bias.

Aggregating opinions

There are two basic forms of aggregation: numerical and behavioural aggregation. Behavioural aggregation techniques include the Delphi approach and variations (Cooke 1991, Ayyub 2001), negotiation, consensus, closure, resolution, and voting or preference voting (Engelhardt and Caplan 1986, Valverde 2001). In general, behavioural aggregation does not guarantee complete consensus and it is not always sensible to seek it. Disagreement cannot necessarily be resolved and consensus may mask important, legitimate differences of opinion. One notable example of a procedure to retain and communicate differences of expert opinion was made by Lunney *et al.* (1996) who displayed the spread of opinions relative to decision thresholds and noted areas of disagreement among experts for each species they evaluated.

It is always useful to eliminate arbitrary elements of disagreement (such as ambiguity and underspecificity). But in the spirit of honest and transparent assessments, differences of opinion about scientific detail, as well as about substantial ethical issues, can be made transparent, without compromising the ability of a group to reach a decision.

Numerical aggregation uses quantitative strategies to arrive at a combined estimate. If the information is probabilistic, then the tools of formal statistics including Bayesian analysis are appropriate. Many approaches including evidence theory and fuzzy sets (see Walley 1991, Klir and Weirman 1998, Sentz and Ferson 2002, Ferson *et al.* 2003) have been developed to cope with non-probabilistic uncertainty.

All aggregation techniques have to deal with the question of weighting the opinions of different experts. Some may be more prone than others to bias and misinterpretation of data, some people have more to lose than others, and so on. For example, some of the opinions in Figure 5 were contributed by people who routinely made more errors than others. Experts may be weighted by consistency, frequency of errors and concordance with known values (Cooke 1991). There have been no explicit applications of weighting strategies in listing decisions for conservation biology.

Akçakaya *et al.* (2000) used non-probabilistic uncertainty methods to combine parameters for the IUCN (1994) criteria. The result of the analysis was a best estimate and bounds that encapsulate uncertainty about parameters and dependencies between the elements of the rule sets (Figure 6).

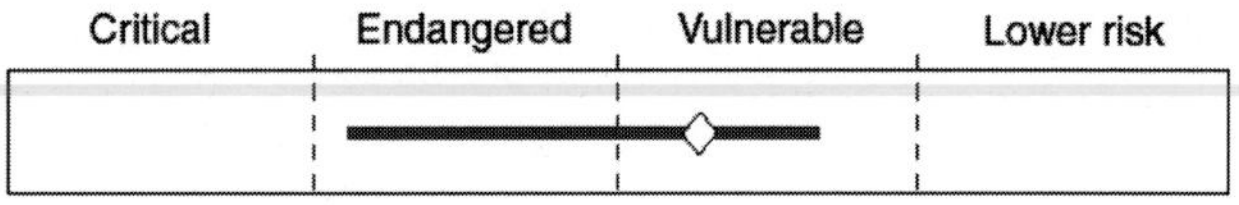

Figure 6. Example output from the method developed by *Akçakaya et al.* (2000). The species is most likely to be Vulnerable, but it is possible that it is Endangered.

Probabilities representing the beliefs of experts may be combined as weighted linear combinations of opinions (Cooke 1991, Vose 1996, Valverde 2001). Numerical averages of individual judgments seem to perform slightly better than group judgments reached by behavioural consensus when the focus is on unambiguous, value free, and sharply defined parameter estimates (Gigone and Hastie 1997, Clemen and Winkler 1999). A few methods have been developed that combine elements of numerical aggregation and behavioural consensus (see Lehrer and Wagner 1981, Cooke 1991, Lehrer 1997). Aggregation methods have not been explored to any great extent in conservation biology and might provide a rich source of ideas for improving the way expert opinions are combined and used in decisions.

Making better use of expert opinion in listing decisions

Overconfidence can be substantially improved by asking people to consider the reasons why they may be wrong (Morgan and Henrion 1990). Performance is enhanced when experts possess appropriate models and are trained to translate subjective assessments into numerical estimates. Experts perform better when they use a consistent format to communicate, when they are trained to provide unbiased estimates and when their judgement is reinforced by immediate, unambiguous feedback (Wright *et al.* 2002). Guidelines for elicitation may be tailored for specific circumstances.

Other strategies might improve the quality of expert judgements. They include a systematic search for motivational biases (Morgan and Henrion 1990), testing expert understanding against known standard problems, making experts aware of potential sources of bias, providing them with training, and monitoring sources of bias during the elicitation process (Meyer and Booker 1990). Cooke (1991) provided some numerical methods for achieving these goals.

Science encourages obedience to scientific authority through its claims of objectivity and expert status. Expert groups, in particular, can be difficult to challenge. Walton (1997) pointed out that scientific argument may be seen as rigid in the sense that it is beyond refutation by non-experts, generating what Walton called a culture of technical control. In conservation biology, the appeal to expert opinion should be a fallible but legitimate strategy (Walton 1997). Anyone with a stake in the outcome should be able to criticize an expert's opinion and be taken seriously. If it is broadly accepted that experts often are advocates of a theory and that they have a stake in the outcome, methods are required that balance their opinions. In addition, honest differences of opinion among experts, those that remain after linguistic uncertainties are eliminated, should be retained and communicated.

Often, experts are the only available source of knowledge to support decisions that would otherwise be made in a vacuum. Group opinions are likely to be better than individual opinions and they provide an efficient and convenient way of compiling and aggregating a broad range of knowledge. But there may be a great deal to be learned from other technical disciplines and from the psychology and sociology of risk perception and decision theory. Some advances have already been made in local assessment protocols (Lunney *et al.* 1996) that would have broad utility. Table 4 summarises some of the potential avenues for improvement. At least some could be implemented almost immediately, with very little cost. Others will take time. Alternative approaches need to be evaluated. However, implementation could substantially improve the reliability, transparency, utility and credibility of listing decisions.

Table 4. Recommendations for expert opinion in listing decisions

Recommendations for providers of expert opinion
• consider why your opinions may be wrong
• use appropriate models to organize ideas and evaluate the adequacy of available data
• seek training to translate subjective assessments into numerical estimates
• use a consistent format to communicate
• seek immediate, unambiguous feedback on your judgements
• seek situations in which your opinions can be tested independently
• ensure you are aware of potential sources of bias
• be aware of overconfidence and practice to reduce it
• be aware of sources of linguistic uncertainty and seek to eliminate them
• declare your interests (your stake in the outcome)
• provide opinions in a form that can be verified as easily as possible
Recommendations for facilitators
• select experts to encompass a broad range of social and technical strata
• decide or negotiate weighting and aggregation strategies before the elicitation process
• consider alternative elicitation strategies (one on one, small groups, workshops)
• create systems that define terms and attributes as clearly and unambiguously as possible
• ensure the elicitation system actively eliminates linguistic uncertainty
• conduct a systematic search for motivational biases
• ensure experts are aware of potential sources of bias
• measure and correct for overconfidence
• use structured rules to assist decisions
• provide experts with immediate, unambiguous feedback on judgements
• anticipate and manage strong personalities, entrenched positions, anchoring, resistance to change, covert opinions, pressure for conformity, and power differences between people
Recommendations for users of expert opinion
• ensure experts encompass a broad range of social and technical strata
• understand terms, parameters and attributes of the system used by experts
• attribute opinions to individuals
• ensure all those who share the burden of the risk of decisions have the opportunity to cross-examine the information and expert opinions
• request that expert performances against standards be made public
• ensure differences of opinion among experts are communicated in lists
• anticipate overconfidence
• evaluate aggregation and weighting schemes for their effect on the consensus position

Acknowledgements

I am indebted to Natalie Baran and Tracey Regan for their permission to use some of the results of their theses at the School of Botany, University of Melbourne. I am grateful to Kuniko Yamada for her assistance in preparing this manuscript, and to David Keith, Dan Lunney, Pat Hutchings and one anonymous reviewer for their comments. Background information in this paper is drawn from a chapter of a book by the author entitled 'Risks and Decisions for Conservation and Environmental Management', to appear through Cambridge University Press. Most of this work was conducted while I was a sabbatical fellow at the National Center for Ecological Analysis and Synthesis at Santa Barbara, California.

References

Akçakaya, H. R., Ferson, S., Burgman, M. A., Keith, D. A., Mace, G. M. and Todd, C. R. 2000. Making consistent IUCN classifications under uncertainty. *Conservation Biology* **14:** 1001–1013.

ALRC. 1985. Evidence. Australian Law Reform Commission Report 26.Volume 1. Australian Government Publishing Service, Canberra. http://www.austlii.edu.au/au/other/alrc/publications/reports/26/

Anonymous. 1995. Biodiversity: the UK Steering Group report. Volume 2. Action Plans, HMSO, London.

Ayyub, B. M. 2001. *Elicitation of expert opinions for uncertainty and risks.* CRC Press, Boca Raton.

Baran, N, 2000. Effective survey methods for detecting plants. MSc Thesis, School of Botany, University of Melbourne.

Bier, V. 2002. Implications of the research on overconfidence for challenge problem solution strategies. Paper presented to the SANDIA Conference on Uncertainty. Albuquerque: New Mexico, August 2002.

Bier, V. M., Haimes, Y. Y., Lambert, J. H., Matalas, N. C. and Zimmerman, R. 1999. A survey of approaches for assessing and managing the risks of extremes. *Risk Analysis* **19:** 83-94.

Briggs, J. D. and Leigh, J. H. 1996. Rare or threatened Australian plants. CSIRO, Melbourne.

Burgman, M. A. 2002. Are listed threatened plant species actually at risk? *Australian Journal of Botany* **50:** 1-13.

Burgman, M. A. 2004. *Risks and decisions for conservation and environmental management.* Cambridge University Press, Cambridge (in press).

Campbell, L. M. 2002. Science and sustainable use: views of marine turtle conservation experts. *Ecological Applications* **12:** 1229-1246.

Christensen-Szalanski, J. and Bushyhead, J. 1981. Physicians use of probabilistic information in a real clinical setting. *Journal of Experimental Psychology: Human Perception and Performance* **7:** 928-935.

Clemen, R. T. and Winkler, R. L. 1999. Combining probability distributions from experts in risk analysis. *Risk Analysis* **19:** 187-203.

Cooke, R. M. 1991. *Experts in uncertainty: opinion and subjective probability in science.* Oxford University Press, Oxford.

Engelhardt, H. and Caplan, H. 1986. Patterns of controversy and closure: the interplay of knowledge, values, and political forces. Pp. 1-23 in *Scientific controversies: case studies in the resolution and closure of disputes in science and technology*, edited by H. Engelhardt and H. Caplan. Cambridge University Press, New York.

Ferson, S., Kreinovich, V., Ginzburg, L., Myers, D. and Sentz, K. 2003. Constructing probability boxes and Dempster-Shafer structures. SAND2002-4015. Sandia National Laboratories, Albuquerque: New Mexico.

Fischhoff, B. 1995. Risk perception and communication unplugged: twenty years of progress. *Risk Analysis* **15:** 137-145.

Fischhoff, B., Slovic, P. and Lichtenstein, S. 1982. Lay foibles and expert fables in judgements about risk. *American Statistician* **36,** 240-255.

Freckelton, I. 1995. The challenge of junk psychiatry, psychology and science: the evolving role of the forensic expert. Pp. 58-59 in *Tomorrow's law,* edited by H. Selby. Federation Press, Sydney.

Freudenburg, W. R. 1996. Risky thinking: irrational fears about risk and society. *Annals AAPSS* **545:** 44-53.

Freudenburg, W. R 1999. Tools for understanding the socioeconomic and political settings for environmental decision making. Pp. 94-125. in *Tools to aid environmental decision making*, editied by V. H. Dale and M. R. English. Springer, New York.

Gigone, D. and Hastie, R. 1997. Proper analysis of the accuracy of group judgments. *Psychological Bulletin* **121:** 149-167.

Gigerenzer, G. 2002. *Calculated risks: how to know when numbers deceive you.* Simon and Schuster, New York.

Groombridge, B. (Ed) 1994. Biodiversity data sourcebook. World Conservation Monitoring Centre, Biodiversity Series No 1. World Conservation Press, Cambridge.

Hart, A. 1986. *Knowledge acquisition for expert systems.* McGraw-Hill, New York.

IUCN. 1994. IUCN red list of threatened animals. International Union for the Conservation of Nature, Gland: Switzerland.

IUCN. 2001. IUCN Red List Categories. IUCN Species Survival Commission, The World Conservation Union, Gland: Switzerland.

Kahneman, D. and Tversky, A. 1979. Prospect theory: an analysis of decision under risk. *Econometrica* **47:** 263-291.

Kahneman, D. and Tversky, A. 1984. Choices, values, and frames. *American Psychologist* **39:** 342-347.

Keith, D. A. and Burgman, M. A. 2004. The Lazarus effect: can the dynamics of extinct species lists tell us anything about the status of biodiversity? *Biological Conservation* **117:** 41-48.

Kerr, R. 1996. A new way to ask the experts: rating radioactive waste risks. *Science* **274:** 913-914.

Klir, G. and Wierman, M. J. 1998. *Uncertainty-based information: elements of generalized information theory.* Physica-Verlag, Heidelberg.

Krinitzsky, E. L. 1993. Earthquake probability in engineering – Part 1: The use and misuse of expert opinion. *Engineering Geology* **33:** 257-288.

Lawson, J. D. 1900. *The law of expert and opinion evidence* (2nd edition). T. H. Flood, Chicago.

Lehrer, K. 1997. Consensus, negotiation and mediation. Pp. 3-14. in *Consensus under fuzziness,* edited by J. Kacprzyk, H. Nurmi and M. Fedrizzi. Kluwer, Boston.

Lehrer, K. and Wagner, C. 1981. *Rational consensus in science and society.* Reidel, Dordecht.

Lunney, D., Curtin, A., Ayers, D., Cogger, H. G. and Dickman, C. R. 1996. An ecological approach to identifying the endangered fauna of New South Wales. *Pacific Conservation Biology* **2:** 212-231.

May, R. M., Lawton, J. H. and Stork, N. E. 1995. Assessing extinction rates. Pp. 1-24. in *Extinction rates,* edited by J. H. Lawton and R. M. May. Oxford University Press, Oxford.

Meyer, M. A. and Booker, J. M. 1990. Eliciting and analyzing expert judgment: a practical guide. Office of Nuclear Regulatory Research, Division of Systems Research, US Nuclear Regulatory Commission, Washington, DC.

Morgan, M. G. and M. Henrion. 1990. *Uncertainty: a guide to dealing with uncertainty in quantitative risk and policy analysis.* Cambridge University Press, Cambridge.

Morgan, M. G., Fischhoff, B., Lave, L. and Fischbeck, P. 1996. A proposal for ranking risk within Federal agencies. Pp. 111-147. in *Comparing environmental risks*, edited by Davies, J. C. Resources for the Future, Washington, DC.

Plous, S. 1993. *The psychology of judgment and decision making.* McGraw-Hill, New York.

Possingham, H. P. Andelman, S. J., Burgman, M. A., Medellín, R. A., Master, L. L. and Keith, D. A. 2002a. Limits to the use of threatened species lists. *Trends in Ecology and Evolution* **17:** 503-507.

Regan, H. M., Colyvan, M. and Burgman, M. A. 2002. A taxonomy and treatment of uncertainty for ecology and conservation biology. *Ecological Applications* **12:** 618-628.

Regan, T. J., Burgman, M. A., McCarthy, M. A., Mace, G. M., Master, L. L., Keith, D. A. and Andelman, S. J. 2004. The consistency of extinction risk classifications. *Conservation Biology* (in review).

Ruckelshaus, M. H., Levin, P., Johnson, J. B., and Kareiva, P. M. 2002. The Pacific Salmon wars: what science brings to the challenge of recovering species. *Annual Review of Ecology and Systematics* **33:** 665-706.

Sentz, K. and Ferson, S. 2002. Combination of evidence in Dempster-Shafer theory. SAND Report, SAND2002-0835. Sandia National Laboratories, Albuquerque, New Mexico.

Slovic, P. Fischhoff, B. and Lichtenstein, S. 1984. Perception and acceptability of risk from energy systems. Pp. 115-135. in *Public reactions to nuclear power: are there critical masses?* edited by W. R. Freundenburg and E. A. Rosa. AAAS/Westview, Boulder, Colorado.

Slovic, P., Monahan, J. and MacGregor, D. G. 2000. Violence risk assessment and risk communication: the effects of using actual cases, providing instruction, and employing probability versus frequency formats. *Law and Human Behavior* **24:** 271-296.

Trumbo, C. W. and McComas, K. A. 2003. The function of credibility in information processing for risk perception. *Risk Analysis* **23:** 343-353.

Tufte, E. 1997. *Visual explanations: images and quantities, evidence and narrative.* Graphics Press, Cheshire, Connecticut.

Valverde, L. J. 2001. Expert judgment resolution in technically-intensive policy disputes. Pp. 221-238. in *Assessment and management of environmental risks*, edited by I. Linkov and J. Palma-Oliveira. Kluwer Academic, Dordrecht, The Netherlands.

Vose, D. 1996. *Quantitative risk analysis: a guide to Monte Carlo simulation modelling.* Wiley, Chichester.

Walley, P. 1991. *Statistical reasoning with imprecise probabilities.* Chapman and Hall, London.

Walley, P. and DeCooman, G. 2001. A behavioral model for linguistic uncertainty. *Information Sciences* **134:** 1-37.

Walton, D. 1997. *Appeal to expert opinion: arguments from authority.* Pennsylvania State University Press, Pennsylvania.

Wright, G., Bolger, F. and Rowe, G. 2002. An empirical test of the relative validity of expert and lay judgments of risk. *Risk Analysis* **22:** 1107-1122.

(Why) Do we need threatened species legislation?

D. Farrier and R. J. Whelan

Institute for Conservation Biology and Law, University of Wollongong

Wollongong NSW 2522

Email: dfarrier@uow.edu.au[1]

ABSTRACT

This paper focuses on the relationship between threatened species listing and command regulation designed to protect habitat on privately owned land. After outlining the historical development of threatened species regulation in NSW from its early origins in legislation designed to control hunting, it explores the role of approval processes and prosecution in the implementation of command regulation in the threatened species context. It examines the extent to which compensation is payable where land use restrictions are imposed, comparing the approach taken in other Australian states, and exploring the case for compensation in the context of the current debate about property rights, landholder duty of care and stewardship payments. Finally, the paper considers the potential of the NSW environmental planning system to protect threatened species habitat in targeted areas and, in particular, explores the relationship between environmental planning instruments and recovery plans. A case is made for a strategic approach to the use of command regulation, in which government takes the initiative rather than reacting on a piecemeal basis to individual project proposals. The Minister for the Environment, advised by what has recently become the NSW Department of Environment and Conservation[2], should play a key role in the strategic planning process.

Key words: historical origins; listing; command regulation; prosecution; compensation; environmental planning.

The origins of threatened species legislation in NSW[3]

In his 1991 decision in *Corkill v Forestry Commission of NSW*,[4] Justice Paul Stein of the New South Wales Land and Environment Court had to determine the meaning of sections 98 and 99 of the *National Parks and Wildlife Act* 1974 (NPWA), as they then stood, which made it an offence to "take" protected and endangered fauna without a licence. He held that "take" should be interpreted as going beyond activities causing direct physical injury to fauna to include "conduct which modifies habitat in a significant fashion thus placing the species of fauna under threat by adversely affecting essential behavioural patterns relating to feeding, breeding or nesting". "Take" included habitat destruction or degradation which adversely affected a threatened or protected species "leading either immediately or over time to a reduced population".[5] The crucial factor leading to this conclusion was that the legislation defined "take" to include "disturb".[6]

---

[1] The authors would like to thank two anonymous referees, Judie Rawling, an ecological consultant, and Alison Schumacher and Dan Lunney of the Department of Environment and Conservation, for their comments.

[2] Following recent administrative changes, the National Parks and Wildlife Service has been incorporated into a new Department of Environment and Conservation. References in legislation to the Service and the Director-General of National Parks and Wildlife are now to be interpreted as references to that Department and its Director-General. See *Public Sector Employment and Management (Environment and Conservation) Order* 2003, cl 6, *Government Gazette* No 152, 24 September 2003. The Service will, however, continue to have a separate identity. See http://80-www.nationalparks.nsw.gov.au.ezproxy.uow.edu.au:2048/npws.nsf/Content/new_department. Those employees of the Service who worked on protected area management are now to be found in the Parks Services Division of the Department. Consideration is being given to locating the Service's off-reserve regulatory functions in the Environment Protection and Regulation Division.

[3] See Jarman and Brock (2004).

[4] *Corkill v Forestry Commission of NSW* (1991) 73 LGRA 126.

[5] Ibid at 139-140. On appeal, the Court of Appeal felt it unnecessary to go into this issue in depth because it was at least clear that the proposed logging programme of the Forestry Commission necessarily involved the taking and/or killing of protected fauna: *Forestry Commission of NSW v Corkill* (1991) 73 LGRA 247 at 250.

[6] NPWA s 5.

Pp 30 - 48 in *Threatened species legislation: is it just an Act?* 2004, edited by Pat Hutchings, Daniel Lunney and Chris Dickman. Royal Zoological Society of New South Wales, Mosman, NSW, Australia.

This was a pivotal moment in the evolution of NSW threatened species legislation from its humble beginnings at the end of the nineteenth century. The early legislative initiatives were designed to protect "game" birds and other animals during breeding seasons so that they could be hunted at other times of the year. "Game" was not restricted to indigenous species. The first piece of legislation, enacted in 1866, was entitled *An Act to provide for the preservation of Imported Game and during the breeding season of Native Game*.[7] Legislation in 1879[8] proclaimed that it was "expedient to encourage the importation and breeding of Game not indigenous to the Colony of New South Wales and also to prevent the destruction of Native Game during the breeding season". It made it an offence to "shoot capture kill take or destroy" any listed native or imported game during the closed season.[9] Pheasant, partridge, grouse and deer were identified as game, but the list also included the Sooty Oyster-catcher which has now graduated from being a game species to a vulnerable species,[10] the "Dragoon-bird"[11] and the "Regent-bird".[12] The list could be extended to include any animals "known to be destroyers of snakes, vermin or insects which are injurious to vegetation". The threats to these species were not insignificant: the legislators felt it necessary to forbid the use of firearms with a barrel length greater than six feet or with a bore exceeding one inch in diameter.[13]

The word "take" which was to play a decisive role in the *Corkill* decision over a century later, appeared in the 1879 legislation, but in a context where it could hardly have been argued that it included the destruction of habitat.[14] It disappeared in the subsequent version of the legislation: the *Birds Protection Act* of 1881.[15] The prohibition only covered those who "willfully kill capture or injure".[16] This legislation claimed to have the same objectives as the 1879 legislation. However, it also protected listed "song birds", including the Indian Minah,[17] the Starling and some native birds, including the "Great Kingfisher – commonly known as the Laughing Jackass".[18] This represents the first move towards conservation other than for hunting, because song birds were protected throughout the year, not only during closed seasons. But this protection ceased after five years.[19] It was renewed from 1893, but again only for another five years, and extended to a broader range of native birds, including Lyre Birds *Menura* and the Emu *Dromaius Novae Hollandiae*,[20] which some feared were in danger of extermination (Stubbs 2001, 33).

The formula of "kill capture or injure" was again used in the *Birds Protection Act* of 1893,[21] the *Birds Protection Act* 1901 and the *Native Animals Protection Act* 1903. By 1901, the Indian Minah had been naturalised, listed as an "Australian Bird". Third parties were encouraged to become involved in the enforcement process, with one half of the penalty going to those providing information leading to conviction. The other half went to the Zoological Society.[22] Walker (1991, 19) reports that the Society received seven pounds eleven shillings and three pence in 1903.

In 1918, significant modifications were made to the provisions relating to protected birds and mammals. These changes signaled a movement in the legislation away from a narrow focus on conserving game and songbirds to a much broader focus on protecting fauna (Stubbs 2001, 51). It was at this point that the prohibition on "take or kill", defined to include "disturb", [23] emerged,[24] although its significance was not to be fully realised until 1991.

---

[7] 29 Vic 22.

[8] 42 Vic 10: *An Act to secure the protection of certain birds and animals.*

[9] Ibid s 2.

[10] TSCA Schedule 2, Part 1.

[11] This is probably a reference to the noisy pitta (*Pitta versicolor*). See the correspondence at http://menura.cse.unsw.edu.au:64800/2001/09/msg00200.html.

[12] This is probably a reference to the regent bowerbird (*Sericulus chrysocephalus*). Scientific names were not provided in the Act.

[13] 42 Vic 10, s 3.

[14] Ibid s 2.

[15] 45 Vic 29. However, the *Local Government (Validation and Amendment) Act* 1922 provided that ordinances could be made "to regulate the taking, use, sale or destruction of fauna …".

[16] Ibid ss 5, 6.

[17] The spelling is now "Mynah".

[18] 45 Vic 29, Schedule 3. Scientific names were not provided in the legislation.

[19] Ibid s 5.

[20] *Birds Protection Act* of 1893, s 6.

[21] 56 Vic 18.

[22] *Birds Protection Act* of 1881, s 14; *Birds Protection Act* of 1893, s 15; *Birds Protection Act* 1901, s 15; *Native Animals Protection Act* 1903, s 9. This provision disappeared with the enactment of the 1918 legislation.

[23] It was defined to *mean* "hunting, shooting, killing poisoning, netting, snaring, spearing, pursuing, taking, disturbing, or injuring". Under the *National Parks and Wildlife Act* 1974, interpreted in *Corkill*, "take or kill" was defined to *include* "hunt, shoot, poison, net, snare, spear, pursue, capture, disturb, lure or injure": s 5(1).

[24] *Birds and Animals Protection Act* 1918, ss 6, 3. This broad definition continued under subsequent legislation: *Fauna Protection Act* 1948, s 4; *National Parks and Wildlife Act* 1974, s 5(1).

Only if an open season was declared could protected species be taken.[25] The 1918 legislation also abandoned the approach of specifically listing protected fauna. All birds and mammals were defined as protected *unless they were listed*.[26] The list identified *unprotected* species, a practice that has continued to this day.[27]

"Rare" fauna first appeared on the scene in the *Fauna Protection Act* 1948. There was no provision for rare fauna to be listed in the legislation. Rare fauna were simply to be proclaimed by the Governor from time to time, without the benefit of a definition. The consequences of designating a protected species as rare were limited. The penalty for taking/killing was increased, from a fine to a prison sentence of six months. Licences could only be issued in very particular circumstances – for scientific specimens and nuisance animals.[28]

The decision in *Corkill* ushered in what was perceived to be a radical new approach to the protection of rare fauna,[29] although it was one that had lain nascent in the law since 1918. It became an offence in NSW to carry out land use activities that significantly modified the habitat of endangered fauna unless this was permitted under the terms of a licence. This was confirmed in amendments made to the NPWA by the *Endangered Fauna (Interim Protection) Act* 1991. Prohibited "take" of listed fauna was defined to include "significant modification of the habitat of the fauna which is likely to adversely affect its essential behavioural patterns".[30]

In spite of what was, in the popular understanding, an ostensibly dramatic expansion in land use regulation, there were, at this stage in the law's development, significant limits on its overall impact. In the first place, the regulatory system only covered *significant* habitat modification, although no guidance was provided as to what was significant. Secondly, although frogs as a group were covered for the first time, invertebrates and plants were not covered. While the *Wild Flowers and Native Plants Protection Act* 1927 made it an offence to pick a native plant/flower identified by proclamation as protected, and "pick" included "destroy", the operation of this offence was significantly restricted because it did not apply to plants grown on private land which were picked with the consent of the landholder.[31] Private landholders could authorise picking of protected native plants from their land for sale.[32] As a result of amendments in 1945,[33] however, sales of protected native plants grown on private land, other than Christmas Bush *Ceratopetalum gummiferum*, were outlawed except where the landholder was licensed to grow them for sale.[34] The basic objective of the provisions was now to regulate commercial exploitation of particular native plants. Private landholders could "pick" threatened plants, including clearing and destroying them, provided that the plants in question were not subject to commercial exploitation and were not picked for that purpose. These provisions continue in operation under the NPWA.[35]

The *Threatened Species Conservation Act* 1995 (TSCA) removed one of the constraints on the operation of the threatened species regime by extending the listing process to cover invertebrates and plants: all species of animals and plants can now be listed as threatened; so can ecological communities, defined as "an assemblage of species occupying a particular area", and populations.[36] Another constraint was removed by abandoning the requirement that *significant* habitat damage be caused before an offence is committed.[37] Section 118D of the NPWA provides, expansively, that a person must not "do anything that causes damage to any habitat ... of a threatened species, an endangered population or an endangered ecological community if the person knows that the land concerned is habitat of that kind". Under section 118A(2), it is an offence to "pick" a threatened plant species or a plant that is part of an endangered population or ecological community, with "pick" defined to include "destroy" and "injure".[38]

---

[25] *Birds and Animals Protection Act* 1918, s 6.

[26] Ibid s 3.

[27] See *Fauna Protection Act* 1948, subsequently repealed by the *National Parks and Wildlife Act* 1974, which continues to define "protected fauna" as fauna not named in Schedule 11: s 5(1).

[28] *Fauna Protection* Act 1948, ss 20, 25, 26. These restrictions on the grant of licences disappeared in the NPWA: ss 99(2), 120.

[29] The decision in *Corkill* was even more dramatic than this because it covered the habitat not only of endangered but protected fauna. This aspect of the decision was quickly neutralised by the *Endangered Fauna (Interim) Protection Act* 1991, Schedule 1(8), which inserted s 98(5) into the NPWA, making it clear that incidental takings of protected fauna in the course of carrying out activity under some other approval, such as a development consent, were exempt from regulation.

[30] NPWA s 5(1).

[31] *Wild Flowers and Native Plants Protection Act* 1927, ss 4(1), 2.

[32] Ibid s 5.

[33] *Wild Flowers and Native Plants Protection (Amendment) Act* 1945.

[34] *Wild Flowers and Native Plants Protection Act* 1927, s 5, as amended.

[35] The list of protected native plants set out in Schedule 13 now includes those used in the cut-flower industry, along with orchids, ferns and waratahs.

[36] TSCA Part 2.

[37] *Donnelly v Delta Gold Pty Ltd* [2001] NSWLEC 55. In spite of drawing this distinction, however, Bignold J, in finding that the activity in question did not damage habitat, relied principally on evidence which had been advanced in the case relating to *significant* impact: [292]-[296].

[38] NPWA s 5.

The decision in *Corkill*, and its subsequent confirmation in the *Endangered Fauna (Interim Protection) Act* 1991, placed threatened species legislation at the forefront of clearing regulation, because vegetation clearance poses a major, and perhaps the biggest, threat to biodiversity (EPA 1997, para 4.1.3; NPWS 1999, p 6; EPAA 2003, para 6.2). The TSCA further consolidated its role in this context by removing some of the remaining constraints on its operation. But in August 1995, State Environmental Planning Policy 46 had been introduced, extending controls on broadacre land clearing throughout the State, and this was followed by the *Native Vegetation Conservation Act* 1997 (NVCA) which came into operation at the beginning of 1998 (Farrier *et al.* 1999: 361-370).[39] As we shall see, the two strands of regulation are closely integrated, but significant issues remain about whether, from one perspective, integration has gone far enough, and from another, whether an appropriate balance has been achieved.

Listing as a trigger for command regulation

From a landholder's perspective, the core element of threatened species legislation – the provision for listing species as threatened – standing alone, should be uncontroversial. Listing is not, however, an end in itself. Nor is it designed solely to alert people in the community of the need to take voluntary action at their own expense. Listing is controversial because it triggers various forms of regulation which involve allocation of community resources and/or the imposition of legal constraints on behaviour. These may include recovery planning, conservation agreements, protective orders, species impact statements and criminal offence provisions.

Some of these do not pose a threat to landholders because they depend on a consensus being reached, for example, voluntary conservation agreements. The effect of more invasive regulatory provisions, such as special directions to particular landholders restricting land use, are frequently softened in practice by requiring agencies to take into account socio-economic impact as well as the conservation needs of listed species, and to pay compensation (see below). Criminal offence provisions, on the other hand, are controversial where they restrict the activities of private landholders generally, without offering compensation. The effect is aggravated where they flow *directly* from a decision to list a species made exclusively on scientific grounds, and the list is continually expanding as new information about the conservation status of species comes to light. In broad terms, this is the kind of regulatory system that is in place in NSW. The question is whether we are expecting too much of it in terms of delivering conservation outcomes.

Not all Australian jurisdictions use listing as a direct trigger for criminalising activities that are likely to have an adverse impact on threatened species habitat. In Victoria, for example, the offence provisions are not triggered until critical habitat has first been designated,[40] and this has occurred only once during the life of the legislation (Moorrees 2003).[41] The Minister has the power to make interim conservation orders regulating private land use,[42] lasting for up to two years,[43] but they can only be made to conserve the *critical habitat* of a species that has been at least nominated for listing.[44] In practice, no orders have been made (Raff 1998, 659). In Tasmania, it is not clear whether listing directly triggers offence provisions.[45] The Minister does, however, have the power to make 30-day interim protection orders, to conserve habitat on private land where activities threaten taxa that have been listed or "accepted" for listing by the Scientific Advisory Committee.[46] As in Victoria,[47] social and economic consequences must be considered before an order is made[48] and compensation is payable for financial loss.[49]

---

[39] The *Native Vegetation Act* 2003 will eventually repeal (s 52) and replace the *Native Vegetation Conservation Act* 1997, but, as at May 2004, had still not been proclaimed to commence.

[40] *Flora and Fauna Guarantee Act* 1988 (Vic), ss 46-48 and 3(1): definition of "take". Protected flora on private land which is not critical habitat can be taken by or with the consent of the landholder. See also *Wildlife Act* 1975, ss 41-43, relating to offences of taking endangered, notable and protected animals. "Take" is not defined.

[41] This declaration was later rescinded.

[42] *Flora and Fauna Guarantee Act* 1988 (Vic), s 27(b), (c).

[43] Ibid s 32.

[44] Ibid s 26(1)(c), (d).

[45] As in a number of other jurisdictions, this depends on the precise meaning to be given to the concept of "take". "Take" is defined broadly to include " kill, injure, catch, damage, destroy and collect": *Threatened Species Protection Act* 1995 (Tas), s 3(1). However, the offence provision draws a distinction between "disturb" and "take", with the former offence only applying on land subject to interim protection orders and land management agreements: s 51(1). This suggests that the offence was not meant to cover habitat destruction except where there is proof that particular individuals are injured, damaged or destroyed. In 2002, however, the legislation was amended to put it beyond doubt that those carrying out forestry under a certified forest practices plan, among other things, do not need a permit to "take" species unless specifically required: s 51(3). This lends considerable force to the argument that where habitat is destroyed other than by activities covered by this provision, habitat destruction does constitute a "take". See also *Nature Conservation Act* 2002 (Tas), ss 3(5),(7); 26-27. For the position in other jurisdictions, see *National Parks and Wildlife Act* 1972 (SA), ss 5, 47, 51; *Nature Conservation Act* 1992 (Q), ss 88, 89 and Schedule; *Wildlife Conservation Act* 1950 (WA), ss 16(1), 23F, 6(1); McDonald et al (2003).

[46] For interim conservation orders in Queensland, see *Nature Conservation Act* 1992 (Q), ss 102, 103, 105, 108.

[47] *Flora and Fauna Guarantee Act* 1988 (Vic), ss 26(5), 43.

[48] *Threatened Species Protection Act* 1995 (Tas), s 32.

[49] Ibid s 45.

This paper assesses the role of threatened species listing in triggering regulation consisting of demands made by government, enforced by government agencies, potentially leading to the imposition of sanctions against those in breach. Economists have pejoratively labelled it *command and control regulation*, to distinguish it from economic instruments, such as pollution charges, tradeable permits and subsidies. It will be referred to here as *command regulation* in that it takes the form of top-down orders issued by government. Whether it succeeds in *controlling* behaviour in such a way as to achieve desired conservation outcomes is quite a different matter.

Criminal offences, such as those found in sections 118D and 118A of the NPWA, are the stereotypes of command regulation. However, command regulation in the environmental context contains important features that distinguish it from traditional criminal offences. Consequently, we need to question assumptions about how it should be enforced in this context insofar as these are based on how it has traditionally been enforced in other contexts. These distinguishing features and their ramifications will now be explored.

Command regulation and approvals processes

The objective of most command regulation in the environmental context is not to outlaw specified behaviour entirely, but to herd those proposing to carry out particular activities into assessment and approvals/licensing processes. This is inevitable to some extent because the acceptability of proposals depends on the particular configuration of site and activity. For those who get to the stage of making an application, the emphasis within these processes is likely to be not so much on determining whether any approval should be given, but on facilitating the activity by allowing it to go ahead subject to conditions that attempt to ameliorate environmental impact. The theory is that proposals that are completely unacceptable are deterred in preliminary discussions with the approval agency, before they become formal applications.

Under the NSW legislation, the approval required to avoid committing an offence under sections 118D or 118A is, in most situations, development consent under Part 4 of the *Environmental Planning and Assessment Act* 1979 (EPAA).[50] If, for example, a proposed development is permissible with consent under the zoning tables of a local environmental plan, the consent authority will usually be a local council. The Minister for Natural Resources also acts as a consent authority under Part 4 when deciding whether to approve the clearance of native vegetation under the NVCA.[51] Where Part 4 procedures are followed and development consent is granted, a separate licence from the Department of Environment and Conservation (formerly the National Parks and Wildlife Service) is not required.

What has happened here is that the current legislation has largely abandoned the separate licensing system under the auspices of what was then the National Parks and Wildlife Service that operated in relation to activities involving habitat destruction following the decision in *Corkill* and, subsequently, under the amendments contained in the *Endangered Fauna (Interim Protection) Act* 1991. Threatened species issues are now taken into account in the course of other approvals/licensing processes. Usually they are dealt with under EPAA Part 4. The separate licensing system under TSCA Part 6 now only operates at the margins, to cover activities otherwise prohibited by sections 118D and 118A, for which neither development consent under Part 4, nor an approval or licence under any other legislation is required.[52]

Forestry on private land, for example, is frequently exempted from the need to obtain Part 4 development consent from the Minister for Natural Resources for clearing native vegetation under the NVCA.[53] Nevertheless, in some local government areas, particularly on the NSW south coast, local environmental plans require development consent to be obtained for private forestry in rural/non-urban zones, but from the local council rather than the Minister. If development consent is obtained, then activities otherwise prohibited by NPWA sections 118D and 118A are allowed if they are essential to carrying out the consent.[54]

In other local government areas, particularly on the north coast, most of the equivalent plans do not require development consent for private forestry (Kelly and Prest 2000). However, it is an offence to cause waters to be polluted without an environmental protection licence, and this would include diffuse run-off resulting from private forestry.[55] If private forestry was to be licensed

---

[50] NPWA s 118D(2)(b)(i), for example, provides that it is a defence to show that the activity was essential to carrying out development in accordance with development consent. See also s 118A(3)(b)(i).

[51] This will continue to be the position when the *Native Vegetation Act* 2003 is brought into operation: ss 13-14. Clearing is also permissible under a property vegetation plan approved by the Minister: ss 12(1)(b), 27.

[52] NPWA ss 118D(2), 118A(3).

[53] The exemption covers "private native forestry", defined as the "clearing of native vegetation in a native forest in the course of its being selectively logged on a sustainable basis or managed for forestry purposes (timber production) (NVCA Schedule 4, cl 3; State Environmental Planning Policy 46, Schedule 3). It does not apply to private forestry on State protected land (primarily, steeply sloping land and land along river banks: NVCA ss 4(1), 7) and in the Western Division. In his second reading speech on what is now the *Native Vegetation Act* 2003, the Minister indicated that private forestry would be regulated through property vegetation plans. However, these can only authorise the clearing of remnant native vegetation and protected regrowth if the proposed clearing will "improve or maintain environmental outcomes" (s 29). This phrase is not defined in the legislation, but the Minister indicated his belief that "[a] sustainable forestry operation should, by definition, maintain or improve the long-term condition of the forest through sound silvicultural practices" (*Hansard*, Legislative Assembly, 12th November 2003).

[54] NPWA ss 118D(2)(b)(i), 118A(3)(b)(i).

[55] *Protection of the Environment Operations Act* 1997, ss 120(2), 122, and Dictionary.

under these pollution control provisions after being assessed under the environmental assessment provisions of EPAA Part 5, then activities otherwise prohibited by sections 118D and 118A would be allowed to the extent to which they are essential to carrying out the licensed activity.[56] In practice, however, the former Environment Protection Authority (now the Department of Environment and Conservation) did not encourage applications for environmental protection licences in these circumstances. Consequently, in this situation, a separate licence from the Department of Environment and Conservation (DEC) under TSCA Part 6 would be required to authorise activities which would otherwise amount to an offence under sections 118D or 118A, but only where there is likely to be a significant effect on listed species, populations or ecological communities.[57]

Situations in which a separate licence is required from the DEC will, however, be exceptional. Usually, development consent under EPAA Part 4 will be required, or, alternatively, the proposal will be subject to assessment and approval under Part 5 of that Act. Where a proposal is processed under Part 4 or Part 5, the DEC is reduced to playing a back-stop regulatory role in exceptional circumstances: the Director-General can only veto or attach conditions to proposals which are referred to it by local councils or other approval agencies on the grounds that they are *likely to significantly affect* threatened species, populations or ecological communities.[58] In this event, a species impact statement must be provided.[59] The DEC loses even this limited regulatory role in two situations. In the first place, where the Land and Environment Court upholds an appeal under EPAA Part 4 by an applicant for development consent, it does not need the concurrence of the Director-General before deciding to grant consent, nor does it have to incorporate conditions attached to any concurrence which has been given.[60] Secondly, where a Minister is the decision-maker under EPAA Part 4 or 5, the DEC simply has to be *consulted*.[61] This is the position, for example, under the NVCA because it is the Minister for Natural Resources rather than the Department who formally makes decisions under this legislation on whether or not to give development consent for proposals to clear native vegetation.[62]

While the theory is that the DEC will get to know about situations where there is likely to be a significant effect, the gatekeeper of this preliminary decision, frequently a local council, may not have the expertise to make it, and the DEC has no power to call in proposals so that it can make the determination on likely significant effect itself. In these circumstances, the DEC is largely marginalised as a decision-maker. Where a gatekeeper determines that a proposal is *not* likely to significantly affect threatened species, populations or ecological communities, then, except in those limited situations where the Land and Environment Court is successfully invited to overturn this decision,[63] assessment of the proposal's threatened species implications and the final decision on whether proposed destruction of habitat should proceed is left in its hands - a decision-maker with limited expertise, or, in the case of some local councils, none at all (Douglas 1999, 137-141).

In practice, few proposals get through to the DEC for its consideration. The practice appears to be for the gatekeeper to encourage proponents to put in place ameliorative measures so that any likely significant effect is removed (Saxon 1997, 90).[64] The scientific basis of these measures (for example, translocation and corridors) may be quite questionable (Farrier *et al.* 2002; Kelly and Prest 2000, 596-598).

There are not only issues about the expertise of decision-makers on the front-line in relation to threatened species issues but also about the culture in which some of them operate. Where local councils are involved in decisions on whether to give consent to development under EPAA Part 4, there are inbuilt systemic biases in favour of development, which many councils will find it difficult to resist. So, for example, Andrew Kelly has pointed to the vital interest that local councils have in loosening

---

[56] NPWA ss 118D(2)(b)(iii), 118A(3)(b)(iii).

[57] TSCA s 95(2). In situations where there is not likely to be such an effect, a certificate to this effect must be issued, and this will again constitute a defence to proceedings under NPWA ss 118A or 118D: NPWA ss 118A(3)(a1), 118D(2)(a1).

[58] EPAA, s 79B(3),(8). The Director-General also has concurrence powers where critical habitat has been designated (see text below at notes 133-136).

[59] Ibid s 78A(8)(b).

[60] *Land and Environment Court Act* 1979, s 39(6). The Court will, however, take into account the views of the Department in evidence, provided that this does not involve undue delay: *Australand v Sutherland Shire Council* [2004] NSWLEC 9.

[61] EPAA s 79B(3). In these circumstances, the recommendations of the Director-General must nevertheless be forwarded to the Minister responsible for the final decision, and if the latter does not accept them, (s)he must give reasons and these must be made available for public inspection at the DEC's head office: EPAA s 79B(6),(7).

[62] NVCA s 14.

[63] *Timbarra Protection Coalition v Ross Mining NL* (1999) 46 NSWLR 55.

[64] *Smyth v Nambucca S.C.* (1999) 105 LGERA 65. However, the Land and Environment Court draws a distinction between measures which are part of the proposal, which can be taken into account in determining likely significant effect, and those introduced as a condition attached to a development consent, which cannot: *Drummoyne M.C. v Maritime Services Board* (1991) 72 LGERA 186; *Byron Shire Businesses for the Future Inc v Byron Council* (1994) 84 LGERA 434; *Westport Marina Development Pty Ltc v Concord Council* (2000) 109 LGERA 451; *Silverwater Estate Pty Ltd v Auburn Council* [2001] NSWLEC 60. But see the comments of Bignold J in *Donnelly v Delta Gold* [2001] NSWLEC 55 on the difficulties of drawing such a distinction. See also *Transport Action Group Against Motorways Inc v Roads and Traffic Authority* (1999) 104 LGERA 133 at 152.

subdivision restrictions, motivated by their considerable dependence for income on rates. Unimproved land values for rateable rural properties are increased when rural land is rezoned for residential purposes. In addition, the number of rateable properties will be increased by the creation of new allotments through subdivision approval (Kelly and Farrier 1996, 385). The result is that the rating system operates as a perverse incentive in the development control process.

On top of this, where land has been zoned under an environmental planning instrument so as to make specified development, such as rural residential subdivision, permissible with consent on a project-by-project basis, there is inevitably a built-in presumption that some configuration of development of that particular kind will be given approval. The focus, then, is on the details of the development rather than on whether it should be allowed to go ahead at all. In considering the details, particularly constraints stemming from threatened species considerations, local councils will inevitably have in mind the need to keep development economically viable, and seem likely to avoid attaching stringent conditions that, from the proponent's point of view, might constitute a de facto refusal. At this stage in the process, it is usually far too late to start carrying out surveys to detect the presence of threatened species and to use these as a basis for refusing development consent altogether. The focus at this point is inevitably on the environmental impact of the particular configuration of the proposed development, not on whether it should be allowed to go ahead at all. This is an issue which ought logically to have been determined when the earlier decision about the appropriate zoning of the land was made.

Given the significance of the approvals process, command regulation in the environmental context lacks many of the characteristics with which it has been associated, in the context of traditional criminal offences. What is formally a system of command regulation relies much more in practice on bargains being struck rather than orders issued. This stems in part from the fact that the decision-maker's brief is not to protect the interests of threatened species at all costs, but to balance the socio-economic interests of the proponent and the broader community against the community's interest in threatened species conservation. Even in those few situations where the DEC becomes involved in the decision-making process, the Director-General must still take social and economic considerations into account in deciding whether or not to grant concurrence.[65] We have seen that local council decision-makers may have a direct self-interest in seeing a project go ahead, while in other agencies the recency of environmental management responsibilities may clash with a long-established agency culture supportive of the interests of landholders, one in which the focus in the past has been on facilitating access to natural resources rather than conserving them.

Command regulation and prosecution

In core areas of criminal law, offenders are likely to be prosecuted and punished if they are caught. Where command regulation has been used to regulate behaviour impacting adversely on the environment, this close association breaks down. Prosecution and punishment are the exception rather than the rule (Hawkins 1984; Bartel 2003; Grabosky and Braithwaite 1992).

In part, this stems from the fact that, as argued above, the function of command regulation in the environmental context has traditionally been not to outlaw specified activities outright but to herd them into approvals processes, where decisions are made on the merits of particular proposals. The criminal law is perceived as no more than a backstop to deal with those who operate outside the approvals system altogether. The expectation is that they will be few and far between. The operating assumptions are that most of those carrying out activities which require approval will want to play within the rules, that their activities are to be encouraged in general terms because of the socio-economic contribution their proposal will make, and that they should be facilitated when they are carried out under conditions which minimise environmental impact, as distinct from preventing it altogether. These assumptions may be valid for the most part in an urban development context, where there is a long tradition of command regulation of this kind and where major developers specialise in bringing forward development proposals, and consequently have an ongoing interest in maintaining good relationships with regulators. However, they may not ring so true where agricultural land clearing is involved. Even in the context of urban development, anecdotal evidence suggests that criminal law offers no guarantee that, once approval has been secured, proponents will comply with its terms and, in particular, conditions designed to minimise environmental impact (Rawling 2004). It seems likely that more vigorous enforcement is needed at this stage.

A more vigorous approach to enforcement need not, however, mean an increased emphasis on prosecution. Command regulation in the environmental context can generally be enforced not only through the criminal process but also via civil law enforcement mechanisms. Under NSW legislation, these allow not only enforcement agencies but also any member of the community to bring proceedings to restrain breaches and remedy them.[66] Under the *Environment Protection and Biodiversity Conservation Act* 1999 (Cth) (EPBC), even financial penalties can be awarded in civil proceedings brought by the Commonwealth environment Minister.[67]

In civil proceedings, the objective is prevention and remediation rather than retributive or deterrent punishment. Consequently, safeguards designed to protect the innocent from conviction traditionally associated with

[65] EPAA s 79B(5)(h).

[66] EPAA s 123; NPWA s 176A; TSCA s 147.

[67] See, for example, EPBC ss 12, 16, 18, 20, 481-486.

criminal prosecutions can be watered down. Liability need only be proved on *a balance of probabilities*, for example, rather than *beyond a reasonable doubt*. Under NSW legislation, the trade-off is that the ultimate sanction is limited to an order requiring that environmentally harmful activities cease and the environment restored to its previous condition.[68] Yet, in practice, the financial impact of a restoration order may be considerably greater than a fine imposed in criminal proceedings.[69]

Ayres and Braithwaite take a further step in breaking down the assumption that prosecution must inevitably follow breach of command regulation (Ayres and Braithwaite 1992). They start from the position that business regulation is designed to achieve particular objectives, and that the purpose of enforcement is to gain compliance rather than to meet out retributive punishment. Within this scenario, they produce a persuasive rationale for resorting to prosecution only on a very selective basis. They argue that an enforcement strategy designed to change behaviour is much more likely to be effective if it is based on a mix of punishment and persuasion: a "tit-for-tat" strategy. While a firm is cooperating, the enforcement agency should refrain from deterrent responses, particularly prosecution (Ayres and Braithwaite 1992, 5, 21). Most breaches can effectively be dealt with and compliance achieved by informal warnings, or, if this fails, formal notices. These responses are at lower levels of what they call an *enforcement pyramid*, with prosecution close to[70] the pinnacle. The strong expectation is that in most cases, compliance will be secured through these less intrusive interventions towards the base of the pyramid. Only the very few, represented by the tapering off of the pyramid as it moves towards its pinnacle, will resist by continuing to breach the law: for these prosecution is ultimately appropriate (Ayres and Braithwaite 1992, Chap 2).

The argument is that prosecution and punishment will lose much of their impact if they are threatened too soon and too often. Punishment should be "in the background until there is no choice but to move it to the foreground" (Ayres and Braithwaite 1992, 47). At the same time, it must be perceived as inexorable for those who will not cooperate and adjust their behaviour following intervention at the lower levels of the enforcement pyramid:

> "*A paradox of the pyramid is that to the extent that we can absolutely guarantee a commitment to escalate if steps are not taken to prevent the recurrence of lawbreaking, then escalation beyond the lower levels of the pyramid will rarely occur. This is the image of invincibility...*" (Braithwaite 2002, 34).

While Ayres and Braithwaite's conclusions about appropriate enforcement strategies, and particularly the role of prosecution, are likely to be very relevant in some environmental contexts, such as industrial point source pollution control, they rest on a number of assumptions that limit their relevance in the present context. In the first place, they work best where what we are seeking is compliance through positive action to improve environmental performance over the long-term. Secondly, they rest on a fundamental assumption that there is an ongoing relationship between regulatory agency and regulated population, which can tolerate relatively minor breaches in pursuit of long-term improvement. Finally, Ayres and Braithwaite's analysis assumes that the consequences of breach are not irreversible, or at least, that, because of a lack of scientific evidence, we can convince ourselves that they rapidly dissipate, or are a very, very small component contributing to a cumulative impact.

Regulation designed to protect the habitat of threatened species does not fit the scenario envisaged by Ayres and Braithwaite. Where, for example, remnant vegetation is completely cleared, whether for agriculture or residential subdivision, it will take many years to regenerate, will be highly vulnerable to weed invasion, and may never achieve its original state. Clearing of this degree is a dramatic, one-off event that completely removes any opportunity for improved environmental performance through restoration and management. It leaves no room for a continuing relationship between offender and regulatory agency, unless it is formed around restoration. Development in urban contexts has traditionally been treated by the planning system as a one-off event: development is approved once and for all at the outset rather than being licensed on a periodic basis. This approach has undoubtedly gone too far and what has been a system of development control needs to pay much greater attention to ongoing management. There would clearly be room for an enforcement strategy based on the regulatory pyramid where, for example, a condition requiring the retention of habitat is complied with, but there are subsequently breaches of conditions requiring it to be restored and managed. Outside of this scenario, however, we must look to alternative rationales for prosecution policy in the threatened species context.

Constraints on prosecution set by offence definitions

Offence definitions set parameters for prosecution. Under the NSW legislation, "routine agricultural activities" are exempted from command regulation regardless of how much damage they cause to listed species, populations or ecological communities.[71] In part, this reflects a strong sentiment, long espoused in "existing use" provisions in planning legislation, that investment in established land-use activities which complied with regulatory requirements at the time when they were established, should not be undermined by having new regulatory burdens placed upon

[68] EPAA s 124; NPWA s 176A; TSCA s 147.

[69] See, for example, *Al Oshlack v Iron Gates Pty Ltd* [1997] NSWLEC 89.

[70] Licence revocation is identified as being at the highest point of the pyramid: Ayres and Braithwaite (1992), 35.

[71] TSCA s 113A.

them. The Commonwealth's *Environment Protection and Biodiversity Conservation Act* 1999, for example, provides that the continuance of a use which was lawful immediately before the legislation came into operation, and which is not enlarged, expanded or intensified,[72] cannot be subjected to command regulation under that legislation. However, the concept of "routine agricultural activities" in the NSW legislation is not confined to uses already existing at the time the legislation came into operation. Does this, then, mean that any activity that is "routine" in the sense that it is customarily regarded as an agricultural activity within that industry is exempted from regulation, even though it represents a completely new undertaking so far as the particular piece of land is concerned? Or does the activity have to be a related offshoot of a specific, already established and ongoing agricultural operation? Could it seriously be suggested that it would be a routine agricultural activity, for example, for a grain farmer to bring goats on to land and set them loose on remnant vegetation that comprises threatened species habitat?[73]

Secondly, under the offence definition in NPWA section 118D, before taking criminal proceedings the DEC must be in a position to prove beyond a reasonable doubt that the landholder accused of damaging habitat *knew* that the land in question was habitat of a threatened species, an endangered population or an endangered ecological community. This means showing actual knowledge or belief, such as where the landholder has been told by an officer of the DEC that the area comprises habitat of that kind.[74] It may extend to situations where the landholder realised that it was likely that the area comprised habitat, in the sense that there was a significant or real chance that this was the case,[75] but an argument that the accused *ought* to have realised that relevant habitat was involved, and was therefore *negligent*, would fail.[76]

It is apparent from this that the primary objective of this offence is retributive punishment of those who deliberately flout the law and deterrence of those who are tempted to do so.[77] It is not designed to encourage landholders to act more cautiously and carefully. This has a crucial bearing on the enforcement strategy pursued. It would seem to follow that prosecution should be a weapon of immediate rather than last resort. The problem is that it is going to be extremely difficult to catch those who "shoot, shovel and shut up". In addition, where a prosecution is undertaken in these circumstances, the enforcement agency cannot afford to fail as a result of evidentiary problems, because an unsuccessful prosecution is likely to result in a significant loss of credibility with the regulated population (Hawkins 1984, 201), leading to reduced compliance in the future.

In some circumstances, it may be possible to prosecute under NPWA section 118A. The offence under section 118A is narrower than the offence under section 118D in terms of the species it protects. The section only covers situations where a threatened plant species or a plant, which is part of an endangered population or ecological community, is destroyed or injured. Habitat of threatened fauna would only be protected to the extent that it comprises listed flora. On the other hand, the provision is silent on the question of the extent to which the prosecution must prove the landholder's awareness of these facts. In these circumstances, in interpreting the legislation, the courts may decide that Parliament, rather than requiring proof of knowledge, intended that liability should be *strict*.[78] This has been the approach taken, for example, in relation to the offence of polluting waters under the *Protection of the Environment (Operations) Act* 1997.[79] This is also the position under Queensland's *Nature Conservation Act* 1992. It is an offence to destroy or injure ("take") a plant listed as presumed extinct, endangered, vulnerable or rare without a licence,[80] with a defence under the *Queensland Criminal Code* where a landholder acts under an honest and reasonable mistake of fact.[81]

---

[72] *Environment Protection and Biodiversity Conservation Act* 1999, s 523(2). See also s 522B.

[73] Whatever the position under the TSCA/NPWA, this would appear to amount to the "clearing" of native vegetation under the NVCA, unless it falls within an exemption. "Clearing" is defined expansively to include "substantially damaging or injuring native vegetation in any other way": NVCA s 5(1). Under the *Native Vegetation Act* 2003, when brought into operation, clearing for "routine agricultural management activities" will be exempt from regulation: s 22. However, these are defined narrowly to cover such things as pest control, firewood collection, commercial harvesting of native vegetation, and construction and maintenance of rural infrastructure: s 11. Under the new legislation, there is also an exemption allowing clearing for the continuation of existing cultivation, grazing or "rotational farming practices", essentially an existing use exemption: ss 9(3), 23. But this does not allow clearing of remnant native vegetation, only regrowth.

[74] *Raad* [1983] 3 NSWLR 344; *Hall* (1985) 81 Cr App R 260.

[75] *Bahri Kural* (1987) 162 CLR 502; *Alan Saad* (1987) 70 ALR 667.

[76] Compare the fault requirements for this offence with those for the offence of damaging *critical* habitat under section 118C. Section 118C(2) provides that if a map of the declared critical habitat has been published in the Government Gazette, it is not necessary for the prosecution to prove knowledge. If a map has not been published, then "the prosecution must prove that the person knew that the habitat was declared as critical habitat or that the person knew, or ought to have known, that it was critical habitat": s 118C(3).

[77] *Director-General of National Parks and Wildlife v Wilkinson* [2002] NSWLEC 171.

[78] See *VAW (Kurri Kurri) Pty Ltd v Scientific Committee* (Established under s 127 of the *Threatened Species Conservation Act* 1995) (2003) 128 LGERA 419 at 462, per Hodgson JA; *Hornsby Shire Council v Vitone Developments Pty Ltd* [2003] NSWLEC 272 at [114], per McClellan J.

[79] *Hunter Water Board v State Rail Authority of NSW* [No1] (1992) 75 LGRA 15; *Tiger Nominees Pty Ltd v SPCC* (1992) 75 LGRA 71.

[80] *Nature Conservation Act* 1992 (Q), s 89 and Schedule. It is also an offence to take a common plant: *Nature Conservation (Protected Plants) Conservation Plan*, s 7. However takings carried out by the landholder are exempt: *Nature Conservation (Protected Plants) Conservation Plan*, s 41(1)(b). See also s 97 in relation to the taking of non-listed plants in critical habitat and areas of major interest.

[81] *Criminal Code (Q)* s 24.

Where liability is strict, defendants must produce some evidence[82] that they were acting under a reasonable mistake of fact if they are to escape conviction. One possible line of argument would be that they did not realise that a particular listed species was present on the site,[83] and that this belief was a reasonable one. This involves them showing that they actively considered the issue of whether listed plants would be destroyed or injured, and, in addition, that their conclusion that this was not the case was a *reasonable* one.[84] In this scenario, the essential issue is what precautionary steps must be taken before any mistake that results is treated as being a reasonable one. It might be argued, for example, that a reasonable landholder would have carried out a species survey before carrying out the activity, and that therefore the accused ought to have done so.[85]

We might expect enforcement agencies to take a more selective approach to prosecution where liability is strict. This could involve an initial focus, not on prosecution but on educating landholders as to the more precise nature of the precautionary measures that it would be reasonable for them to take before clearing native vegetation. Subsequently, strategic prosecutions could be brought against those who did not make any effort to act reasonably.

Property rights, compensation and landholder duty of care

Compensation consists of backward-looking payments to landholders for lost expectations, based on any reduction in land value attributable to the land use restrictions imposed. Private landholders claim "property rights" in relation to land in an attempt to ward off command regulation of land use without compensation. In the current environment, government's right to regulate is usually, though reluctantly, conceded. The issue is principally about compensation.

It is crucial to distinguish compensation from incentives paid to landholders. Compensation is not designed to induce landholders to change their behaviour. Stewardship payments, on the other hand, acknowledge the difficulties caused for farmers by the imposition of restrictions on their land use, but they pay not for the restrictions themselves but for active conservation management of the land. Where restrictions on land use have resulted in loss of income, they provide farmers with an alternative source of income (Farrier 1995a; Farrier 1995b).

Section 51(xxxi) of the Commonwealth Constitution provides for compensation where there is an "acquisition of property" by the Commonwealth. Although the law in this area is increasingly confused, the preferable view is that it does not require compensation to be paid in situations where the Commonwealth simply regulates land use, even where the restrictions are very significant. The basic principle is that for section 51(xxxi) to come into play, there must be more than simply a taking (destruction of a right). The Commonwealth or others must obtain a benefit. Where the Commonwealth simply regulates the use of land, the argument is that it does not gain any proprietary benefit (acquisition) even if it is to be regarded as having taken property. However, the particular High Court decision in which this position was most clearly stated[86] involved Commonwealth regulation of land owned by a State, and the legal politics may well be different where regulation of private land use is involved. The *Environment Protection and Biodiversity Conservation Act* 1999 covers the possibility that its land use controls could be held to constitute an acquisition under the Australian Constitution by providing that, should this be the case, the Commonwealth must pay "a reasonable amount of compensation".[87]

The current understanding of the constitutional position in Australia can be contrasted with the position in the USA, where it has been held that the equivalent constitutional provision requires compensation to be paid in *some* situations where significant restrictions are imposed by government on land use. In practice, however, it has proved very difficult to define these situations clearly, apart from extreme circumstances when land use restrictions deprive the landholder of all economically beneficial or productive use of the land.[88]

The question of the scope of section 51(xxxi) of the Commonwealth Constitution is, however, largely an academic one. This is because most land use regulation stems from State Governments. It is State Governments rather than the Commonwealth that have historically allocated and managed land in Australia. Section 51(xxxi) does not apply to State legislation. Nor is there anything in the Constitutions of the States that guarantees compensation where development, or even current land use for that matter, is regulated in the public interest.

Where land use restrictions are imposed in the interests of species conservation, State practice differs in relation to the provision of compensation. There is no provision for compensation in NSW.[89] In Queensland, Victoria and

---

82 The burden of proof ultimately rests on the prosecution to prove beyond a reasonable doubt that the accused was not acting under a reasonable mistake of fact: *He Kaw Teh* (1985) 15 A Crim R 203.

83 The conservation status of a species that the defendant knows to be present would be a question of *law* rather than fact, and would not provide the basis for a defence based on a reasonable mistake of *fact*.

84 *State Rail Authority v Hunter District Water Board* (1992) 65 A Crim R 101.

85 See the comment by McClellan J in *Hornsby Shire Council v Vitone Developments Pty Ltd* [2003] NSWLEC 272 at [114].

86 *Commonwealth v Tasmania* (1983) 158 CLR 1. But see Sperling (1999).

87 *Environment Protection and Biodiversity Conservation Act* 1999, s 519.

88 *Lucas v South Carolina Coastal Council* 112 S. Ct. 2886 (1992).

89 In exceptional circumstances, the landholder can require government to purchase the land under the terms of the *Land Acquisition (Just Terms Compensation) Act* 1991, outlined below in note 123.

Tasmania, compensation is available where short-term orders are put in place.[90] In Western Australia, compensation is provided for a period of five years when consent to take "rare" flora is refused, but not rare fauna. After this, consent must be given unless the land is acquired.[91]

Another interesting variation is found in Queensland provisions that allow private land to be declared a nature refuge. Normally, nature refuges are created by a conservation agreement reached between Minister and landholder.[92] But where agreement cannot be reached on a nature refuge proposal, and the Minister considers that the area contains an area of major interest or critical habitat, a procedure can be instituted to make a compulsory declaration,[93] and to attach binding land use covenants.[94] Where compulsory procedures are used, compensation is payable for loss, *but only where this flows from restrictions imposed on the current use to which land is being put*.[95] In other words, it is not provided to the extent that the land's development potential is restricted. One possible rationale for this "half-way house" is that landholders are being offered an alternative to a compulsory declaration – a conservation agreement – in which they might be able to negotiate better terms in relation to land use restrictions, compensation or stewardship payments. By comparison, those whose land is subject to an interim conservation/protection order have no option but to submit to the restrictions on land use that it imposes.

The stance currently taken by some farmers' organisations is not that compensation should be paid for all restrictions imposed on land use. The demand is that it should be paid where land use is restricted *over and above a limited duty of care*. The duty of care reflects the restrictions on land use which farmers are prepared to comply with without compensation. Beyond this, they claim property rights relating to land use and development, for the withdrawal of which they expect compensation.

The NSW Farmers' Association has produced the following definition of the landholder duty of care in a recent *Discussion Paper* (Keogh 2002):

> *"The owner of land used for farming has a duty of care for the land.*
>
> 1. *Any action taken on the land should not result in harm to another person or the property of another person, provided always that such harm is foreseeable, able to be predicted on the basis of current knowledge, occurs within reasonable proximity, and it is reasonable for the owner to take action to avoid such harm.*
> 2. *The land should be managed in a manner that ensures the sustained productive health of the land, by taking all reasonable steps to:*
>
> - *avoid causing or contributing to land degradation*
> - *conserve soil*
> - *protect water resources*
> - *eradicate or to the best extent possible control pest animal and weed species*
>
> 3. *Action should be taken by owners of farm land to avoid causing noise or air pollution, but only to the extent that it is reasonable to do so given current agricultural practices"*

What this definition is trying to do is to differentiate between landholder activities that *deliver a community benefit*, for which compensation should be paid, and those that *cause harm to the community*. At first sight, land use restrictions designed to protect threatened species look as if they are designed to provide a public benefit. However, it could equally be argued that loss of biological diversity, with potentially important values – as a source of industrial chemicals and pharmaceuticals, for example – represents a public harm.

Difficulties in apportioning landholder and community responsibilities would arise under this definition where land use restrictions have multiple justifications. For example, a restriction on the clearing of native vegetation may be imposed to prevent land degradation (soil erosion, salinity), to conserve habitat of threatened species and to prevent the release of carbon into the atmosphere. In determining the amount of compensation payable, the contribution made by land use regulation to each of these objectives would have to be apportioned among them. In some situations (where, for example, clearing approval is refused because of a high risk of salinity, which would fall clearly within the landholder duty of care) little, if any, compensation would be payable even though an inevitable consequence might be that habitat is preserved at the same time.

It is possible to contemplate extreme cases where the lines between compensable and non-compensable restrictions are easier to draw, and where there is likely to be extensive agreement on the appropriateness of compensation. For example, there is a very persuasive argument that compensation should be paid where land use restrictions effectively close down a well-established use, as distinct from modifying it, or deprive the land of all economically viable use.

Apart from these equity considerations, there are also very powerful pragmatic reasons from an enforcement perspective for ensuring that landowners see the discovery of a listed species as a blessing rather than a curse. Otherwise, they may be tempted to "shoot, shovel and shut up". Yet this is not necessarily an argument for one-off compensation. Ongoing stewardship payments, as part of a management agreement, may provide a much better conservation outcome.

---

[90] *Nature Conservation Act* 1992 (Q), s 108; *Flora and Fauna Guarantee Act* 1988 (Vic), s 43 ; *Threatened Species Protection Act* 1995 (Tas), s 45.

[91] *Wildlife Conservation Act* 1950 (WA), s 23F(7)-(9).

[92] *Nature Conservation Act* 1992 (Q), s 45.

[93] Ibid s 49.

[94] Ibid s 51(3) and Schedule.

[95] Ibid s 67.

Tasmania's Forest Practices Code, 2000, under which forest practices plans required for forestry activities are assessed, imposes a duty of care on landholders. This goes beyond the NSW Farmers' Association version by not only protecting soil and water values but also providing for the "reservation of significant cultural values". However, this is to be achieved not by identifying the values of particular areas but by setting aside arbitrary percentages of land: up to 5% of land to be totally excluded from forestry operations, or 10% where the protection of significant cultural values is compatible with partial harvesting. Anything beyond this is "deemed to be for the community benefit" and is to be secured only on a voluntary basis or through the payment of compensation.[96] This is reflected in provisions in the *Nature Conservation Act* 2002, which allow land owners to apply for compensation if their forest practices plans are amended or refused in order to protect rare or endangered flora or fauna.[97] In return, the landowner must be prepared to enter into a conservation covenant, with terms fixed by an arbitrator where agreement cannot be reached with the Minister.[98] In determining the amount of compensation, the emphasis is on the lost value and potential of the land,[99] although arrangements can be made for additional management payments under the terms of the covenant.[100] If compensation is refused or not paid, certification of a forest practices plan cannot be denied on the ground that it fails to address adequately threatened species issues previously identified, nor can it be amended.[101] The whole scheme and the desired conservation outcomes can be defeated, therefore, if budgets do not make adequate provision for compensation.

Delivering command regulation through strategic plans

The debate is no longer about whether command regulation should be used to protect threatened species habitat on private land. It is clear that, in the short term at least, adequate provision will not be made for biodiversity conservation in the open market place, and that arrangements made on an entirely voluntary basis between government and individual landholders (conservation agreements, land acquisition) will only operate at the margins. Currently, the issue is about the extent to which the blow of command regulation should be softened by monetary payments, and the precise form that these should take.

Another matter that requires urgent attention, however, is the way in which command regulation is delivered. We have seen that in NSW, under the TSCA and EPAA, it is currently triggered directly by listing decisions based on purely scientific criteria, although the impact of this on proponents is substantially mitigated by the operation of approvals processes that are weighted towards facilitating proposals. One of the major problems is that regulation is not precisely targeted at those areas where it is clear that development will have an impact on threatened species habitat. It applies wherever habitat of threatened species, endangered ecological communities or endangered populations occurs, but responsibility for determining location is not allocated: there is nothing in the legislation requiring habitat to be identified upon listing, only a provision for *discretionary* identification of *critical* habitat, which in practice is rarely used.[102] At the same time, the criminal offence provisions do not clearly allocate responsibility to individual landholders. The messages are mixed. As we have seen, the offence under NPWA section 118D, with its requirement for proof of knowledge, assumes that there is some mechanism for conveying to particular landholders that their land does indeed comprise protected habitat. The offence under section 118A, on the other hand is silent on the question of whether actual knowledge must be proved or whether proof of negligence will suffice.

The problems are particularly acute where ecological communities are involved. In *VAW (Kurri Kurri) Pty Ltd v Scientific Committee* (Established under s 127 of the Threatened Species Conservation Act 1995),[103] a decision concerned with the validity of a listing determination, Spigelman CJ made it clear that, while the listing had to be "sufficiently flexible so as to enable the protection of communities through periods of seasonal and climatic variation", ecological communities had to be defined by the Committee with "reasonable certainty", because listing triggers offence provisions.

---

96 *Forest Practices Code* 2000, Section D, p 52.

97 *Nature Conservation Act* 2002 (Tas), ss 33, 40.

98 Ibid s 40.

99 Ibid s 42.

100 Ibid s 34(3)(a).

101 Ibid s 44(8).

102 Once critical habitat has been designated, activities on the area can be controlled by regulation made under the legislation (TSCA s 51). However, only two critical habitat designations have been made: the habitat of the Mitchell's rainforest snail *Thersites mitchella* in Stotts Island Nature Reserve and the habitat of the little penguin *Eudyptula minor* population in Sydney's North Harbour. The latter explicitly excludes formed backyards and residential areas but does include some private land that is not in use because it is steep and rocky. Regulations have been made in relation to the critical habitat of the little penguin (*Threatened Species Conservation Regulation* 2002). Companion animals are prohibited and there are prohibitions on interfering with burrows and approaching within 5 metres of a penguin on land. A recommendation has been made that the critical habitat of the Bomaderry zieria *Zieria baeurlenii*, be designated within the Bomaderry Bushland. The area in question is public land. See http://www.nationalparks.nsw.gov.au/npws.nsf/Content/Critical+habitat+protection+by+doctype.

103 (2003) 128 LGERA 419.

This meant that a listing determination has to "enable a citizen to decide whether a specific location falls within it". But this did not require maps identifying where the community was to be found, or even the identification of the minimum number of species required to occur together for the community to exist.[104] The difficulties which this creates were recently explored in *Hornsby Shire Council v Vitone Developments Pty Ltd*.[105] The issue was whether a listed ecological community was so badly degraded that it no longer existed. The Court was faced with the conflicting views of two experts, one of whom argued that account should be taken of the prospect of "assisted natural regeneration" in deciding whether the community was still present.[106] McClellan J accepted that the community had been present in the past. He indicated that, in his view, the listing determination had failed to clearly address the issue of restoration, other than by making it clear that not all of the species identified as characterising the community needed to be present at any one time: this would depend on the disturbance history of the site. He concluded that the extent to which active regeneration could be taken into account would depend on the circumstances of each case, but that here, the community no longer existed.[107]

While the issue in the *Vitone* decision was whether a species impact statement should have been prepared and considered because of the likely effect on a listed community, McClellan J specifically drew attention to the problems created for landholders who have to comply with offence provisions. One solution he offered was to rely on what was "the fundamental mechanism of the planning process": " a catalogue containing the parcels of land where the relevant communities have been identified". There are significant tensions here between the flexibility demanded by ecology in identifying ecological communities, and the criminal law's insistence on a degree of certainty. If this cannot be resolved, they may have to part company.

Another problem with the way in which command regulation to protect threatened species habitat is currently delivered in NSW is that it fails to set development parameters well ahead of specific proposals. It defers difficult decisions to the point when a head of steam has built up, with landholders committed to particular projects. The emphasis at this point is on mitigating environmental impact on a particular site rather than addressing it at a landscape level. The issue is how a detailed proposal can be made appropriate for a site, not what the site is appropriate for. As we have seen, this is accentuated where land has been zoned to allow projects of the general kind proposed to go ahead, albeit with consent. The existing delivery system also defers scientific input, in the shape of the "eight part test" (to determine whether a proposal is likely to significantly affect threatened species, populations or ecological communities, or their habitats) and species impact statements (Farrier *et al.* 2002, Whelan *et al.* 2004), until it is too late to carry out the research, let alone for the results to have a substantial bearing on the decision. Any research that is carried out must be paid for by the developer, and will inevitably focus on the specific site. It cannot adequately inform the issue of cumulative impact.

If we are to get serious about conserving threatened species and their habitats, we need to start making relatively firm commitments in strategic plans rather than deferring the issue to the exercise of broad discretions at a stage when a particular proposal is on the table and the focus is on this rather than the conservation significance of the site in the broader landscape. Significant habitat needs to be identified and protected in planning instruments, upon or soon after listing, rather than being left to the vagaries of eight-part tests and species impact statements, and last minute decisions. Decisions made well in advance in strategic plans will be more palatable because they hose down aspirations before they become expectations. Those who do not take heed will find it difficult to deny culpability based on lack of awareness, opening up the way for implementation of a vigorous enforcement policy.

The land use planning system, set up in New South Wales under Part 3 of the *Environmental Planning and Assessment Act* 1979 (EPAA), provides a mechanism for targeted protection of threatened species habitat (Saxon 1997, 93-94; Douglas 1999, 143-144). This legislation has very broad objectives. One of them is to encourage "the protection of the environment, including the protection and conservation of native plants and animals, including threatened species, populations, ecological communities and their habitats".[108] Environmental planning instruments – local environmental plans, regional environmental plans and State environmental planning policies – can be made for these purposes.[109] They can be used as a delivery mechanism for command regulation of development proposals that impact adversely on threatened species habitat.

Commitments to conservation made in environmental planning instruments need to be firm ones. Provisions that simply defer decisions to the exercise of broad discretions by decision-makers in relation to specific project proposals would simply reflect the status quo. Where there is a risk of habitat being irreversibly destroyed, discretion at the stage of project control must be firmly constrained, or even removed by prohibiting development altogether. There are precedents for this in other States in a different but

---

[104] Ibid at 423, per Spigelman CJ. See also Beazley JA at 453; Hodgson JA at 459-462.

[105] [2003] NSWLEC 272.

[106] Ibid at [76].

[107] Ibid at [91]-[99]; [114].

[108] EPAA s 5(a)(vi).

[109] EPAA ss 24, 26(1)(e1).

related context: command regulation of native vegetation clearance.[110] The Queensland legislation, for example,[111] severely curtails the exercise of discretion by decision-makers where a development permit is required,[112] in particular areas identified by vegetation mapping. The *Vegetation Management Act* 1999 provides that, until Regional Vegetation Management Plans are approved, clearing applications relating to freehold land are to be assessed in accordance with the *State Policy for Vegetation Management on Freehold Land*,[113] which incorporates the Code for the clearing of vegetation.[114] So far as biodiversity conservation is concerned, the *State Policy* provides[115] that the specified purposes of the Code are to be achieved "by not clearing" in remnant *endangered* regional ecosystems and declared areas of high nature conservation value[116] except in very limited circumstances (where clearing is necessary to protect the area from a threatening process, or essential for establishing infrastructure, such as a fence or road, and no suitable alternative exists).[117]

To ask for firm commitments in strategic land use plans is not to deny the inevitability of adjustment in line with changing community sentiment. The provisions of strategic plans are by no means immutable. National park reservations in NSW can only be revoked by an Act of Parliament.[118] Plans, on the other hand, can be amended following a statutorily defined process of public exhibition and comment.[119] What we have witnessed in NSW in recent years, however, has been a growing erosion of the commitment to implement plans insofar as they impose constraints on development, with the gradual introduction of mechanisms into both the legislation and planning instruments to counteract the alleged "inflexibility" of the strategic planning system (Farrier *et al.* 1999, 91-93). Chief among these mechanisms has been the replacement of outright prohibitions on development with discretionary approvals processes. In theory, discretions are constrained insofar as they must be exercised in line with detailed planning objectives. In practice, however, these constraints have been largely immobilised by the decision of the Land and Environment Court in *Schaffer Corporation Ltd v Hawkesbury City Council*. Here it was held that it is not necessary to show that development promotes or is ancillary to planning objectives, or even that it is compatible, provided that it is not "antipathetic".[120]

---

[110] In NSW, the *Native Vegetation Act* 2003, which has not yet commenced operation, provides that consent cannot be given for clearance of remnant native vegetation or protected regrowth outside of urban areas (s 5(1)(c) and Schedule 1, Part 3) unless the clearing will "improve or maintain environmental outcomes": ss 8, 14(3), 29. The extent to which this constrains the discretion of the Minister will depend on the precise meaning of "improve or maintain". This is not defined in the legislation, but can be defined in regulations made under the Act: s 15(1)(c). Note that the *clearing* must improve or maintain environmental outcomes, not simply compensate for environmental impacts. This would not appear to authorise an offsets strategy, whereby compensation for environmental harm caused by clearing in one area is provided by restoration/rehabilitation of another area.

[111] See also the *Native Vegetation Act* 1991 (SA). This Act provides that in deciding whether to give consent to an application to clear native vegetation, the Native Vegetation Council must not make a decision which is "seriously at variance with" the principles of clearance of native vegetation: s 29(1). These are set out in Schedule 1 of the Act, delineating a number of instances when native vegetation *should not be cleared.* They include situations where native vegetation has significance as habitat for any type of wildlife, whether or not threatened, comprises a high level of diversity of plant species, includes rare, vulnerable or endangered plants, constitutes a rare, vulnerable or endangered plant community or is significant as a remnant of vegetation in an area which has been extensively cleared. The legislation effectively creates a presumption against clearing in these situations, and takes the further step of forbidding a decision that deviates significantly from this presumption. Even though the Council is required to have regard to the applicant's desire to operate the business as efficiently as possible, this is only to operate as a factor within these overall limits (s 29(3)), representing a significant downgrading of traditional economic considerations in the decision-making process. As a result, the legislation significantly *constrains* the discretion exercised by the Native Vegetation Council. These constraints have been tightened even further by recent amendments which provide that the Council cannot give consent to clearance if the vegetation comprises a stratum which is substantially intact, except to the extent that "harvesting" it will not result in lasting damage, lead to significant soil damage or erosion, or result in any long term loss of biodiversity: s 27(2), (3).

[112] Some activities involving the clearing of native vegetation are exempt from any requirement to obtain development consent. These are clearing to build a single residence; clearing for essential management (safety, infrastructure, garden maintenance) and forest practices (*Integrated Planning Act* 1997 (Q), Schedule 8, cll 3A(a),(b); 13(c)). "Forest practices" are defined in Schedule 10 to cover forest management in plantations, and also in native forests under a native forest management code, provided that the existing forest type is regenerated, trees are not taken for export woodchip, and there is no land degradation. The term does not include clearing for the initial establishment of a plantation.

[113] *State Policy for Vegetation Management on Freehold Land* (November 2002).

[114] *Vegetation Management Act* 1999 (Q), s 20(2); *State Policy for Vegetation Management on Freehold Land*, cl 8.2.2.

[115] *State Policy for Vegetation Management on Freehold Land*, Appendix 2.

[116] *Vegetation Management Act* 1999 (Q), Part 2, Division 4. Areas of high nature conservation value can incorporate an area of regrowth vegetation where it will enhance an endangered regional ecosystem: s 19(1)(d).

[117] Proposals are being explored, at the time of writing, to extend this strong presumption against land clearing to remnant "of concern" regional ecosystems on freehold land, in conjunction with an incentives/compensation package.

[118] NPWA s 37(1)(a).

[119] EPAA Part 3.

[120] (1992) 77 LGRA 21. This decision has most recently been applied in *New Century Developments Pty Limited v Baulkham Hills Shire Council* [2003] NSWLEC 154. See, however, *Dem Gillespies v Warringah Council* [2002] NSWLEC 224, where Bignold J departed from this line of authority and held that "consistent" meant "compatible" in the context of the local environmental plan under consideration in that case.

If plans simply defer decisions to unconstrained discretions exercised at the stage of project control, or they can be changed to accommodate market imperatives without careful deliberation, they cease to be strategic plans. Significant commitments to conservation made in plans should only be reversed after careful consideration, particularly where the results of reversing them – for example, to allow a residential subdivision to go ahead on threatened species habitat – are themselves irreversible. The principles of ecologically sustainable development require us to confirm our allegiance to a process that emphasises cautious decision-making in the face of irreversible development.

The irony is that, at the present time, "commitments" in local environmental plans to *allow development* are being kept at the expense of biodiversity conservation. In the past, in many parts of NSW, areas were rezoned to allow development, particularly residential subdivisions, at a time when requirements for prior environmental assessment were quite inadequate. As a result, the precise configuration of environmental, residential/tourism, recreation and rural residential zones are unlikely to reflect environmental constraints, such as significant vegetation and habitat of threatened species. The law offers no guarantee to landholders that land which has been zoned in the past to allow development will not at a later point be rezoned to restrict or prohibit it ("down-zoning" or "back zoning").[121] Landholders only have a legal right to continue a lawful existing use, not to develop.[122] The only, very limited, concession made is to allow landholders to force purchase by government agencies in limited circumstances where their land has been effectively set aside for public purposes.[123]

Regardless of the law's attitude to down-zoning, there is considerable reluctance to use it in practice. There may well be confusion over the formal legal position, with landholders assuming that a land use allowed under an existing zoning, even one which requires development consent, falls within their "property rights", and local councils fearing actions for negligence.[124] Yet in the light of current levels of scientific uncertainty about the habitat requirements of threatened species,[125] there are compelling arguments for not giving long term guarantees to landholders that development will be allowed to go ahead, or that they will be compensated if it is not. Any suggestion that decisions that zone land for development should be immutable is fundamentally at odds with arguments for adaptive management in the face of scientific uncertainty.

***Intra*generational equity**

The down-zoning scenario discussed above clearly raises equity issues about how the cost of biodiversity conservation should be spread among existing generations. Where, however, ownership of land has not changed hands following the initial rezoning that allows development, and the landholder has not invested in any steps to develop the land, then later down-zoning simply defeats expectations, removing what has been no more than an increase in the paper value of the land. Where the land has changed hands and the increased value attributable to the rezoning has been factored into the purchase price, equity considerations are more compelling. However, the reality is that in many cases there is a strong element of speculation in these dealings, and the purchaser has no intention to develop the land in the short-term. With the vendor's cooperation, purchasers can always protect their position by making a development application *before* they purchase land, but the problem from the developer's point of view is that any consent obtained would lapse within a set period of time.[126]

---

[121] Some restrictions on proposals to down-zone land in draft local environmental plans have been imposed on local councils by the Minister in directions made under EPAA s 117(2). For example, draft plans "shall retain existing provisions enabling a dwelling house to be erected on an existing allotment" (Direction G9(ii)) and "shall not substantially reduce existing zonings of land for industrial development" (Direction G11). However, these are designed to provide for some level of stability in planning over time rather than to protect the interests of private landholders. Moreover, the legislative scheme ultimately allows a final plan to be made that does not comply with the Directions. Besides, in *Smith v Wyong Council* [2003] NSWCA 322, the NSW Court of Appeal held that failure by a council to comply with a section 117 Direction when preparing a draft plan did not make it invalid.

[122] EPAA Part 4, Division 10.

[123] Under the *Land Acquisition (Just Terms Compensation) Act* 1991, ss 21-24, government agencies can be required to purchase land "reserved ... exclusively" for specified public purposes in environmental planning instruments made under the *Environmental Planning and Assessment Act* 1979 (EPAA ss 26(c), 27), where owners will suffer hardship if there is any delay. Land is exclusively reserved not only where it is set aside for use for a particular purpose, but also where it is "expressly set apart ... for use for such a purpose and also for other purposes, but those other purposes do not constitute a reasonable use of the land". Relevant public purposes include "open space", so it may be possible to argue that a down-zoning which effectively meant that land was being set aside for the conservation of threatened species habitat, triggers these provisions. But this would be an extreme case.

[124] The argument of a thwarted developer would be that the council's initial decision to zone the land for development led to the council being liable in tort for negligence for the loss incurred by the developer because it breached its duty of care by failing to carry out the environmental assessment that a reasonable council would have carried out at that stage. However, in *Alex Finlayson Pty Ltd v Armidale City Council* (1994) 84 LGERA 225, 252, Burchett J of the Federal Court held that a duty of care did not arise in relation to a zoning decision because this was a policy rather than an operational matter.

[125] The intractable problem is that the fewer individuals left, the more difficult it is to carry out research on their habitat requirements. This assumes, of course that a close focus on conserving the habitat of threatened species is the most effective way of conserving biodiversity, a crucial issue which is beyond the scope of the present paper.

[126] EPAA s 95.

This does not, however, address the fundamental equity issue raised in a strategic planning scenario where, depending on the conservation significance of particular areas of land, some landholders may be substantially prevented from developing, while others are allowed to go ahead. It is at this point that the issue of compensation is presented most starkly. From one perspective, it could be argued that landholders must take their land as they find it: constraints are imposed on development not only by geological features, such as landslip, but also by biodiversity conservation imperatives. Yet the latter stem from broad community valuations imposed on landholders rather than what should be the self-interest involved in not erecting a building that might fall down. Apart from this, the reality is that disgruntled landholders make poor conservationists. This might not matter too much if conservation required no more than maintenance of the *status quo*. The reality is that it involves active management, and the person in the best position to carry this out is the landholder. Farrier (1995a; 1995b) has argued that, in these circumstances, stewardship payments by government for active conservation management by landholders make far more sense than paying compensation for loss of land value caused by command regulation.

Activities on land that are established and ongoing when new restrictions are brought into operation present particular difficulties. Provisions in the EPAA prevent the use in planning instruments of command regulation to remove or control "nonconforming" or "existing" uses, provided that they are not expanded or intensified.[127] Depending on the circumstances, this could include activities impacting adversely on the habitat of threatened species, such as the removal of dead wood and bush rock, as well as periodic logging.

Provisions which protect existing operations from the demands of new regulatory initiatives are justified by powerful equity arguments, but these are considerably weakened where the aim is not to shut down an activity altogether but to require modifications to be made, in the interests of threatened species conservation for example. The Victorian planning legislation specifically provides for this,[128] and the NSW legislation should be amended to similar effect.

Where threatened species considerations require more fundamental constraints to be imposed on nonconforming uses, including abandoning them altogether, we need to seriously consider paying compensation or to look to alternatives to command regulation. One approach is for government to purchase the land, another for government to make management payments for stewardship or the provision of environmental services, yet another, to require someone who has been allowed to develop land elsewhere to make payments to those who are to abandon their nonconforming use, as part of an offset scheme.

There is no reason why these alternative policy instruments should not be delivered through strategic land use plans. While it is true that environmental planning instruments in NSW have traditionally been associated with command regulation, this stems from planning tradition rather than legislative constraints. The EPAA specifically provides that environmental planning instruments are not restricted to controlling activities[129] but can, more generally, make provision for "protecting and conserving native animals and plants, including threatened species, populations and ecological communities, and their habitats".[130]

The role of the Department of Environment and Conservation

One crucial issue relates to the role, within the strategic planning process, of the NSW Department of Environment and Conservation (DEC). Currently, in NSW, before preparing draft environmental planning instruments, or the environmental studies which sometimes precede them,[131] councils or the Department of Infrastructure, Planning and Natural Resources must consult with the DEC if of the opinion that threatened species, populations or ecological communities, or their habitats could be affected by the plan.[132] This admits the DEC to a consultation process, in which its views must be heard, but not necessarily heeded.

One way in which the DEC can take the initiative under current legal arrangements is to identify the critical habitat of a listed threatened species, population or ecological community.[133] This must then be reflected in the relevant environmental planning instrument.[134] A species impact statement will have to be prepared by those seeking an approval,[135] which means that, in theory, decision-makers will be better informed. In addition, the concurrence of the DEC is required before approval can be given.[136] However, critical habitat designations are seen to be exceptional rather than a crucial component of the ongoing process of strategic land use planning in the course

[127] Ibid ss 106-109.

[128] *Planning and Environment Act* 1987 (Vic), s 6(4A).

[129] EPAA, s 26(1)(b), (f).

[130] Ibid s 26(1)(e1).

[131] Ibid ss 57, 74(2).

[132] Ibid ss 34A, 5C.

[133] TSCA Part 3.

[134] EPAA s 26(2).

[135] Ibid s 78A(8)(b).

[136] Ibid s 79B(3).

of which decisions about appropriate land use are made and reviewed over time. Critical habitat is designated by the Minister for the Environment. It requires proof that the habitat in question is "critical to the survival of the species, population or ecological community".[137] This is hardly precautionary. In practice, only two declarations of critical habitat have been made in NSW at the time of writing, primarily over public land.[138]

In other situations, where critical habitat has not been declared, limited constraints are imposed on the grant of approvals for activities involving habitat destruction. In some situations, where a Minister makes the final decision, for example clearing control under the *Native Vegetation Conservation Act* 1997, the DEC has no power of veto but simply has to be consulted, and then only in situations where there is likely to be a significant effect on threatened species, populations or ecological communities.[139] Where a Minister is not the final decision-maker, the DEC acquires a power of veto through the concurrence process.[140] But we have already seen that this decision involves balancing species conservation against social and economic factors. At this late point in the process, when the whole focus is on a specific proposal by a particular landholder for a particular site, there is little room to manoeuvre when it comes to addressing issues of cumulative impact, trade-offs between socio-economic factors and species conservation, and equity between different landholders.

One possible policy position which might be adopted is that the Minister for the Environment, advised by the DEC, should have a power of veto, not over specific project proposals through the concurrence process, but over provisions in draft environmental planning instruments which are likely to have a significant impact on threatened species habitat. There is a precedent for this. Under the *Water Management Act* 2000, the concurrence of the Minister for the Environment is needed before the Minister for Natural Resources can sign off on a water management plan.[141]

Beyond this, there is a persuasive argument that where a recovery plan has identified specific habitat requirements of threatened species, the Minister for the Environment should be able to take the initiative and put forward proposals for habitat protection in a draft environmental planning instrument, for exhibition and debate.

Links with recovery planning

At present, recovery plans are the poor relations in the broader system of environmental and natural resources planning which exists in NSW under the *Environmental Planning and Assessment Act* 1979, the *Native Vegetation Conservation Act* 1997, the *Water Management Act* 2000 and the *Catchment Management Act* 1989 (Farrier 2002). It is not simply that recovery plans do not bind landholders in the same way that environmental planning instruments, regional vegetation management plans and water management plans do. There is not even a requirement that the provisions of other natural resource management and land use plans should be consistent with the provisions of recovery plans.

On this latter issue, the NSW legislation appears to obfuscate deliberately. At one point it states categorically that Ministers, local councils and public authorities "must not make decisions that are inconsistent with the provisions of a recovery plan".[142] It then goes on, however, to negate this by providing that, where discretion is being exercised, as where a local council is determining the contents of a draft local environmental plan, or the Minister is making a determination on the contents of the final plan, recovery plans must simply be taken into account.[143] In other words, while the measures proposed in recovery plans must be considered, they do not have to be followed.

There is not even a guarantee that a species impact statement will have to be prepared where a proposed activity is inconsistent with a recovery plan. Amendments in 2002 to the so-called "eight part test" only identify consistency with a recovery plan as one of a number of factors to be considered in determining whether there is likely to be a significant effect on threatened species, populations or ecological communities, or their habitats. Even this limited provision has not yet commenced operation.[144]

Recovery plans are given much higher status under the Commonwealth's *Environment Protection and Biodiversity Conservation Act* 1999 (EPBC). Where a plan has been made or adopted under the EPBC, it is binding, but only on Commonwealth agencies and in Commonwealth areas. The Commonwealth must implement recovery plans to the extent to which they apply in Commonwealth areas,[145] and Commonwealth agencies must not take any action that contravenes a plan.[146] On the other hand, when it comes to implementing Commonwealth recovery plans covering land under State jurisdiction, including private land, the Commonwealth's only resort is to seek the *cooperation* of the States.[147]

137 TSCA ss 37(1), 47.

138 See note 102.

139 EPAA s 79B(3).

140 Ibid.

141 WMA ss 41(2).

142 TSCA s 69(1).

143 Ibid s 69(2).

144 *Threatened Species Conservation Amendment Act* 2002, Schedules 1[53] and 2.1[4].

145 EPBC s 269(1).

146 Ibid s 268.

147 Ibid s 269(2).

Where the approval of the Commonwealth Minister for the Environment is required, because a proposed activity is likely to have a significant impact on Commonwealth listed species or ecological communities, Commonwealth recovery plans can bite hard. The EPBC provides that in deciding whether or not to give approval, and if so under what conditions, the Minister "must not act inconsistently" with a recovery or threat abatement plan.[148] Where this leads to an adverse Commonwealth decision, it will override any decisions by State agencies and local councils in favour of allowing a proposal to go ahead.

If, for example, a Commonwealth recovery plan did go as far as to identify areas of habitat that need to be protected, it would have a crucial bearing on the Commonwealth Minister's decision on whether to approve a proposal. But unless landholders make a practice of checking the contents of individual Commonwealth recovery plans, this constraint will only became apparent when a specific project is proposed, and then only if the Commonwealth Minister makes the preliminary decision that the proposal requires Commonwealth approval because it is likely to have a significant impact on listed threatened species or ecological communities.[149] There is no requirement to link Commonwealth recovery planning with strategic land use planning at a State level.

Conclusion

NSW provisions relating to the protection of threatened species habitat betray their roots in a system originally designed to regulate hunting. At a symbolic level, heavy reliance appears to be placed on command regulation triggered directly by threatened species listing, and backed up by the threat of prosecution. In practice, however, this is significantly moderated by the operation of an *ad hoc* approvals process dominated by socio-economic considerations, with conservation outcomes a hoped-for side-effect of ameliorating provisions rather than a primary objective. Decision-making processes are driven by individual development proposals rather than strategic thought, and assessment of cumulative impact is a significant casualty. The government agency with expertise in threatened species conservation and scientific research is marginalised.

An alternative approach involves re-orientating the decision-making process away from project control towards strategic planning. Rather than consistently deferring decisions to a time when particular development proposals are put forward for consideration, the initiative should be taken at a much earlier stage. At this point, commitments can be made to the protection of threatened species habitat by heavily constraining the discretion of those responsible for deciding on approvals, or putting in place absolute prohibitions. The orientating principle should not be that we can, at one and the same time, keep our environmental cake as well as consuming it through development, but that destruction of threatened species habitat is largely irreversible and should only be contemplated, and then carefully, at a strategic level. While these commitments should be subject to ongoing review, adjustments should only be through a deliberate process of plan amendment in which community participation in decision-making is central. The DEC can play a much more effective role at this level than it currently plays through the exercise of its limited concurrence powers within the project control process.

References

Ayres, I. and Braithwaite, J. 1992. *Responsive Regulation: Transcending the Deregulation Debate*. Oxford University Press, New York.

Bartel, R. 2003. Compliance and complicity: an assessment of the success of land clearance legislation in New South Wales. *Environmental and Planning Law Journal* **20**: 116 – 141.

Braithwaite, J. 2002. *Restorative Justice and Responsive Regulation*. Oxford University Press, New York.

Douglas, S. 1999. Local Government and the Threatened Species Conservation Act – The Greatest Potential; The Weakest Link. *Australasian Journal of Natural Resources Law and Policy* **6**: 135-148.

Environment Protection Authority (NSW). 1997. *New South Wales State of the Environment 1997*. EPA, Sydney, NSW.

Environment Protection Authority (NSW). 2003. *New South Wales State of the Environment 2003*. EPA, Sydney, NSW.

Farrier, D. 1995a. Policy instruments for conserving biodiversity on private land. Pp. 337-359 in *Conserving Biodiversity: Threats and Solutions*, edited by R. A. Bradstock, T. D. Auld, D. A. Keith, R. T. Kingsford, D. Lunney and D. P. Sivertsen. Surrey Beatty and Sons, Chipping Norton, NSW.

Farrier, D. 1995b. Conserving biodiversity on private land: incentives for management or compensation for lost expectations. *Harvard Environmental Law Review* **19**:303-408.

Farrier, D., Lyster, R., Pearson L. and Lipman Z. 1999. *The Environmental Law Handbook: Planning and Land Use in New South Wales*. 3rd ed., Redfern Legal Centre Publishing, Sydney, NSW.

Farrier, D. 2002. Fragmented law in fragmented landscapes: the slow evolution of integrated natural resource management legislation in NSW. *Environmental and Planning Law Journal* **19**: 89-108.

Farrier, D., Whelan, R. and Brown, C. 2002. Addressing scientific uncertainty in local government decision-making processes. *Environmental and Planning Law Journal* **19**: 429-444.

Grabosky, P. and Braithwaite, J. 1986. *Of Manners Gentle: Enforcement Strategies of Australian Business Regulatory Agencies*. Oxford University Press, Melbourne, Victoria.

Hawkins, K. 1984. *Environment and Enforcement*. Clarendon Press, Oxford, UK.

Jarman, P.J. and Brock, M.A. 2004. The evolving intent, coverage and use of legislation to protect biodiversity in New South Wales. Pp 1-19 in *Threatened Species Legislation – Is It Just An Act?*, edited by P. Hutchings, D. Lunney and C. Dickman. Royal Zoological Society of NSW, Mosman, NSW.

148 Ibid s 139.

149 Ibid Part 7.

Kelly, A. and Farrier, D. 1996. Local government and biodiversity conservation in New South Wales *Environmental and Planning Law Journal* **13**: 374 –389.

Kelly, A. and Prest, J. 2000. Implementation of threatened species law by local government in New South Wales. *Environmental and Planning Law Journal* **17**: 584-601.

Keogh, M. 2002. *Property rights and farming in Australia: Defining a Duty of Care for Farm Land.* NSW Farmers, Sydney, NSW, June 2002.

McDonald, L., Bradshaw, S. D. and Gardner, A. 2003. Legal protection of fauna habitat in Western Australia. *Environmental and Planning Law Journal* **20**: 95-115.

Moorrees, A. 2003. Manager Threatened Species and Communities, Department of Sustainability and Environment (Vic). Pers. comm. 25 February 2003.

National Parks and Wildlife Service (NSW). 1999. *NSW Biodiversity Strategy*. NPWS, Hurstville, NSW.

Rawling, J. 2004. Case studies and lessons learnt: translocations conducted as amelioration for development. Australian Network for Plant Conservation Translocation Workshop, 4 June 2004, Mt Annan Botanic Garden, Sydney, NSW.

Saxon, M. 1997. "The Threatened Species Conservation Act 1995 – Is the Development Approval Process Working?" Pp 85-95, in *On the Brink: Your Bush, Their Habitat, Our Act: Is the Threatened Species Act Working?* edited by H. Webb. Nature Conservation Council, Sydney, NSW.

Sperling, K. 1999. If caution really mattered. *Environmental and Planning Law Journal* **16**: 425-440.

Stubbs, B. J. 2001. From 'useless brutes' to national treasures: A century of evolving attitudes towards native fauna in New South Wales, 1860s to 1960s. *Environment and History* **7**: 23-56.

Walker, R. B. 1991. Fauna and flora protection in New South Wales, 1866-1948. *Journal of Australian Studies* **28**: 17-28.

Whelan R., Brown C. and Farrier D. 2004. The Precautionary Principle: What is it and how Might it be Applied in Threatened Species Conservation? Pp 49-58 in *Threatened Species Legislation – Is It Just An Act?*, edited by P. Hutchings, D. Lunney and C. Dickman. Royal Zoological Society of NSW, Mosman, NSW.

The precautionary principle: what is it and how might it be applied in threatened species conservation?

R.J. Whelan[1], C.L. Brown and D. Farrier

Institute for Conservation Biology and Law, University of Wollongong

Wollongong, NSW, 2522, Australia

[1] Email: rob@uow.edu.au

ABSTRACT

The Precautionary Principle is incorporated in law and policy via Ecologically Sustainable Development provisions but it is widely misunderstood and frequently ignored. Threatened species typically satisfy key elements of the Precautionary Principle in that serious or irreversible damage would be caused by inappropriate development or management, and scientific uncertainty usually surrounds an understanding of their ecology and prediction of impacts. Where could the New South Wales *Threatened Species Conservation Act* 1995 be precautionary? The listing process could be, but a significant amount of information is needed before a population, species or community can be listed, so nominations of taxa for which little information is available are likely to be unsuccessful. A precautionary approach to listing would be to use the IUCN classification "Data Deficient" as a category, especially if was designed to trigger a process of ecological study to gather relevant data, and a Species Impact Statement in cases where a data deficient taxon was likely to be affected by a specific development proposal. The assessment of likelihood that significant impacts will result from a development could be precautionary if determining authorities assumed a significant impact wherever scientific uncertainty existed. Currently, the application of precaution here is variable. The 'Eight-part-test' is effectively a checklist that is applied to determine whether there is 'likely to be a significant impact', thus triggering a Species Impact Statement, under the NSW threatened species legislation. Eight-part-tests that conclude no significant impact, perhaps with mitigation works, are often accepted despite questions about their validity. A conclusion that there is likely to be a significant impact will not necessarily halt a development proposal, because compromises with socio-economic benefits can be made. We urge a more thorough application of the Precautionary Principle, to force decision-makers to acknowledge this compromise, rather than hiding behind the pretence that lack of knowledge implies no detrimental impact.

Key words: scientific uncertainty, threatened species, biodiversity, conservation, policy

The precautionary principle – definition and misuse

Much has been written of the Precautionary Principle in relation to environmental issues (e.g. Deville and Harding 1997; Harding and Fisher 1999; Calver *et al.* 1999 a, b; O'Riordan *et al.* 2001; Harremoes *et al.* 2002), and it is a topic of debate and discussion (e.g. UPEM 2004). This principle first arose as the "Foresight Principle" in relation to environmental policy in Germany in the 1960s and it was used, according to Stein (2000), as a way of assessing political decisions about the environment. The Precautionary Principle was given international status at the time of the Rio Declaration on Environment and Development, in 1992, as one of the "Guiding Principles" for achieving ecologically sustainable development (ESD). Principle 15 is worded as follows:

> "*In order to protect the environment, the precautionary approach shall be widely applied by States according to their capabilities. Where there are threats of serious or irreversible damage, lack of full scientific certainty shall not be used as a reason for postponing cost-effective measures to prevent environmental degradation.*"

While the Rio Declaration is not binding on Australia under international law, Principle 15 has been adopted as part of the commitment to ESD. Under the *National Strategy for Ecologically Sustainable Development* (1992)[1], the Council of Australian Governments accepted the precautionary principle as one of the "guiding principles" of ESD, and the Intergovernmental Agreement on the Environment (1992)[2], signed by the Australian Government, States, Territories and Local Government, makes specific reference to the principle (Spry 1997, Nagorcka 2003). In 1999, the Productivity Commission reviewed the extent of adoption of ESD in Australian

[1] <http://www.deh.gov.au/esd/national/nsesd/index.html> Accessed May 28th 2004.

[2] <http://www.deh.gov.au/esd/national/nsesd/index.html> Accessed May 28th 2004.

Pp 49 - 58 in *Threatened species legislation: is it just an Act?* 2004, edited by Pat Hutchings, Daniel Lunney and Chris Dickman. Royal Zoological Society of New South Wales, Mosman, NSW, Australia.

Government Agencies (Productivity Commission 1999) and concluded as follows: "...at the broadest level those agencies which have accounted for ESD objectives have either done so explicitly (specific recognition of ESD principles and objectives or use of similar terms) or implicitly (no specific recognition of ESD principles and objectives or similar terms but ESD principles are incorporated in decision making processes). However, very few actually refer to ESD objectives specifically."

In NSW, the concept of ESD has been incorporated into many pieces of legislation bearing on environmental management (Stein 2000) and the policies of a number of State agencies (e.g. Roads and Traffic Authority Environment Policy[3]; Sydney Water ESD Policy[4]; National Parks and Wildlife Service Corporate Plan 2000-2003[5]). The relevant definition is to be found in section 6(2) of the *Protection of the Environment Administration Act* 1991. In relation to the present discussion, this Act provides:

> "... *[E]cologically sustainable development requires the effective integration of economic and environmental considerations in decision-making processes. Ecologically sustainable development can be achieved through the implementation of the following principles and programs:*
>
> *(a) the precautionary principle - namely, that if there are threats of serious or irreversible environmental damage, lack of full scientific certainty should not be used as a reason for postponing measures to prevent environmental degradation. In the application of the precautionary principle, public and private decisions should be guided by:*
>
> *(i) careful evaluation to avoid, wherever practicable, serious or irreversible damage to the environment, and*
>
> *(ii) an assessment of the risk-weighted consequences of various options...*"

It has been argued often that several features of the precautionary principle limit its power. Its content is obscured by the way it is written, with the use of multiple negatives; "**lack** of full scientific certainty should **not** be used as a reason for **postponing** measures to **prevent** environmental degradation" (Nagorcka 2003). Further, various parts of the principle assume some level of objective assessment, such as what constitutes serious or irreversible environmental damage. This may not be achievable. Fisher and Harding (2001) presented a range of reasons explaining why the precautionary principle is only weakly implemented in Australia. Whether as a deliberate political strategy or by ignorance, or both, the precautionary principle is often ignored or undermined by presenting it as "unscientific". Both internationally and in Australia, there are worrying examples of this interpretation, as illustrated in the following examples.

The World Trade Organisation's "Sanitary and Phytosanitary Measures" (SPS) agreement provides guidelines for the limiting of free trade in order to protect human, animal or plant life or health (WTO 1998). A country wishing to bar imports on these grounds must show, using a risk-assessment process, that an unacceptable level of harm will occur and that it cannot be mitigated. Although the SPS agreement does allow for some trade restrictions, as a temporary measure, in cases of scientific uncertainty (Article 5(7)), there is clearly a view that the application of the precautionary principle is somehow unscientific. The following statement by Secretary of Agriculture[6], Ann Veneman, reveals both satisfaction at having prevented the use of the precautionary principle and also a misunderstanding of the place of science in precautionary decision-making when there is uncertain knowledge:

> "*We were successful in precluding the Europeans from including what is called the precautionary principle, so to speak. Now, that principle they wanted in there to be able to use, we believe, could have undermined the whole sanitary and phytosanitary agreement, allowing countries to say: 'well, in the interests of precaution, we're going to take this action', undermining the sound science that's contained in the sanitary and phytosanitary agreement ... That we could not live with, and that was not allowed to go in to the text.*"

Consider the similarity between this reaction to the precautionary principle and the situation surrounding the nomination for listing "Changes to plant-pollinator associations caused by bumblebees, *Bombus* spp." as a Key Threatening Process under the Australian Government's *Environmental Protection and Biodiversity Conservation Act* 1999. The following criteria are considered in judging whether a nomination is eligible for listing:

(i) the process could cause a native species or an ecological community to become eligible for listing in any category; or

(ii) it could cause a listed threatened species or a listed threatened ecological community to become eligible to be listed in another category representing a higher degree of endangerment; or

(iii) it adversely affects 2 or more listed threatened species or 2 or more listed threatened ecological communities.

---

[3] "The RTA will demonstrate due diligence in the provision of its services, manage its work activities in a manner that is consistent with the principles of ecologically sustainable development, and will deliver continuous improvement in environmental performance..." (RTA 1997).

[4] "The purpose of this Policy is to commit Sydney Water to ensuring the principles of Ecologically Sustainable Development (ESD) underpin our decision making processes, and as such, how ESD is incorporated into our activities." (Sydney Water 2003). This policy explicitly refers to the precautionary principle as one part of giving effect to ESD.

[5] "...the conservation of nature be undertaken within the framework of ecologically sustainable development..." (NPWS 2003).

[6] Address by Secretary of Agriculture, Ann Veneman, to the US National Association of Farm Broadcasters. Friday, November 16, 2001. <http://www.usda.gov/news/releases /2001/11/0236.htm> Accessed January 10th 2004.

In recommending against the listing of changes to plant-pollinator associations caused by bumblebees, *Bombus* spp." as a Key Threatening Process, the Scientific Committee emphasised lack of certain knowledge:

> "*Difficulties in assessing this process arise as it is a relatively recent event and the evidence available does not strongly prove a negative effect, only a possible potential to threaten listed species and broader ecological processes. A number of experts state there is insufficient or no evidence against any of the criteria to justify listing, some believing the evidence presented is circumstantial and conflicting. Others believe that although data is lacking there is a clear potential for impacts and that the nomination is comprehensive and balanced. Further views include that waiting for compelling evidence would mean leaving* Bombus terrestris *unchecked and that listing and abatement should occur as a precautionary measure.*"[7]

In considering whether the process could cause a native species or an ecological community to become eligible for listing in any category, the Scientific Committee concluded as follows: "... considers that while potentially invasive, the ecological effects of this process on unlisted native species and ecological communities are not clearly defined or easily predicted, and that there are few quantitative data on actual or potential impacts. The information is considered insufficient to determine whether the threatening process meets this criterion at this time." In considering whether it might drive already listed species to a higher level of listing, the committee concluded: "...the potential for this process to cause the Swift Parrot, Helmeted Honeyeater or Regent Honeyeater to become eligible for listing in a category representing a higher degree of endangerment, is not sufficiently known, nor easily predicted. The information is considered insufficient to determine whether the threatening process meets this criterion at this time." Finally, in relation to adverse effects on already listed species or communities, it concluded: "...although the process may be adversely affecting the listed Swift Parrot, there is, as yet, no evidence of an effect, and the process is not currently adversely affecting the Helmeted Honeyeater or the Regent Honeyeater. The threatening process is not adversely affecting at least 2 listed threatened species and is therefore not eligible under this criterion."

In a reaction similar to that expressed by Veneman, above, the decision by the Minister for the Environment not to list changes to plant-pollinator associations caused by bumblebees as a Key Threatening Process was lauded by the Australian Hydroponics & Greenhouse Association[8]: "Mr Graeme Smith, President of the Australian Hydroponics & Greenhouse Association, welcomed Dr Kemp's decision as cautious good news for the industry, although there is still much work to be done. 'The Minister has taken a sensible approach, and it shows that good science does prevail,' he said." In this case, too, there was a perception that applying the precautionary principle would have been somehow unscientific.

There is a growing literature on the relative merits of the precautionary principle and other ways of accommodating scientific uncertainty, such as environmental risk assessment (e.g. Goklany 2001). We consider that the precautionary principle should not be treated as an alternative approach to risk assessment, because its objective is not to assess the risk. Rather, its purpose is well summarised by Fisher (2004) as follows: "In terms of its substance, the principle is concerned with the process of decision-making or with how decisions are made. ... The focus is upon how the quality of the scientific evidence in relation to setting a standard is factored into the decisionmaking process."

The following example illustrates a misuse of the term precaution in describing policy that was formulated to guide conservation management. The honeybee *Apis mellifera* is an introduced insect that is the basis for an important domestic and export honey industry. Many commercial hives are located so that the bees can harvest nectar and pollen from native vegetation communities, some of them in national parks.[9] A summary of the possible impacts of honeybees is contained in a review commissioned by the Australian Nature Conservation Agency[10], in which Paton (1996) concluded:

> "*At some plants, seed production was reduced when honeybees were frequent floral visitors (e.g.* Callistemon rugulosus*) while at others seed production was enhanced (e.g.* Banksia ornata*). Plant species whose seed production increased were those that received inadequate attention from their native pollinators. Plant-pollinator systems are vulnerable to perturbations like habitat clearance and degradation, and some Australian plants may now depend on honeybees for full pollination because their native pollinators have declined dramatically or even disappeared in some areas. Whether honeybees should be included or excluded from selected areas will depend on which native taxa are to be favoured in those areas. Some plants may benefit by the presence of honeybees while other plants and animals may continue to suffer degradation in their presence.*"

This combination of potential for impact and a high degree of scientific uncertainty about how the potential impacts might be realised in particular situations is surely a good argument for applying the precautionary principle to the formulation of policy. On the face of it, this appears to be

---

[7] *Environment Protection and Biodiversity Conservation Act* 1999 <www.deh.gov.au/ biodiversity/threatened/ktp/unsuccessful/ bumblebees.html> Accessed January 10th 2004.

[8] *Practical Hydroponics and Greenhouses*. Issue 72 (Sept.-Oct. 2003) <www.hydroponics. com.au/current_issue/np72.html> Accessed January 10th 2004.

[9] Mostly as a result of the transfer of Crown Lands and State Forests to the National Park estate, over the past several years.

[10] Now called the Department of Environment & Heritage. <www.deh.gov.au/ biodiversity/invasive/insects/bees/index.html> Accessed January 10th 2004.

the case, because the beekeeping policy[11] adopted by the NPWS, is introduced as adopting "...a precautionary and pragmatic approach to the issue of beekeeping on Service managed land by allowing existing sites to continue but not allowing any new/additional sites." The detail in the policy indicates that it is not, in fact, precautionary at all. One of the objectives of this policy is to provide managers with guidance for deciding on the need to relocate sites. To give effect to this, the Relocation Protocol states "...the Service will examine the need for relocating a site where ... scientific studies have shown that a particular landscape may be susceptible to adverse impacts from foraging honeybees" (s. 34). A precautionary approach would assume some degree of impact (given well-cited indications of the potential for impact but a paucity of scientific studies to quantify the impact in particular circumstances) and therefore relocate hives from conservation areas until studies have been conducted and have shown that there are no adverse impacts. A compromise position, between precaution and pragmatism, would be to initiate (and fund) scientific studies at all sites not being relocated, as an integral part of applying the policy, in order to ensure that relevant information is being collected on which to base a judgment of the urgency of relocation, rather than committing to action only after a relevant study happens to have been done.

Many populations, species or ecological communities that are rare or threatened will typically satisfy the key elements of the precautionary principle: (i) serious or irreversible damage would be caused by inappropriate development or management, and (ii) scientific uncertainty usually surrounds prediction of impacts. It is therefore worthwhile examining the ways in which the NSW *Threatened Species Conservation Act* 1995 already has the potential to be used in a precautionary way, and considering how it might be changed so as to make it more precautionary in practice – in line with the NSW commitment in the National Strategy for Ecologically Sustainable Development (1992) and the Intergovernmental Agreement on the Environment (1992).

Precaution in Listing

Listing is the key foundation for action under the NSW *Threatened Species Conservation Act* 1995. It triggers both reactive processes, which occur in response to a development proposal or other "action" (e.g. Species Impact Statements, Licensing), and also proactive processes aimed at improved management including acquisition of better knowledge (e.g. Recovery Plans, Threat Abatement Plans).

In assessing a nomination for listing of a population, species, or community, the Scientific Committee uses a range of criteria. A species can be listed as endangered if there is a likelihood of it becoming extinct in NSW, through factors threatening its survival or evolutionary development[12] or it is in immediate danger of extinction because numbers of individuals have been reduced to a critical level or its habitats have been "drastically reduced".[13] Listing as vulnerable is warranted if a species is likely to become endangered unless the factors that are affecting its survival or evolutionary development are removed.[14] A population can be classed as endangered if it is facing a high risk of becoming extinct in NSW *and* it is of conservation value because it is disjunct or near the limit of its geographical range[15], likely to be genetically, morphologically or ecologically distinct[16], or it is of significant conservation value for some other reason[17]. Eligibility for listing an endangered ecological community revolves around determining whether the definition of an ecological community ("an assemblage of species occupying a particular area"[18]) is met in a particular case and showing that the community is likely to become extinct if the factors threatening the survival and evolutionary development are not removed.[19] A process may be listed as a key threatening process if it is adversely affecting two or more listed species, populations or ecological communities or if it could cause species, populations or ecological communities not yet threatened to become threatened.[20]

It is safe to assume that not every population, species, community or process in New South Wales has been thoroughly considered for possible nomination, even though there might now be some systematic assessment of the status of species and communities (for example, the process of developing listings under the NSW *Endangered Fauna (Interim Protection) Act 1991* involved a systematic process of expert assessment of all vertebrate fauna in NSW; Lunney *et al.* 1996). Thus, it is almost certain that there are species that are rare, declining, or otherwise under threat of extinction but about which so little is know that they are not yet the subject of nomination. Smith (1997), for example, claimed that some species have become extinct while waiting for listing to take place under the *Endangered Species Act* 1979 (USA).

---

[11] NPWS Beekeeping Policy 2002 <http://www.nationalparks.nsw.gov.au/npws.nsf/ Content/Beekeeping+policy> Accessed January 10th 2004.

[12] TSCA s. 10(a).

[13] TSCA s. 10(b).

[14] TSCA s. 14.

[15] TSCA s. 11(a).

[16] TSCA s. 11(b).

[17] TSCA s. 11(c).

[18] TSCA s 4(1).

[19] TSCA s. 12(a).

[20] TSCA s. 15(a) and (b).

Furthermore, as Farrier (1999) pointed out, criteria for listing typically demand a high level of scientific evidence. In Victoria, for example, under the *Flora and Fauna Guarantee Act* 1988, a taxon or community of flora or fauna is eligible to be listed if it is "in a demonstrable state of decline which is likely to result in extinction". The NSW legislation likewise demands an assessment of likelihood of extinction and of levels of threat to species' survival and evolutionary development that is impossible, given the current state of scientific knowledge. To help the Scientific Committee assess a nomination for the listing of a species[21], a nominator is asked to provide information on the following: past and current distribution of the species in NSW; past and current population size in NSW; information on threats to the species, the intensity of the threat and whether the threat(s) occurs throughout the whole range or parts of the range of the species; details of any recent surveys; in the case of a species which is presumed extinct, details of any surveys for the species in the last 50 years; information on occurrence of the species in National Parks and Wildlife Service reserves or State Forest reserves. Most of the information needed to assess a nomination for listing is lacking (even for some of the species and communities that are currently known to be common and widespread!) and the NSW Scientific Committee makes a considerable effort to find the information that does exist. Although the IUCN categories of conservation status and the decision rules supporting them are generally accepted internationally, Akcakaya *et al.* (2000) argued that there is considerable uncertainty due to limited knowledge, measurement error, environmental variability and vagueness in the definitions of some parameters.

Uncertainty is explicitly addressed in the 2001 revision of the "categories and criteria" in the IUCN Red List of Threatened Species[22]:

> *"The data used to evaluate taxa against the criteria are often estimated with considerable uncertainty. … The way in which this uncertainty is handled can have a strong influence on the results of an evaluation. Details of methods recommended for handling uncertainty are included in Annex 1…"*

Annex 1 of this IUCN document addresses the adoption of precaution in deciding on the conservation category as follows:

> *"… assessors need to consider whether they have a precautionary or evidentiary attitude to risk (known as risk tolerance). A precautionary attitude will classify a taxon as threatened unless it is certain that it is not threatened, whereas an evidentiary attitude will classify a taxon as threatened only when there is strong evidence to support a threatened classification. Assessors should resist an evidentiary attitude and adopt a precautionary but realistic attitude to uncertainty when applying the criteria; for example, by using plausible lower bounds, rather than best estimates in determining population size, especially if it is fluctuating."*

Given the various uncertainties, and the potential for serious or irreversible damage to biodiversity if bad decisions are made, an extreme precautionary approach would be to list all known (extant) species as endangered unless and until there is sufficient evidence available to warrant down-listing to vulnerable or to support removal from a list altogether, as indicated in Annex 1 of the IUCN document. This is probably an unrealistic goal, given political realities and also concern by scientists that the many nominations for de-listing that would inevitably flow from this approach would damage their credibility (e.g. Bonham 1999), and it is logistically unachievable in practice. We therefore argue for a "*precautionary but realistic attitude to uncertainty*" in listing.

One way of using precaution in listing would be to follow the advice given by the IUCN: namely, to use the lowest value in a range of estimates of population size or species range as if they were the real values for numbers and distribution when assessing status. Another way of achieving precaution in the face of uncertainty would be to make explicit use of the IUCN's "Data Deficient" category. This category is defined as follows: "A taxon is Data Deficient when there is inadequate information to make a direct, or indirect, assessment of its risk of extinction based on its distribution and/or population status. A taxon in this category may be well studied, and its biology well known, but appropriate data on abundance and distribution are lacking. Data Deficient is therefore not a category of threat. Listing of taxa in this category indicates that more information is required and acknowledges the possibility that future research will show that threatened classification is appropriate."

We argue that it would be valuable to include a schedule to the TSCA for "Data Deficient" species. Listing on this schedule would trigger a SIS if a listed taxon were to be deemed likely to occur in the area of a development proposal, and the Director General's requirements would include gathering of relevant data to allow the assessment of impact. Listing as data-deficient would also trigger a survey and research program to be completed "as soon as practicable", as is the case with species recovery plans for those taxa listed as endangered,[23] at which time an assessment of a nomination for listing under one of the threatened categories ("vulnerable" or "endangered") would be required.

Notwithstanding our comments above, in relation to the problems of adequately incorporating precaution into the listing of populations, species and perhaps communities, an examination of the recent nominations

---

[21] NPWS 2003. For example, nomination form for listing a species as vulnerable or threatened. <www.nationalparks.nsw.gov.au/npws.nsf/Content/How+to+nominate+a+species+ as+threatened> Accessed January 20th 2004.

[22] The IUCN Red List of Threatened Species: 2001 Categories and Criteria, Version 3.1. <www.redlist.org/info/categories_criteria2001.html> Accessed January 20th 2004.

[23] TSCA s 56(1).

for the listing of Key Threatening Processes under the NSW legislation does indicate some level of precaution. First, the Scientific Committee made a preliminary determination in December 2003 listing the introduction of the 'large earth bumblebee' *Bombus terrestris* as a Key Threatening Process (EPAAss 34A, 5C). Interestingly, this decision was informed by the same body of scientific evidence as the decision by the Australian Government **not** to list under the federal legislation (*EPBC Act* 1999), namely studies on the effects of feral bumblebees in Tasmania (e.g. Semmens 1996; Buttermore 1997; Hingston and McQuillan 1998; Hingston *et al.* 2001; Low 1999). Second, the importation of the 'red imported fire ant' has been listed in NSW, based on general knowledge of the biology of this introduced species, preliminary studies of the impact on biodiversity of colonies that have become established in Brisbane, and climatic modelling suggesting that this species, if imported, could inhabit most of the NSW coastal belt and the more mesic inland areas. Listing its importation therefore represents a precautionary approach. Third, as mentioned above, competition from feral honeybees has been listed as a Key Threatening Process in NSW, once again based on an understanding of the general biology of honeybees, the knowledge that feral colonies are widespread in NSW, and a range of studies suggesting the potential for deleterious ecological effects of interactions between honeybees and native fauna and flora.

It is worth noting that, even if listing of these Key Threatening Processes is viewed as precautionary, in practice there may be a long time lag between listing and action. Listing triggers the preparation of a Threat Abatement Plan but, to date, only two have been completed (Predation by the Plague Minnow; Predation by the Red Fox)[24] out of the 20 Key Threatening Processes so far declared.

Precaution in Assessing the Likelihood of Significant Impacts

Assessing nominations for listing, and some of the planing processes that are triggered by listing (for example, recovery planning, threat abatement planning, and, as we propose, research planning to overcome data deficiency) are conservation actions that can be (and should be) undertaken well in advance of specific development proposals. In addition, there are environmental planning instruments made under the *Environmental Planning and Assessment Act* 1979 (EPAA), which can regulate development likely to affect threatened species, for example by identifying areas that should be zoned for environment protection. The Director-General of the Department of Environment and Conservation[25] must be consulted where proposed environmental planning instruments may affect threatened species, populations or ecological communities, or their habitats. However, Farrier *et al.* (2002) argued that, in practice, species conservation issues usually surface at a point when approval is sought for a particular development proposal. There is often uncertainty in the information available to consent authorities on which they must base an assessment of the likely impact of the proposed development on listed taxa, yet, at this late stage in the process, significant time constraints make it difficult to carry out adequate scientific research into potential environmental impacts.

This difficulty is partly acknowledged in the multi-stage process that must be followed by a consent authority (usually a local Council) on receipt of a development proposal (see Farrier *et al.* 2002):

1. The Council must first decide whether listed species or the habitats of a listed species are likely to be present on the site.

If so…

2. The Council (and ultimately the courts, in the few cases which get that far) must decide whether the proposed development is likely to significantly affect any listed species identified as being present, or their habitats (the so-called "eight part test").[26]

If so…

3. The developer must provide a species impact statement (SIS), complying with the requirements of the Director of National Parks and Wildlife, which must be placed on public exhibition along with the development application.[27]

Then…

4. The Council must consider potential impact on species and their habitats (including any SIS that has been prepared) in making a decision on whether to grant development consent and, if so, what conditions should be attached.[28]

If the Council proposes to give consent…

5. The agreement (concurrence) of the Director-General of National Parks and Wildlife is required in situations where the proposed development is likely to significantly affect threatened species or their habitats.[29]

If the Council refuses to give consent, or cannot give consent because the NPWS refuses to give concurrence…

---

[24] <http://www.nationalparks.nsw.gov.au/npws.nsf/Content/Threat+abatement+plans +by+doctype> Accessed January 10th 2004.

[25] Formerly the Director General of the National Parks and Wildlife Service.

[26] EPAA ss 78A(8)(b), 5A.

[27] EPAA s 78A(8)(b); TSCA s 111; EPAR cll 5(1)(c), 87, 89.

[28] EPAA s 79C.

[29] EPAA s 79B(3)(b). See Farrier (2002).

6. The developer may appeal on the merits of the case to the Land and Environment Court, which, in essence, makes the decision in the place of the council,[30] without needing to get the concurrence of the Director-General of National Parks and Wildlife if it decides to give consent.[31]

Even at the first step of this process, deciding whether listed species or habitats are present at the site of a development proposal, information can be elusive. Brown (2003) described the case study of *Tetratheca glandulosa* (Elaeocarpaceae) in a proposed housing subdivision in Sydney (Aquatic Drive, Allambie Heights), in which difficulties stemmed from the fact that this listed species is cryptic when not flowering. During the development assessment process and subsequent Land and Environment Court merit appeals, there was considerable argument surrounding the question of whether *Tetratheca glandulosa* occurred on the parcel of land in question (see also Farrier *et al.* 2002). Precaution could be exercised at this step if Council officers negotiating with developers assume that a listed species occurs at the site, where there is a reasonable chance of it occurring based on the nature of the vegetation, soils and other site characteristics, even if there are no specific records of presence. Achieving a precautionary approach at this stage requires that relevant Council officers have sufficient ecological experience and training, or at least access to appropriate information, to allow them to make the judgement – and also to have sufficient credibility in negotiating with a proponent of a development about the need to determine whether there is likely to be a significant impact on a listed taxon (population, species, community). We have knowledge of situations in which Council officers do indeed exercise this judgement.

The second step of the process outlined above is the "eight-part-test", which is a check-list designed to help the consent authority decide whether there is likely to be a significant impact on a threatened (listed) entity (population, species, community) as a result of the proposed development.[32] As part of this process, the consent authority can take into account ameliorative measures proposed in the development application. This raises scientific uncertainties at two levels: firstly, assessment of the significance of the impact of the proposal on a listed taxon and the likelihood that the impact will happen; secondly, the likely response of the listed taxon to the amelioration measures. The questions here are of cause-and-effect. We argue that, in practice, there will be little scientific knowledge to provide guidance, partly because the extreme level of complexity in ecological systems makes it likely that cause-and-effect relationships are going to be highly site- and species-specific and hence not readily predicted from previous studies in a specific situation. In any case, there have been few previous studies for most species on any site.

If precaution were to be exercised here, uncertainty should lead to an assumption of a significant impact and assumption that amelioration measures will not mitigate the impact sufficiently, thus leading the proponent into the preparation of a Species Impact Statement (SIS). However, the evidence from Claire Brown's work (Brown 2003) on *Tetratheca glandulosa* and *Darwinia biflora* (Myrtaceae), and some statistics on how few development applications proceed to a SIS and therefore require the concurrence of the Director General of the Department of Environment and Conservation (see Farrier *et al.* 2002), suggest that Councils are likely to accept a conclusion that there is not likely to be a significant impact, perhaps as a result of mitigation works that are assumed to be likely to be effective. For example, in the site of a proposed subdivision, which contained a population of the vulnerable plant species *Darwinia biflora*, the proponents included the establishment of a bushland habitat corridor as a mitigation measure, to avoid the isolation of the population from remnant bushland adjacent to the site.[33] The vegetation in the corridor did not provide suitable habitat for the plant, but the assumption was that it would facilitate pollination of the isolated plants by enabling the movement of insects involved in pollinating the species.

In this case, there was no knowledge about whether the pollinators of the plant would move through the proposed corridor and adequately pollinate the *D. biflora* plants in this otherwise isolated population. Indeed, the identity of the insects that might be responsible for pollination of this plant species is currently unknown! Nevertheless the Council officer dealing with the application was prepared to accept that such a corridor of remnant vegetation would have a mitigating effect.

The acceptance of such a proposition in relation to insect pollination of an isolated plant population is perhaps understandable, because the idea that the retention of a linear strip of habitat between at least two patches will increase animal movement and allow for the persistence of individual populations is attractive, sounds reasonable, has found its way into conservation biology texts, and has been demonstrated for insect pollinators (butterflies) in one study – albeit in the USA (Haddad 1999; Haddad and Baum 1999). However, the limited research that has been carried out has shown that the effectiveness of corridors in enhancing animal movements is highly likely to be species- and landscape-specific (Beier and Noss 1998) and there may even be costs in terms of disease transmission, spread of fire and predation by domestic animals (Simberloff and Cox 1987). We could find no empirical research published on the role of vegetated corridors in facilitating pollinator movements or enhancing the reproductive success of plants in the remnant populations. In this case study, we see scientific 'folk lore' being used in the course of the decision-making, far beyond legitimate limits of current

[30] EPAA s 97(1), (5).

[31] *Land and Environment Court Act* 1979, s 39(6).

[32] EPAA ss 78A(8)(b), 5A

[33] This is relevant because section (d) of the eight-part-test asks whether "an area of known habitat is likely to become isolated from currently interconnecting or proximate areas of habitat for a threatened species…"

knowledge. Brown (2003) has argued that, in many cases, accurate completion of an assessment of whether there is likely to be a significant effect, using the eight-part-test, would need the results of the studies that are conducted to prepare a Species Impact Statement (step 3, above).

The preparation of a Species Impact Statement (SIS) may appear to be outside the realm of the precautionary principle, because it is a data-collection exercise with the potential to overcome scientific uncertainty before a decision is made on a development proposal. However, time limitations and financial constraints on environmental consultants imposed by an imperative for rapid decisions mean that there is typically a lack of data and often debate about the appropriateness of the design of studies. The scientific uncertainty here should lead consultants to apply the precautionary principle in interpreting their results, perhaps by using particular statistical approaches (e.g. Quinn & Keough 2002), but we doubt that this occurs in most cases.

As the Director General of the Department of Environment and Conservation sets requirements for an SIS[34], there is an opportunity for precaution to be exercised at this stage, if there is uncertainty about the evidence for potential impacts, by requiring that sufficient work is done in the SIS to inform the Council's decision on the development application. The SIS must have regard to the principles of ESD in describing any feasible alternatives to the action that are likely to have lesser impact and giving reasons for carrying out the action in the manner proposed.[35] There is an implication that these factors should be salient ones for councils in making a final decision. However, there is no requirement that councils should *apply* the precautionary principle when making a final decision on whether a proposal should be allowed to go ahead, and in deciding whether or not to grant concurrence, the Director-General of National Parks and Wildlife is simply directed to take the precautionary principle into account, not to apply it.[36]

The degree to which precaution might currently be being applied in these three steps of the process (steps 3 to 5, above) is hard to assess, given that the decision-making process is decentralised, with a large number of local councils involved and so few development applications judged to be likely to have a significant impact on threatened taxa, so as to require the preparation of a SIS and the concurrence of the Director General of the NPWS. This would be a fruitful and important topic for further investigation.

Benefits of Better Use of the Precautionary Principle

The foregoing review indicates that the precautionary principle is embedded in policy in NSW (mostly as an implicit part of ESD) and there are many opportunities for the application of the precautionary principle in decision-making within the processes of the TSCA, and in policy. Nagorcka (2003) argued that an examination of the use of the precautionary principle can reveal how much the natural environment is valued by assessing the extent to which decisions that are made forgo socio-economic benefits, which are often certain and quantifiable, in order to achieve some poorly quantifiable and uncertain environmental benefits. We are concerned that opportunities to use the precautionary principle are too often being ignored.

There may be several explanations for this reluctance. First, the precautionary principle is not well understood and is being portrayed as "unscientific", especially in comparison with a risk-assessment process. This is clearly wrong, because the precautionary principle is an approach to decision-making in situations of uncertain scientific knowledge – it is not a method for evaluating a level of risk or predicting likely outcomes (see Farrier *et al.* 2002). Second, individual staff who have a responsibility for decision-making (e.g. Council officers dealing with development applications) may not be sufficiently well trained or experienced to judge the state of ecological knowledge and hence scientific uncertainty in particular situations. As a result, they may be too easily convinced that there will not be a significant impact on threatened taxa and that proposed mitigation works will be effective. Third, there is an exaggerated fear as to the role the precautionary principle might play in halting development, arising from a fundamental misunderstanding of the way in which it operates. The precautionary principle does not require environmental impact to be given a higher weighting in decision-making processes than socio-economic considerations. Rather, it addresses the issue of scientific uncertainty about the environmental impact of activities. It simply helps decision-makers to determine what that environmental impact should be presumed to be, in a context where scientific research is scant or ambiguous. When the decision-maker comes to make the final decision, socio-economic considerations relating to a particular activity may be considered to outweigh environmental impact, however significant that impact is taken to be, using either direct evidence of impact or inferring the impact by applying the precautionary principle. The precautionary principle should make it difficult for a decision-maker to dismiss potential environmental impact on the grounds of lack of clear evidence. In particular, the principle should operate to outlaw the use of economic considerations to justify sloppy "science" at earlier stages in the decision-making process – in the context of eight-part tests of significant effect and SISs. In situations where there is sufficient scientific evidence to trigger the operation of the precautionary principle, there should be no question of the presumed environmental impact being rebutted by, for example, the results of species surveys carried out at the wrong time of year because proponents are not prepared to devote the time and resources necessary to do a proper job.

---

[34] TSCA s111 (2).

[35] TSCA s 110(2)(h).

[36] EPAA s 79B(5).

We conclude that proper understanding of the precautionary principle and its proper application in decision-making processes will lead to greater understanding of potential environmental impacts and greater transparency in decision-making. Decision-makers will be forced to acknowledge that they are approving proposals even where the impact on threatened species or their habitats is presumptively significant. Adoption of the precautionary principle prevents the assumption that scientific uncertainty is indicative of minimal environmental impact, which admits the possibility of presenting a compromise decision as a win-win situation for development and the environment. Thus, the decision on whether the eight-part-test indicates the likelihood of a significant impact would trigger a SIS whenever there is uncertainty about threatened taxa involved. Likewise, the NPWS Beekeeping Policy would assume that there are going to be significant impacts of licensed beekeeping sites and require relocation, unless a judgement of socio-economic benefit is perceived to outweigh these detrimental environmental effects. Finally, the statements on the NPWS web site on how threatened species affect developers and landowners[37] would be modified. It currently states: "In most cases, a development can be modified to accommodate the interests of both the landholder and the NPWS, to ensure the continued survival of the threatened species, population or ecological community." Given the level of scientific uncertainty, and consequent adoption of the precautionary principle in making judgements of potential impacts, it could state instead: "Where there is some scientific evidence that there may be serious or irreversible harm to threatened species, populations or ecological communities from proposed development, but the evidence is not conclusive, it must be assumed that there will be threats to their survival, unless adequate evidence is produced to undermine this assumption. However, development applications may still be approved when social or economic benefits, or both, are judged to be more important than the potential damage to threatened taxa."

References

Akcakaya, H., Ferson, S., Burgman, M., Keith, D., Mace, G. and Todd, C. 2000. Making consistent IUCN classifications under uncertainty. *Conservation Biology*. **14**: 1001-1013.

Beier, P. and Noss, R.F. 1998. Do habitat corridors provide connectivity? *Conservation Biology* **12**: 1241-1252.

Bonham, A.J. 1999. *Angolypta* crawls off the list. *Invertebrata* **14**: 5. <http://www.qvmag.tased.edu.au/zoology/invertebrata/printarchive/ printpdf/inv14.pdf> Accessed January 20th 2004.

Brown, C.L. 2003. *Can Legislation Deliver Conservation? An Assessment of the Threatened Species Conservation Act 1995 (NSW) Using Two Threatened Plant Species as Case Studies*. PhD Thesis, Faculty of Law and Department of Biological Sciences, University of Wollongong.

Buttermore, R.E. 1997. Observations of successful *Bombus terrestris* (L.) (Hymenoptera: Apidae) colonies in southern Tasmania. *Australian Journal of Entomology* **36**: 251-254.

Calver, M.C., Bradley, J.S., and Wright, I.W. 1999a. Towards scientific contributions in applying the precautionary principle: an example from southwestern Australia. *Pacific Conservation Biology* **5**: 63-72.

Calver, M.C., Bradley, J.S., and Wright, I.W. 1999b. Environmental management: the precautionary principle and null hypothesis. *Pacific Conservation Biology* **5**: 78-82.

Deville, A. and Harding, R. 1997. *Applying the Precautionary Principle*. Federation Press, Sydney.

Farrier, D. 1999. Factoring biodiversity conservation into decision-making processes: the role of the Precautionary Principle. Pp 99-121 in *Perspectives on the Precautionary Principle*, edited by R. Harding and L. Fisher. Federation Press, Sydney.

Farrier, D., Whelan, R.J. and Brown, C.L. 2002. Addressing scientific uncertainty in local government decision-making processes. *Environmental and Planning Law Journal* **19**: 1-15.

Fisher, E. and Harding, R. 2001. The Precautionary Principle in Australia. Chapter 9 in *Reinterpreting the Precautionary Principle*. O'Riordan, T. *et al.* (editors). Cameron May, London.

Fisher, E. 2004. Precaution, uncertainty and principles of good administration. International Symposium: Uncertainty and Precaution in Environmental Management. Copenhagen, Denmark. <upem.er.dtu.dk/files/L%Fisher.pdf> Accessed July 20th 2004.

Goklany, I.M. 2001. *The Precautionary Principle: A Critical Appraisal of Environmental Risk Assessment*. Cato Institute, Washington D.C.

Haddad, N.M. 1999. Corridor and distance effects on interpatch movements: a landscape experiment with butterflies. *Ecological Applications* **9**: 612-622.

Haddad, N.M. and Baum, K.A. 1999. An experimental test of corridor effects on butterfly densities. *Ecological Applications* **9**: 623-633.

Harding, R. and Fisher, E. (eds) 1999. *Perspectives on the Precautionary Principle*. Federation Press, Sydney.

Harremoes, P., Gee, D., MacGarvin, M., Stirling, A., Keys, J., Wynne, B. and Guedes Vas, S. (eds) 2002. *The Precautionary Principle in the 20th Century: Late Lessons from Early Warnings*. Earthscan, London.

Hingston, A.B. and McQuillan, P.B. 1998. Does the recently introduced bumblebee *Bombus terrestris* (Apidae) threaten Australian ecosystems? *Australian Journal of Ecology* **23**: 539-549.

Hingston, A.B., Marsden-Smedley, J., Driscoll, D.A., Corbett, S., Fenton, J., Anderson, R., Plowman, C., Mowling, F., Jenkin, M., Matsui, K., Bonham, K.J., Ilowski, M., McQuillan, P.B., Yaxley, B., Reid, T., Storey, D., Poole, L., Mallick, S.A., Fitzgerald, N., Kirkpatrick, J.B., Febey, J., Harwood, A.G., Michaels, K.F., Russell, M.J., Black, P.S., Emmerson, L., Visoiu, M., Morgan, J., Breen, S., Gates, S., Bantich, M.N., Desmarchelier, J.M. 2001. Extent of invasion of Tasmanian native vegetation by the exotic bumblebee *Bombus terrestris* (Apoidea: Apidae). *Austral Ecology* **27**: 162-172.

Low, T. 1999. *Feral Future*. Viking, Melbourne.

Lunney, D., Curtin, A., Ayers, D., Cogger, H.G. and Dickman, C.R. 1996. An ecological approach to identifying the endangered fauna of New South Wales. *Pacific Conservation Biology* **2**: 212-231.

[37] <www.nationalparks.nsw.gov.au/npws.nsf/Content/How+do+threatened+species+ affect+developers+and+landowners> Accessed January 21st 2004.

Nagorcka, F. 2003. Saying what you mean and meaning what you say: precaution, science and the importance of language. *Environmental and Planning Law Journal* **20**: 211-222.

NPWS 2003. NSW National Parks and Wildlife Service Corporate Plan 2000-2003. NPWS, Hurstville. <www.nationalparks.nsw.gov.au/npws.nsf/ Content/Corporate+Plan+20002003> Accessed December 15th 2003.

O'Riordan, T., Cameron, J. and Jordan, A. (eds) 2001. *Reinterpreting the Precautionary Principle.* Cameron May, London.

Paton, D.C. 1996. *Overview of Feral and Managed Honeybees in Australia.* Australian Nature Conservation Agency, Canberra.

Productivity Commission 1999. *Implementation of Ecologically Sustainable Development by Commonwealth Departments and Agencies*, Report No. 5, AusInfo, Canberra.

Quinn, G., and Keough, M. 2002. *Experimental Design and Data Analysis for Ecologists.* Cambridge University Press, Cambridge.

RTA 1997. The NSW Roads and Traffic Authority – Environment Policy. <www.rta.nsw.gov.au/environment/downloads/rtaenviropolicy.pdf> Accessed January 20th 2004.

Semmens, T.D. 1996. Flower visitation by the bumble bee *Bombus terrestris* (L.) (Hymenoptera: Apidae) in Tasmania. *Australian Entomologist*. **23**: 33-35.

Simberloff, D. and Cox, J. 1987. Consequences and costs of conservation corridors. *Conservation Biology* **1**:63-71.

Smith, J. 1997. Skinning cats, putting tigers in tanks and bringing up baby: a critique of the Threatened Species Conservation Act 1995 (NSW). *Environmental and Planning Law Journal* **14**: 17-37.

Spry, M. 1997. Hinchinbrook Island and the Precautionary Principle in Australian environmental law. Research Note 4, 1997-98. *Department of the Parliamentary Library, Parliament of Australia.* <www.aph.gov.au/library/ pubs/rn/1997-98/98rn04.htm> Accessed December 15th 2003.

Stein, P.L. 2000. Are decision-makers too cautious with the precautionary principle? *Environmental and Planning Law Journal* **17**: 3-23.

Sydney Water 2003. Ecologically Sustainable Management Policy. <www.sydneywater.com.au/html/about_us/esd1.cfm> Accessed December 15th 2003.

UPEM 2004. International Symposium: Uncertainty and Precaution in Environmental Management. Copenhagen, Denmark, June 7-9 2004. <upem.er.dtu.dk/programme.htm> Accessed June 12th 2004.

WTO 1998. Understanding the WTO Agreement on Sanitary and Phytosanitary Measures. <www.wto.org/englans/tratop_e/sps_e/spsund_e.htm> Accessed January 10th 2004.

Opening a can of words. The importance of definitions in the *NSW Threatened Species Conservation Act*

Paul Adam*

School of Biological, Earth and Environmental Sciences

University of New South Wales NSW 2052

p.adam@unsw.edu.au

ABSTRACT

The *NSW Threatened Species Conservation Act 1995* provides for the listing on schedules of threatened species, populations, ecological communities and of key threatening processes. The definitions of these concepts are discussed. The implicit component parts of the definition of species are analysed and differences between legal and science based approaches are discussed. Issues arising from application of the concept of endangered ecological community are discussed in the light of the interpretations provided by the Court of Appeal. Listing is a means to an end, and not an end in itself. The importance of listing to the promotion of the development of conservation measures for threatened species is discussed. The potential for the implementation of Threat Abatement Plans to play a major role in biodiversity conservation is stressed.

Key words: threatened species, endangered ecological communities, endangered populations, key threatening processes, definition of concepts.

Introduction

The *NSW Threatened Species Conservation Act 1995* is an important component of the legislation which governs the development and application of biodiversity conservation policy in New South Wales.

The development of conservation legislation in the State is reviewed by Jarman and Brock (2004) and Farrier and Whelan (2004). The *Threatened Species Conservation Act* was, in the most immediate sense, the replacement of the *Endangered Fauna (Interim Protection) Act 1991*, which had a sunset clause effective at the end of 1995. However, it was also the opportunity to give effect to obligations to conserve biodiversity arising from the Rio Biodiversity Convention of 1992. As such it introduced a number of new features into the State's legislation. Unlike previous legislation it is not taxonomically restrained although matters covered by the *Fisheries Management Act 1994* are excluded. When the Threatened Species Conservation Bill was first introduced into Parliament in late 1995 it was all encompassing, but during its passage through the legislature it was amended by the Government to exclude fisheries matters. This created a number of apparent anomalies, for example some dragonflies are eligible for listing under the *Threatened Species Conservation Act* as they have non-aquatic larvae but the majority would be categorized as fish. While fisheries and terrestrial resource management are separated under many jurisdictions, separation for purposes of determining conservation status of elements of biodiversity is possibly unique. The Commonwealth *Environment Protection and Biodiversity Conservation Act 1999* does not differentiate between 'fish' and 'others' in its listing process. The vertebrate fauna on the Schedules of the *Endangered Fauna (Interim Protection) Act* were transferred to the Schedules of the *Threatened Species Conservation Act*. The initial Schedules also included vascular plants but for the first time the way was open for formal recognition of the conservation status of organisms such as fungi, lichens and invertebrates as well as the traditional objects of concern. Importantly ecological communities are recognized and also, a category unique to New South Wales, populations. The Act also provides for public involvement through nomination of matters to be listed to an independent Scientific Committee, and through submissions on preliminary determinations.

The Act deals extensively with concepts which are in everyday use by ecologists. However, they are not always used with precision and textbooks do not offer a uniformity of definitions. Additionally society at large has some understanding of the concepts, adding an additional range of nuances of meaning. It is important, therefore, that those who have need to interpret and apply the Act are aware of the definitions which are provided in (s)4 of the Act, rather than assuming some perceived common usage. Not all terms are defined, and hence there may be circumstances were several interpretations could be offered and where resolution may require a determination in the Courts. (An example of an instance where the meaning of

*The views expressed in this paper are those of the author and not of the Scientific Committee established by the Threatened Species Conservation Act.

Pp 59 - 67 in *Threatened species legislation: is it just an Act?* 2004, edited by Pat Hutchings, Daniel Lunney and Chris Dickman. Royal Zoological Society of New South Wales, Mosman, NSW, Australia.

a concept in general usage amongst ecologists could not be agreed upon by expert witnesses arose in Slack-Smith and Another-v-Director General of the Department of Land and Water Conservation [2003] NSWLEC.189, a case under the *Native Vegetation Conservation Act 1997*, where there was failure to come to agreement on whether or not a particular species was predominant in the vegetation in dispute.) The definitions provided in the *Threatened Species Conservation Act* are generally similar to those in (s)528 of the *Commonwealth Environment Protection and Biodiversity Conservation Act*, but there are some differences.

The context

The context in which listing occurs, and the subsequent importance of listings, can be best appreciated in the light of the objectives of the Act((s)3) which are:

(a) to conserve biological diversity and promote ecologically sustainable development, and

(b) to prevent the extinction and promote the recovery of threatened species, populations and ecological communities, and

(c)to protect the critical habitat of those threatened species, populations and ecological communities that are endangered, and

(d) to eliminate or manage certain processes that threaten the survival or evolutionary development of threatened species, populations and ecological communities, and

(e) to ensure that the impact of any action affecting threatened species, populations and ecological communities is properly assessed, and

(f) to encourage the conservation of threatened species, populations and ecological communities by the adoption of measures involving co-operative management.

The first objective of the Act is the conservation of 'biological diversity'. Although the Act is entitled the *Threatened Species Conservation Act*, threatened species are only a subset of biological diversity. While there are sound reasons for promoting conservation of threatened species, a reductionist approach which separates threatened species and their management from biodiversity as a whole will not necessarily advance biodiversity conservation nor, in the long term, save the species concerned. Recognition of endangered ecological communities and key threatening processes, and the implementation of community recovery plans and threat abatement plans, have the potential to assist conservation of biological diversity (including the many as yet unknown and undescribed components of biodiversity) to a much greater extent.

Species

The definition of species in the Act is as follows:

species *of animal or plant includes any defined sub-species and taxon below a sub-species and any recognizable variant of a sub-species or taxon.*

This does not adopt any particular concept of species, although the Commonwealth has explicitly adopted the biological species concept in the definition provided in the *Environment Protection and Biodiversity Conservation Act.*

Species means a group of biological entities that:

(a) interbreed to produce fertile offspring; or

(b) possess common characteristics derived from a common gene pool;

and includes:

(c) a sub-species; and

(ca) for the purposes of Part 13A-a distinct population of such biological entities; and

(d) except for the purposes of Part 13A-a distinct population of such biological entities that the Minister has determined, under section 517, to be a species for the purposes of this Act.

Although many textbooks indicate that the biological species concept is the preferred basis for defining species the concept is, in the vast majority of cases, an hypothesis rather than a reflection of the process through which taxonomic decisions were made. Many species have been described on the basis of long dead specimens rather than observation in the field. It is difficult, for example, to discern much about the mating behaviour of insects pinned in a display case! On the basis of experience and intuition, taxonomists would expect that what are regarded as 'good' species on morphological or other characters would behave as 'biological' species. In some cases this has been confirmed subsequently, but in most it remains an assumption.

Importantly, entities below species rank can, for the purposes of the Act, be treated as if they were species. This is an important mechanism for capturing components of the genetic variation level of biodiversity. Also, importantly, the entities need not have been formally named to be listed, provided they are 'recognisable'.

To fully understand the scope of the species definition the meanings of animal and plant also need to be considered. [The binary division into plant and animals is understandable to the general public, but does not accord with the currently scientifically accepted three domain classification e.g. Campbell and Reece 2002].

animal *means any animal-life that is indigenous to New South Wales or is known to periodically or occasionally migrate to New South Wales, whether vertebrate or invertebrate and in any stage of biological development, but does not include:*

(a) humans, or

(b) fish within the meaning of Part 7A of the Fisheries Management Act 1994.

This allows migrant species to be considered for listing, although whether a distinction can be drawn between vagrant and 'occasionally' migrant species may be debatable. New South Wales for purposes of the Act extends 3 nautical miles to sea, and hence marine mammals, reptiles and birds become potentially eligible for listing.

The definition of plant is similar:

plant *means any plant-life that is indigenous to New South Wales, whether vascular or non-vascular and in any stage of biological development, and includes fungi and lichens, but does not include marine vegetation within the meaning of Part 7A of the Fisheries Management Act 1994.*

The exclusion of marine vegetation precludes listing of marine algae, seagrasses and mangroves, but not freshwater algae and macrophytes. A separate listing process, overseen by the Fisheries Scientific Committee, is now provided for in the *Fisheries Management Act*.

For both plants and animals the definition encompasses 'any stage of biological development'. This is extremely important for ensuring long term conservation, but does create practical difficulties, both for making determinations for listing and subsequently in the assessment of sites and the significance of possible impacts of development on species. The importance of protecting birds' eggs from collection and destruction has long been recognized in legislation in many jurisdictions. Invertebrate larval stages, seeds and spores have received much less attention in legislation and site assessment. Nevertheless the existence of a seedbank or larvae in the soil may permit individual species or whole communities to regenerate after disturbance. Just because a site has been cleared above ground does not automatically mean that all conservation values are lost, and that threatened species, populations and communities might not still exist. Including "all stages of the life cycle" within the concept of the species, while making good biological sense, may create some interesting dilemmas for conservation.

One of the iconic species for European plant conservation is the Killarney fern *Trichomanes speciosum* (Ratcliffe *et al.* 1993). This beautiful filmy fern was always regarded as an uncommon plant of the most oceanic parts of western Europe, and is thought to have become very rare as a result of excessive collection during the Victorian era (Ratcliffe *et al.* 1993, Moore 1998). However, the object of concern was the sporophytic generation – the stage the public would regard as a fern. As with all ferns *Trichomanes* shows alternation of generations, with sexual reproduction occurring in the haploid gametophyte. In most ferns the gametophyte is short lived, but in *Trichomanes* it is persistent and more desiccation and frost tolerant than the sporophyte (Moore 1998). It was known that in some North American species of *Trichomanes*, gametophytes occurred at sites geographically distant from known locations of sporophytes. The gametophytes of *T. speciosum* in Europe were long over looked, or, if noticed, dismissed as algae, but were recorded for the first time, in abundance, in 1989. Subsequently they were shown to be widespread in eastern England and as far east in Europe as the Czech Republic (Rumsey *et al.* 1998). How should the conservation status of this species be assessed?

We know of no example in Australia where the ecology and distribution of different life cycle stages are so different, but we should be aware of the possibility of such cases.

The final piece of the jigsaw of species' definition is indigenous.

A reference in this Act to animal-life or plant-life indigenous to New South Wales is a reference to animal-life or plant-life of a species that was established in New South Wales before European settlement.

Clearly we do not want to use scarce conservation resources to promote introduced animals and weeds and further threaten native biodiversity. The cut off of European settlement has, however, brought the dingo *Canis lupus dingo* within the purview of the Act even though it is a relatively recent introduction (Dickman and Lunney 2001). However, it also creates an arbitrary deadline at which 'natural' processes are deemed to have ceased. Despite the huge increase in human assisted dispersals, some species have undoubtedly spread and become established in new locations without obvious human assistance since 1788. Even if such a species were otherwise to satisfy the criteria for listing as a threatened species it would be ineligible. (Most documented examples of recent natural spread are birds but there is no reason to suppose that the phenomenon does not occur in other taxa. One example of a species which has recently spread over much of the globe, including Australia, is the cattle egret *Bubucus ibis* L. – see Handtke and Mauersberg 1977). Unfortunately we do not have a comprehensive account of the biota of NSW in 1788 (nor do we have one for 2003). When a species is newly discovered in NSW, and is thought to be threatened, the Scientific Committee must decide whether it is indigenous or a recent immigrant before listing can occur. If what has been discovered is a new species of tree whose population includes large mature individuals then the decision is straight forward, but for most taxa there has to be an element of expert opinion rather than absolute certainty. To date, the question of the indigenous status of species listed has not been seriously challenged.

Eligibility for listing

Having dissected the meaning of species (in the context of the Act), the next issue is eligibility for listing. The criteria for listing of endangered and vulnerable species are provided in s10 and s14 of the Act respectively.

A species is eligible to be listed as an endangered species if, in the opinion of the Scientific Committee:

(a) it is likely to become extinct in nature in New South Wales unless the circumstances and factors threatening its survival or evolutionary development cease to operate, or

(b) its numbers have been reduced to such a critical level, or its habitats have been so drastically reduced, that it is in immediate danger of extinction, or

(c) it might already be extinct, but is not presumed extinct.

A species is eligible to be listed as a vulnerable species if, in the opinion of the Scientific Committee, the species is likely to become endangered in New South Wales unless the circumstances and factors threatening its survival or evolutionary development cease to operate.

While the Scientific Committee may be informed by the results of other assessments of conservation status (for example application of the IUCN Scheme – see Burgman

2004), the ultimate test is whether the Committee is satisfied that the criteria specified in the Act are met. Given current knowledge of distribution and abundance (past and present) it is extremely unlikely that the Scientific Committee would be able to assess the status of most invertebrates or cryptogamic plants. For those few species in these categories which can be assessed, and which are found to satisfy the criteria, listing is important, not only as it will provide an impetus for targeted conservation measures but as a means of getting the message across to the broader public that all components of biodiversity are of value. For the majority of invertebrates and lower plants, however, any improvement of conservation status through the *Threatened Species Conservation Act* will arise through their occurrence in listed Endangered Ecological Communities.

Populations

A population is defined in the Act as

a group of organisms, all of the same species, occupying a particular area.

and

A population is eligible to be listed as an endangered population if, in the opinion of the Scientific Committee, it is facing a high risk of becoming extinct in nature in New South Wales and it is of conservation value at the State or regional level for one or more of the following reasons:

(a) it is disjunct or near the limit of its geographic range,

(b) it is or is likely to be genetically, morphologically or ecologically distinct,

(c) it is otherwise of significant conservation value.

In its particular formulation the ability to list populations is unique to NSW. In some circumstances populations can be listed under the *Commonwealth Environment Protection and Biodiversity Conservation Act*, but as if they were species, rather than in a category of their own. The ability to recognize populations is a valuable feature; it allows listing of distinctive subspecific variants and hence addresses in part the conservation of the genetic level of biodiversity and also provides acknowledgement that species become extinct on a population by population basis so that protection of populations is one measure to reduce extinction risk at species level. In some localities, most particularly North Head, endangered populations have become the focus of local conservation effort (Banks 2004, Lambert 2004). A more comprehensive analysis of issues associated with endangered populations is provided by Baker (2004).

Ecological Communities

The provision to list endangered ecological communities potentially provides a very powerful tool for biodiversity conservation. It has proved to be one of the more controversial aspects of the Act, partly because of unfamiliarity by many with the concept, and also because of claims that community listing has a greater potential to obstruct development than does listing species.

Ecology textbooks provide a variety of definitions of 'community' but for purposes of the Act:

ecological community *means an assemblage of species occupying a particular area.*

This definition does not constrain the type of species making up an assemblage. There is a widespread view that equates ecological community with vegetation type, but this is not a requirement of the legislation. The majority of the communities listed to date have been based on plant communities, which reflects the availability of data.

Communities can, by analogy, be considered as Russian Dolls, in both a physical and conceptual sense.

In a physical sense, communities consist of sets of working parts. Each of these sets will, in its own right, satisfy the definition of an ecological community. For example the Hygrocybeae Community of Lane Cove Bushland Park has been listed as an Endangered Ecological Community, even though this fungal assemblage is dependent on the vascular plant community in which it occurs.

Most community classifications are hierarchical, and the entities at any level in the hierarchy meet the definition of ecological community. Thus, depending on our purpose, we could recognize rainforest in NSW as a single entity, as four (subtropical, warm temperate, cool temperate and dry) or as a much larger number of types nested within the higher levels in the hierarchy. For purposes of listing, what level in the hierarchy is appropriate? The Act provides no guidance and the Scientific Committee is required to exercise judgement on a case by case basis.

What is an endangered ecological community?

Section 12 of the Act states:

An ecological community is eligible to be listed as an endangered ecological community if, in the opinion of the Scientific Committee:

(a) it is likely to become extinct in nature in New South Wales unless the circumstances and factors threatening its survival cease to operate, or

(b) it might already be extinct.

There is a number of consequences of this section. Firstly, any listing is of communities and not of individual sites (although clearly stands of plant communities occupy sites, but sites are not identified in the Schedules). There is no direct equivalent in NSW legislation of the concept of Site of Special Scientific Interest embodied in UK conservation and planning law (Evans 1997). A site may be of exceptional interest because, for example, it supports the largest known population of a particular species, but this could not be the basis for listing the site as an Endangered Ecological Community. It is possible that protection could be afforded through reaching a Voluntary Conservation Agreement or via heritage legislation, but it is not within the scope of the *Threatened Species Conservation Act*. Secondly the criteria make no reference to the condition of examples of the community. In relation to Commonwealth listing, Purdie (2003) has recently argued that "Very highly disturbed remnants are generally not considered to be part of the communities listed under the EPBC Act", while noting that "Although highly disturbed areas may not be part of the listed ecological community they can still be very important, especially where very little of the vegetation remains in good condition."

I would argue that there needs to be a clear separation of the processes of identifying a particular stand as being of an endangered ecological community and the assessment of the significance of that stand. The *Threatened Species Conservation Act* amended the *Environmental Planning and Assessment Act 1989* by requiring that possible impacts of proposals on items listed on the Schedules of the *Threatened Species Conservation Act* be assessed during the planning process. The first stage of this assessment is the identification of possible significant impacts through the so-called 8-part test, subsequently there may be a requirement to examine these possible impacts in more detail in a Species Impact Statement (a Species Impact Statement can be prepared for populations and ecological communities as well as for species). If the condition of a stand were used to deny the occurrence of an endangered ecological community then the 8-part test would not be conducted. In my opinion this would be contrary to the intent of the Act. My view in this regard has received support from the Land and Environment Court. In Plumb-v-Penrith City Council and Anor [2002] NSWLEC223, Chief Justice Pearlman decided that Cumberland Plain Woodland (an endangered ecological community) occurred on the site in contention, but the question of the occurrence of the community (which had been debated) should be answered separately from that of its importance, and in that regard determined that an 8-part test would lead to a conclusion of no significant impact and hence a Species Impact Statement had not be required. As Purdie (2003) acknowledged, there may be circumstances when a highly degraded remnant of a community is of great significance, but it is difficult to encapsulate these as a universal rules embodied in the legislation. Rather the legislation has established a regime to allow each situation to be examined on a case by case basis, and this regime needs to be followed rather than evaded.

What does extinction of a community mean?

In order to be listed as endangered, a community must be in danger of extinction (or be already extinct).

For a species, the concept of extinction is straight forward. A species becomes extinct when the last individual dies – while in practice it may be very difficult to be certain when this occurs, there is no doubt as to the measure of extinction. For communities it is not so simple. If the sites of all the former stands of a community no longer support the community, nor provide opportunities for regeneration, then the community could clearly be said to be extinct. At the time of listing, the Eastern Suburbs Banksia Scrub (see Lambert 2004) was clearly close to this state. However, for most communities the threat is not one of being covered by concrete but of environmental degradation (changed nutrient status, hydrology, invasion by weeds and feral animals) leading to loss of components of the original community, or preventing regeneration. If regeneration cannot occur, then the community is effectively extinct; even if sites remain vegetated, the integrity of the community has been lost.

Issues associated with community listing

Community listings have attracted a great deal of controversy and criticism, far more than has been generated by species and population listings. One of the reasons for this is that we lack an agreed community typology appropriate for the listing of endangered ecological communities, even if we restrict consideration to communities based on vegetation types. There are numerous approaches to vegetation classification; the appropriateness of any particular method depending on the use to which it is put.

Historically approaches to vegetation classification in Australia have utilized different attributes at different levels in the hierarchy. The upper level units have been defined structurally or physiognomically, the lower level units floristically, although often only canopy or dominant species are considered (see for example Beadle 1981). These approaches are appropriate for preparing large area vegetation maps, and for addressing a variety of natural resource management issues. The definition of community required by the Act gives absolute predominance to an "assemblage" of species, so in the case of plant communities this would require a floristic approach. This is appropriate in the context of an objective to promote biodiversity conservation. While in some instances there may be concordance between structural and floristic classifications this is not always the case. There is a long tradition in Europe of comprehensive floristic classifications (e.g. Ellenberg 1978, Westhoff and den Held 1969), but these classifications took many years to develop. The National Vegetation Classification of Britain (Rodwell 1991-2000) took nearly a quarter of a century to complete (despite the land area and size of flora being a fraction of that of NSW). Interestingly Rodwell (1991-2000) aimed to provide descriptions of communities, and while maps of the location of quadrats are provided there is no overall vegetation map. Clearly the listing of endangered ecological communities in NSW cannot await the completion of a comprehensive floristically based vegetation classification. The Scientific Committee will have to make assessments for communities originally defined on other than floristic criteria. The work required to obtain sufficient data, on a case by case basis, to characterize the assemblage of species occurring in a community may be considerable.

Criticisms of community listings have been many and varied but the main issues of contention have involved:

- the nature of the 'assemblage of species'
- the nature of the 'particular area'
- the assessment of likelihood of extinction.

Many of the issues have been addressed at length by the Court of Appeal in VAW (Kurri Kurri) Pty. Ltd.-v-Scientific Committee (Established under s127 of the Threatened Species Conservation Act 1995) [2003] NSWCA 297. (Henceforth referred to as VAW).

The total assemblage of species comprising any ecological community is unknown and, for all practical purposes, unknowable. A community will contain a range of taxa from mammals to viruses, many of which will have been uncollected and unnamed. Nevertheless, the Final Determination for the community will include a list of species whose presence can be used to recognize and delineate occurrences of the community. Measures taken to protect and conserve the community will benefit all the organisms within it, not just those included in the determination. Actions not specifically identified in the determinations, which might harm components of a community, for example spraying insecticide on a stand of an endangered ecological community for which the identifying species are vascular plants, would constitute damage to the community.

The lists of species provided in determinations are, in most cases, considerably less than the total number of species which have been recorded for any community. The lists are provided for a number of reasons. Firstly they provide evidence that the community has been defined by an assemblage of species, and that the assemblage is different from that in other communities. Secondly "the terms of the Scientific Committee's final determination must enable a citizen to decide whether a specific location falls within it" (Chief Justice Spigelman in VAW). The list of species must enable the community to be identified. Unfortunately, not every citizen is able to identify species. In the case of fungi or lichens very few ecologists would be confident of their identifications, but in general, competent consultants and informed members of the public should be able to recognize and identify the species concerned. Few of the species mentioned in the determination will be restricted to a single community, it is the co-occurrence of species forming the assemblage that determines the distinctiveness of a community (see Westhoff and van der Maarel 1973). Nor is it a requirement that any of the species listed be themselves threatened. Some communities may contain no currently listed species, in other cases threatened species are so rare that they may occur in only a few stands of the community so that their inclusion on the list would not assist in the identification of most examples of a community.

A frequently-raised issue is how many of the species listed in determinations have to be present in order for a particular community to occur. In Commonwealth of Australia-v-Randwick City Council [2001] NSWLEC 79, the Commonwealth argued in relation to Eastern Suburbs Banksia Scrub that, "to qualify as ESBS, all 46 species of the Final Determination must be present." However, "The Court rejects this construction of the Final Determination since it is unlikely that any ecological community would contain all 46 species." Chief Justice Spigelman in VAW found that the preparation of a final determination "does not necessitate the enumeration of the minimum number of species that must be found together to constitute the community" and also "that the wording of final determinations must be sufficiently flexible as to enable the protection of communities through periods of seasonal and climatic fluctuations."

The number of species from the determination, and the total number of recordable species, in a stand may vary for a number of reasons, including size of stand, disturbance history and local environmental conditions. It would surely defeat the object of the Act if an example of an endangered ecological community could not be regarded as being part of the community immediately after a fire, when the species listed in the determination could not be identified as being present.

Listings of species apply throughout the State. Although a determination for a species may indicate a restricted known distribution, should that species subsequently be found in different locations then it is still an endangered or vulnerable species (as the case may be) in these new localities and all the consequences of being listed automatically apply.

In the case of endangered ecological communities and endangered populations the situation is different. The listing only applies to 'a particular area' as defined in the final determination. If a final determination understates the extent and location of occurrences, the occurrences beyond the boundary of the particular area specified in the determination are not included in the listing and it would be necessary to undertake a formal revision to create a new determination altering the boundary of 'the particular area'.

In an ecological context the particular area occupied by a community is its habitat, and it would be possible to define this in terms of biophysical factors with varying degrees of precision. Determinations for communities include information on habitat but, although the Act does not include any definition of 'particular area', it is clear that merely to describe the distribution in terms of habitat is insufficient and that area requires a spatial geographic description.

The question then arises as to how restrictive the account of the geographic distribution of a community should be. Determinations tend to specify occurrence in bioregions, supplemented by local government areas in which the community is known to occur. In some cases the habitat of a community is so restricted in distribution that 'the particular area' is of very limited extent (for example Genowlan Point *Allocasuarina nana* heathland or the Hygrocybeae Community of Lane Cove Bushland Park), in other cases the 'particular area' may encompass one or more bioregions (for example White Box Yellow Box Blakely's Red Gum Woodland).

It is of the nature of communities, just as it is of species, that extent of distributions vary – not all endangered species will be restricted to single localities, and not all endangered ecological communities will occur as a single patch. Examination of the overseas literature demonstrates that in this regard there is nothing unusual about the distribution patterns of communities in NSW.

(For example in the UK, *Erica vagans – Ulex europaeus* heath is restricted to the Lizard peninsula in Cornwall where it is associated with soils derived from a particular form of serpentine (Rodwell 1991), while *Calluna vulgaris – Scilla verna* heath is found on coastal headlands from Lands End to the Shetlands although, because of the specialized nature of the habitat, the total area is not large (Rodwell 1991)).

A frequent demand from property owners and local councils is that maps should be available of the distribution of listed communities, or that they should be defined by cadastral boundaries. In some cases maps may be available, but often published maps are inadequate to determine whether a particular community occurs on a development site. Vegetation maps at scales of 1:100 000 or 1:250 000 cannot accurately depict the occurrence of small stands of communities, and in any event maps are representative of the situation at the time their supporting data were collected. Changes in distribution occur over time so that maps would need to be checked by ground truthing to ascertain their continuing relevance. Chief Justice Spigelman found in VAW that there was no requirement in Determinations for "the provision of maps indicating where the community may be found".

That a community is likely to become extinct is an opinion to be reached by the Scientific Committee. Clearly there must be grounds for reaching this opinion, but the fact that others might disagree does not invalidate the Committee's opinion. In most instances submissions questioning the opinion have a narrower concept of what might constitute extinction than that discussed earlier in this paper.

Threatening Processes

Species, populations and communities are threatened in most cases because of the operation of particular processes, many as a direct or indirect result of human actions. The identification and listing by the Scientific Committee of Key Threatening Processes results in a obligation on the Government to develop and implement a Threat Abatement Plan. Successfully identifying and addressing Key Threatening Processes should provide a more effective and efficient means of achieving biodiversity conservation than implementing programs on a species by species (or community by community) basis without an understanding of the causes of decline to threatened status.

A threatening process is eligible to be listed as a key threatening process if, in the opinion of the Scientific Committee, it:

(a) adversely affects 2 or more threatened species, populations or ecological communities, or

(b) could cause species, populations or ecological communities that are not threatened to become threatened.

The requirement for listing of a Key Threatening Process is that it adversely affects species populations or communities. Depending on the particular entities threatened, a key threatening process may be widespread or localized in its operation. Although most listed key threatening processes have been state wide one, 'Predation by the Ship Rat *Rattus rattus* on Lord Howe Island', is relevant only to a small part of the State.

Provisional listing

The final aspect of listing which needs to be discussed is provisional listing:

A species is eligible to be provisionally listed in Schedule 1 as an endangered species if, in the opinion of the Scientific Committee:

(a) the species

(i) although not previously known to have existed in New South Wales, is believed on current knowledge to be indigenous to New South Wales, or

(ii) was presumed to be extinct in New South Wales but has been rediscovered, and

(b) the species is not listed in Part 1 of Schedule 1.

The effect of provisional listing is that a species is immediately entered on the schedules, without the processes of provisional and final determination. Subsequent to provisional listing, which lapses after a year, the 'normal' avenue of advertising a provisional determination to seek submissions and then the making of a final determination must be followed. Only species can be subject to provisional listing, there is no mechanism for the emergency listing of populations or communities.

Discussion

The *Threatened Species Conservation Act 1995* has been operating since the start of 1996. During that time a large number of nominations have been assessed by the Scientific Committee (see Dickman 2004). The diversity of sources of nominations is a measure of the success of the Act in tapping the wealth of community knowledge of biodiversity. A process that relied only upon existing "official" sources of information would have been unlikely to have resulted in determinations for fungi, a lichen community or most of the now-listed invertebrates.

The schedules are far from comprehensive, and for many groups of cryptogamic plants and invertebrates, the depth of our ignorance is such that it will be very many years before we can make judgements as to which individual species are threatened. If these species are to be conserved, we cannot wait for listing and the implementation of individual species recovery plans; if the *Threatened Species Conservation Act* is to work in these cases it will be through Threat Abatement Plans addressing Key Threatening Processes (see Leys 2004), or recovery plains for Endangered Ecological Communities (see Auld and Tozer 2004).

Even for what are considered the better known taxonomic groups, vertebrates and flowering plants there will be many changes to the schedules, some arising from the discovery of species new to NSW, others from a reassessment of the status of long known species. Amongst the flowering plants the distribution of threatened species is uneven both taxonomically and geographically (Mokany and Adam 2000, Adam 2002). Whether the patterns are 'real', reflect under recording or result from inadequate assessment of data remains to be determined.

Burgman (2004) has argued that the decision-making process leads to uncertainty as to the status of species, given differences of opinion between experts. The degree of uncertainty may, however, depend on the complexity of the assessment process. Under the *Threatened Species Conservation Act* the number of categories of threat status is limited (non threatened, vulnerable, endangered for species; non threatened, endangered for populations and ecological communities, the greater number of categories in the schemes discussed by Burgman may promote variation in opinions. In the context of a legislative scheme, simplicity is desirable. A more complex scheme may be appropriate in prioritization of management actions, but it is debatable whether such a scheme should be embodied in legislation.

The mechanisms for reviewing the schedules, including through nomination for amendment or delisting, provide means by which perceived uncertainties in the listings.

The great majority of listings by the Scientific Committee have attracted little controversy. Some ecological community determinations have been extremely controversial, but a number has passed through the process to listing with little public comment. The whole concept of ecological communities is still regarded with suspicion in some quarters. In many cases there will be a need to determine the application of listings in the field on a case by case basis and there may be residual uncertainty as to the exact boundaries of stands of particular communities. The Court of Appeal in VAW recognized there was an inherent 'uncertainty' associated with community determinations but that this need not be such as to invalidate determinations. [Chief Justice Spigelman "The intricacy of all ecological communities means that some indeterminateness is bound to arise from the form of expression used to describe them". Justice of Appeal Hodgson "It is plain therefore that a certain amount of vagueness and uncertainty in the determination of boundaries of ecological communities is an inevitable result of formulations chosen by the legislature. It must follow, in my opinion, that a certain amount of vagueness and imprecision will not necessarily cause a determination that a named ecological community is an endangered ecological community to be regarded as void or invalid because of uncertainty or lack of definition]. This does remove from the Scientific Committee the obligation to make its determinations as clear, precise and unambiguous as possible, but it does mean that the absolute certainty demanded by some critics is unrealistic and based on misunderstandings as to the nature of ecological communities.

The listing process is an essential step in developing conservation measures for species populations and ecological communities, but of itself does not guarantee successful conservation outcomes. The real measure of the success of the *Threatened Species Conservation Act* – whether it is indeed more than just an act – will be in the reduction in threats to biodiversity and in the success of *in situ* measures to improve the conservation status, in the wild, of elements of biodiversity.

References

Adam, P. 2002. Rarity, rare plant species and the New South Wales. *Threatened Species Conservation Act – conservation opportunities and challenges*. *Cunninghamia* **7**: 651-669.

Auld, T.D. and Tozer, M. 2004. Endangered ecological communities and landscape conservation in NSW: successes and failures in the Sydney Basin. Pp 94-101 in *Threatened Species Legislation : is it just an Act?* edited by P. Hutchings, D. Lunney and C. Dickman. Royal Zoological Society of NSW, Mosman, NSW.

Baker, J. 2004. Endangered populations: the concept in practice. Pp 82-87 in Threat*ened Species Legislation : is it just an Act?* edited by P. Hutchings, D. Lunney and C. Dickman. Royal Zoological Society of NSW, Mosman, NSW.

Banks, P.B. 2004. Population viability analysis in urban wildlife management: modeling management options for Sydney's quarantined bandicoots. Pp 70-77 in *Urban wildlife: more than meets the eye*, edited by D. Lunney and S. Burgin. Royal Zoological Society of NSW, Mosman, NSW.

Beadle, N.C.W. 1981. *The vegetation of Australia*. Gustav Fischer, Stuttgart.

Burgman, M.A. 2004. Expert frailties in conservation risk assessment and listing decisions. Pp 20-29 in *Threatened Species Legislation : is it just an Act?* edited by P. Hutchings, D. Lunney and C. Dickman. Royal Zoological Society of NSW, Mosman, NSW.

Campbell, N.A. and Reece, J.A. 2002. *Biology*. 6th edn. Benjamin Cummings, San Francisco.

Dickman, C.R. 2004. The Scientific Committee under the NSW Threatened Species Conservation Act1995: seven years of debate. Pp 68-81 in *Threatened Species Legislation : is it just an Act?* Edited by P. Hutchings, D. Lunney and C. Dickman. Royal Zoological Society of NSW, Mosman, NSW.

Dickman, C.R. and Lunney, D. (eds) 2001. *A symposium on the dingo*. Royal Zoological Society of NSW, Mosman, NSW.

Ellenberg, H. 1978. *Vegetation Mitteleuropas mit den Alpen*. 2nd edn. Ulmer, Stuttgart.

Evans, D. 1997. A *history of nature conservation in Britain*. 2nd edn. Routledge, London.

Farrier, D. and Whelan, R. 2004. (Why) Do We Need Threatened Species Legislation? Pp 30-48 in *Threatened Species Legislation : is it just an Act?* Edited by P. Hutchings, D. Lunney and C. Dickman. Royal Zoological Society of NSW, Mosman, NSW.

Handtke, K. and Mauersberger, G. 1977. Die Ausbreitung des Kuhreihers, *Bubulcus ibis* (Linneaus). *Mitteilungen Zoologisches Museum in Berlin (Suppl.) Band 3. Ann. Orn.* **1**: 1-78.

Jarman, P. and Brock, M. 2004. The evolving intent and coverage of legislation to protect biodiversity in New South Wales. Pp 1-19 in *Threatened Species Legislation : is it just an Act?* edited by P. Hutchings, D. Lunney and C. Dickman. Royal Zoological Society of NSW, Mosman, NSW.

Lambert, J. 2004. Threatened species legislation: Does it work for local communities or Local Government? Pp 158-162 in *Threatened Species Legislation : is it just an Act?* Edited by P. Hutchings, D. Lunney and C. Dickman. Royal Zoological Society of NSW, Mosman, NSW.

Leys, A. 2004. Threat abatment plans: strategic pest management for biodiversity conservation. Pp 102-106 in *Threatened Species Legislation : is it just an Act?* Edited by P. Hutchings, D. Lunney and C. Dickman. Royal Zoological Society of NSW, Mosman, NSW.

Mokany, K. and Adam, P. 2002. The biogeographical attributes of the threatened flora of New South Wales. *Cunninghamia* **6**: 873-892.

Moore, P.D. 1998. Plant extinction: frondless ferns lie low to survive. *Nature* **392**: 661-662.

Purdie, R. 2003. Threatened ecological communities. *Australasian Plant Conservation* **12(2):** 10-12.

Ratcliffe, D.A., Birks, H.J.B. and Birks, H.H. 1993. The ecology and conservation of the Killarney fern *Trichomanes speciosum* Willd. in Britain and Ireland. *Biological Conservation* **66**: 321-347.

Rodwell, J.S. (ed.) 1991. *British plant communities vol.2. Mires and heaths.* Cambridge University Press, Cambridge.

Rodwell, J.S. (ed.) 1991-2000. *British plant communities vols.1-5.* Cambridge University Press, Cambridge.

Rumsey, F.J., Jermy, A.C., and Sheffield, E. 1998. The independent gametophytic stage of *Trichomanes speciosum* Willd. (Hymenophyllaceae) the Killarney Fern and its distribution in the British Isles. *Watsonia* **22**: 1-19.

Westhoff, V. and den Held, A.J. 1969. *Plantengemeenschappen in Nederland.* Thieme: Zutphen.

Westhoff, V. and van der Maarel, E. 1973. The Braun-Blanquet approach. Pp.619-726 in *Handbook of vegetation science V. Ordination and classification of vegetation.* edited by R.H. Whittaker. Dr. Junk, The Hague.

Postscript 24 September 2004.

At the time of going to press the *Threatened Species Legislation Amendment Bill* 2004 was before Parliament. The bill will amend the *Threatened Species Conservation Act* 1995 and the *Fisheries Management Act* 1994. The amendments, if passed, will substantially alter the operation of the *Threatened Species Conservation Act.* The Bill will make provision for biodiversity certification, by the Minister for the Environment, of the native vegetation reform package and of environmental planning instruments, EPIs. (EPIs include a range of plans, but the most important in this context will be local government Local Environmental Plans, LEPs). The Minister may grant biodiversity certification where an EPI seeks to promote conservation of threatened species. Where certification is granted, development or an activity permitted by the EPI will be deemed not likely to significantly affect threatened species, thus removing the need to address the significance of possible impacts for threatened species and the need for preparation of a species impact statement. Importantly certification would potentially remain in force regardless of changes in the Schedules of the TSCA. The government has put in place new bodies which will be the major players in the application and policing of natural resource management with the establishment of thirteen independent Catchment Management Authorities and the Natural Resources Commission. The existing requirements for the preparation of recovery plans and threat abatement plans will be replaced by new provisions for the Director-General of the Department of Environment and Conservation to prepare and adopt a Threatened Species Priorities Action Statement, which will set priorities for the recovery of threatened species and for the abatement of key threatening processes. The Bill seeks to establish new categories of critically endangered species and ecological communities and foreshadows regulations which will set criteria to be met for listing under the various schedules. Provision is also to be made for an extra step in the listing process, by which the Scientific Committee would inform the Minister of an intent to make a final determination, and the Minister could refer the matter back to the Committee for further consideration. The ultimate decision would, however, remain with the Scientific Committee.

The intent of the most of the changes is said to be to give greater certainty in land use planning. Whether this will be the practical outcome, and whether it can be achieved at the same time as satisfying the core objective of the TSCA of conserving biodiversity remains to be seen. A few years hence a future symposium may be required to critically analyse the outcome of the new statutory regime.

The Scientific Committee under the NSW *Threatened Species Conservation Act 1995*: seven years of debate

Chris R. Dickman

Chair, Scientific Committee, 1996-2002

Institute of Wildlife Research, School of Biological Sciences, University of Sydney, NSW 2006

Email: cdickman@bio.usyd.edu.au

ABSTRACT

As specified by the NSW *Threatened Species Conservation Act 1995*, the Scientific Committee is responsible for listing threatened species, populations, ecological communities and key threatening processes in schedules of the Act, as well as for several review and advisory tasks. In this paper, I outline the composition, structure and functioning of the NSW Scientific Committee during its first seven years of operation (1996-2002), and then describe the Committee's approach in evaluating five nominations of different entities for listing. Each generated debate about different aspects of the listing process, the amount and quality of information required for the Scientific Committee to be able to make a determination, and about how criteria for listing different entities in the Act should be interpreted. From the debates, the Committee was able to develop robust and effective protocols for review of nominations and assessment of status. These protocols allowed the Scientific Committee to process some 600 nominations for listing and delisting of species, populations, communities and key threatening processes during its first seven years of operation, and should continue to assist the Committee to function in future.

Key words: Scientific Committee, threatened species, populations, ecological communities, threatening processes.

Introduction

Passage of the *Threatened Species Conservation Act* in New South Wales in 1995 provided for the establishment of a Scientific Committee to carry out and oversee a number of statutory obligations. Among these obligations were to maintain lists of threatened species, endangered populations and ecological communities in Schedules 1 and 2 of the Act, and to list key threatening processes in Schedule 3. Additional obligations were to review drafts of joint management agreements between the NSW Department of Environment and Conservation, or DEC (formerly NSW National Parks and Wildlife Service), and other public authorities for conservation of threatened taxa, populations or ecological communities, to provide comment on draft recovery and threat abatement plans and advice on the prioritization of plans, and to advise both the Director-General of DEC and the Minister for the Environment on biodiversity conservation in general. A Scientific Committee was established in 1996 and has functioned continuously to the present.

Effective operation of the Scientific Committee is important for several reasons. Firstly, the listing of threatened species, endangered populations and endangered ecological communities confers recognition of status and some protection of these entities from further threats. It also triggers a requirement by DEC to prepare a recovery plan within a specified time period. Listing of key threatening processes likewise confers recognition of important threats to biodiversity and leads to the preparation of threat abatement plans by DEC in consultation with other stakeholders. Secondly, critical input by the Scientific Committee into draft recovery plans, threat abatement plans, joint management agreements and other documents assists in ensuring that conservation management is underpinned by sound principles. Finally, because of the obligation to provide relevant advice to the Minister for the Environment and Director-General of DEC, the Scientific Committee is potentially able to flag emerging problems and issues of general conservation concern before they become serious. The Scientific Committee is independent of both the Minister and the Director-General and its decisions on listing are final; hence its roles in offering sound advice and commenting on policy, as well as maintaining lists of threatened entities and key threatening processes, are particularly important.

Because of the manifold functions of the Scientific Committee and the significance of its operations, it is of paramount importance that the Committee not only carries out its duties but is also seen to work effectively. A parliamentary review of the Threatened Species Conservation Act two years after its inception vindicated both the independence and work of the Scientific Committee (Anon 1997). However, criticisms have been levelled that the operations of the Scientific Committee are not accountable (Lim 1997), that they "stretch the capacity of science further than is widely recognised" (Cardew 1997), and that they do not consider land use planning (Jerogin 1997). Decisions of the Scientific

Pp 68 - 81 in *Threatened species legislation: is it just an Act?* 2004, edited by Pat Hutchings, Daniel Lunney and Chris Dickman. Royal Zoological Society of New South Wales, Mosman, NSW, Australia.

Committee have also been reported negatively and sensationally in the press (e.g., "Urban bush seized"; Benson 1997), or challenged in the NSW Land and Environment Court (VAW (Kurri Kurri) Pty Ltd versus the Scientific Committee 2002-03) and NSW Court of Appeal (VAW (Kurri Kurri) Pty Ltd versus the Scientific Committee 2003). Although the criticisms have been addressed (Adam *et al.* 1997*a*, *b*; Smith 1997) and legal challenges dismissed, it is still crucial that the Scientific Committee is seen as impartial, objective and effective in carrying out its operations.

In this paper, I outline the functioning of the Scientific Committee in its first seven years of operation (1996-2002), when I had the privilege of being Chairperson. I shall briefly describe the composition of the Committee, how it carried out its tasks, and then focus on the approaches taken by the Committee in evaluating and finally listing several contentious species, populations, communities and key threatening processes in the schedules of the *Threatened Species Conservation Act*. While this account is largely historical, it has strong contemporary bearing because almost all of the entities listed from 1996-2002 remain in the schedules of the Act, and because debates on procedures and approaches led to a robust framework that should assist the operations of both the current and future committees.

Scientific Committee: composition

Section 129 of the *Threatened Species Conservation Act 1995*, as enacted in 1996, specified membership of the Scientific Committee as follows:

(1) The Scientific Committee is to consist of 10 members appointed by the Minister.

(2) Of the members of the Scientific Committee:

(a) two are to be scientists employed by the National Parks and Wildlife Service nominated by the Director-General,

(b) one is to be a scientist employed by a public authority, having expertise in one or more of the areas of study referred to in subsection (3), selected by the Minister,

(c) one is to be a scientist nominated by the Commonwealth Scientific and Industrial Research Organization,

(d) one is to be a scientist employed and nominated by the Australian Museum Trust,

(e) one is to be a scientist employed and nominated by the Royal Botanic Gardens and Domain Trust,

(f) one is to be a scientist nominated by the Ecological Society of Australia,

(g) one is to be a scientist nominated by the Entomological Society of Australia,

(h) one is to be a scientist who is employed by a tertiary educational institution and who is selected by the Minister,

(i) one is to be a scientist having expertise in agricultural science and natural resource management who is selected by the Minister.

3. A person appointed as a member of the Scientific Committee is to have expertise in one or more of the following areas of study:

(a) vertebrate biology,

(b) invertebrate biology,

(c) plant biology,

(d) terrestrial ecology,

(e) plant community ecology,

(f) limnology,

(g) marine ecology,

(h) genetics of small populations,

(i) population dynamics (including population viability analysis or evolutionary ecology).

Members of the original and subsequent Scientific Committees were selected following an extensive nomination process involving the various societies, organisations and authorities listed in section 129, thus ensuring that the Committee had broad expertise in biodiversity conservation and resource management. Lim (1997) commented that the Act did not require representation from local government, the union movement or the manufacturing and mining industries, but apparently confused the scientific role of the Committee with economic, cultural and social implications that arise when developments are proposed. Following amendments to the Threatened Species Conservation Act in 2002, membership of the Scientific Committee has now been expanded to 11.

Scientific Committee: structure and functioning

Any large, diverse body of people faced with a continuing and heavy workload needs an efficient administrative structure to function effectively, and the Scientific Committee is no exception. From its first meeting in May 1996, the Committee met for at least one full day a month and, by December 2002, had sat on 76 occasions. Much of the business of meetings was concerned with evaluating nominations for listing, delisting or changing the status of entities in the schedules of the *Threatened Species Conservation Act*; procedures for this have been described elsewhere (Dickman 1997). However, meetings also provided venues for discussing protocols, refining determinations, reviewing public submissions on preliminary determinations and other documents such as draft recovery plans, debating contentious issues, and developing proposals for listing within the Committee itself. All members were asked at the beginning of meetings to declare any pecuniary or non-pecuniary interests that may have affected their judgments on issues before the Committee, and to bring any issues of general interest to the table. In addition, all members of the Scientific Committee carried out substantial research between formal meetings, reviewing papers, seeking expert opinion and information on particular issues, preparing draft determinations and assessing comments received after advertising of preliminary determinations.

Under sections 22 and 24 of the *Threatened Species Conservation Act*, the Scientific Committee is required to notify the nominator, Minister for the Environment and Director-General of the NSW National Parks and Wildlife Service (DEC) when preliminary and final determinations are made. Notice of any decision must also be published in a statewide newspaper (the *Sydney Morning Herald* was selected), the *Government Gazette*, and any local newspapers that circulate in areas that may be affected by a determination. However, in response to early criticism that many potentially interested parties did not get to hear of the Scientific Committee's determinations until too late, letters of notification were also sent routinely to many government and non-government organisations, interested individuals, scientific advisory committees in Victoria and the Australian Capital Territory, the Commonwealth Threatened Species Scientific Committee, to all local councils that may be affected by a determination, and in some instances landholders, if they could be identified. Most members of the Scientific Committee, especially the Chair and Deputy Chair, were called upon at times to comment on determinations in print, radio or television interviews, or to present seminars on Committee functions to interested societies and organizations (e.g., Ecological Society of Australia, Royal Australian Planning Institute, Royal Zoological Society of New South Wales). As Chair, I estimate that these various activities occupied 2-2½ days a week on average; they could not have been less for the Committee's Deputy Chair during the period 1996-2002, Paul Adam.

While members of the Scientific Committee worked diligently to carry out their tasks, little could have been achieved without the dedicated, full-time support of Executive Officers, Kylie Mayne (1996) and Sue Chate (1997 to the present) and their assistants Jan Bloomfield and Kelly Taylor. In addition to organizing meetings and correspondence, liaising with nominators, submission writers, officers of DEC and other government departments, as well as the Committee itself, executive support has extended to fielding many inquiries from the public as well as to coping with the occasional abusive phone call. Most of the Committee's meetings have been held at the DEC Head Office in Hurstville, but others have been hosted by regional offices in Coffs Harbour, Dubbo and Queanbeyan. DEC has provided considerable further support to the Scientific Committee by providing funds for dedicated research assistants, access to legal advice on the prosecution of Committee duties, a workshop in Dubbo on potentially endangered ecological communities in the state's far west, and briefings on current and emerging issues. Integration of these people and tasks has been essential to the efficient functioning of the Scientific Committee since its inception.

Seven years of debate

Despite the considerable expertise within the Scientific Committee and much support to ensure its effective functioning, many issues that came before the Committee were contentious and resolved only after much debate. For example, the definitions of several important terms in the *Threatened Species Conservation Act*, such as 'vulnerable', 'endangered', 'population' and 'community' leave much scope for interpretation (Adam 2004), while concepts such as community extinction are difficult to delineate; is a community extinct when one species is lost, or should some (pre-defined) set of species disappear before extinction can be ascribed? For species or other entities that are cryptic, poorly-defined or known, what is the minimal kind or amount of information required to make a decision on status? What can be done when there is residual error or uncertainty about the information that is available (Akçakaya *et al.* 2000; Burgman 2004), and what role can the precautionary principle (Calver *et al.* 1999; Whelan *et al.* 2004) play in deliberations?

In general, the Scientific Committee used as much information as was available from different sources to inform its decision-making process. We sought expert opinion on virtually all nominations received. The primary, peer-reviewed literature was consulted frequently but, where appropriate, licence records, unpublished reports and other 'grey' literature documents were also used. Much use was made, in addition, of distributional data provided by DEC, State Forests of NSW, the Australian Museum, Royal Botanic Gardens and Herbarium and other authorities. Consultants were engaged to review certain taxonomic groups for which information was scattered, such as marine mammals, birds and reptiles, and ecological communities in western New South Wales. These resulted in detailed written reports, with recommendations or options for listing, and sometimes also in presentations to the Committee. With certain very contentious nominations, such as for the listing of the Dingo *Canis lupus dingo* and the Grey-headed Flying-fox *Pteropus poliocephalus* as threatened species, the Committee co-organised public fora (with the Royal Zoological Society of NSW and Australasian Bat Society) to gather all available information prior to making any decisions.

On yet other occasions, members of the Scientific Committee undertook their own reviews or on-site investigations of nominated entities. When available information and raw data had been collated, it was sometimes subjected to primary analysis by the Committee itself (e.g., ordinations were run on quadrat-based floristic samples to delineate ecological communities). As far as possible, overall data sets were evaluated using robust, repeatable methods such as modified IUCN criteria (Keith 1998) or point-scoring approaches (Millsap *et al.* 1990; Lunney *et al.* 2000). When all available information had been collated and reviewed, the Committee's next task was to judge whether a nominated entity met the criteria specified by the *Threatened Species Conservation Act.*

In the section below, I describe five case studies of species, populations, ecological communities and key threatening processes that generated a lot of debate at meetings of the Scientific Committee, and outline the rationales that led to each being listed.

Endangered species

Under section 10 of the *Threatened Species Conservation Act*, "A species is eligible to be listed as an endangered species if, in the opinion of the Scientific Committee:

(a) it is likely to become extinct in nature in New South Wales unless the circumstances and factors threatening its survival or evolutionary development cease to operate, or

(b) its numbers have been reduced to such a critical level, or its habitats have been so drastically reduced, that it is in immediate danger of extinction, or

(c) it might already be extinct, but is not presumed extinct."

***Nurus brevis* Motschulsky, 1865.** Nomination of this large carabid beetle was made on the grounds that the species appears to have a restricted distribution and is in very low numbers in New South Wales. Before proceeding to a determination, however, the Scientific Committee debated whether the beetle had been sufficiently sampled to be confident of its true status, whether it could be confused with other carabids (in particular, *Nurus atlas* had also been nominated at the same time for listing as a threatened species), whether it was known to have declined, or faced any current threats. Inquiries indicated that the beetle is a large (5 cm), highly distinctive predator with black and metallic green reflections that should allow ready detection in the field. It had been collected many times during the late nineteenth and early twentieth centuries, and recent records were scant despite targeted searches. As one monitored population had become extinct in the previous decade, only two further disjunct populations were known, loss of the species' apparently preferred rainforest habitat was continuing, and the species was at risk of collectors, *Nurus brevis* was considered to clearly satisfy criterion 10(b) above, and was listed as endangered on 16 March 2001 (Appendix 1).

Vulnerable species

Under section 14 of the *Threatened Species Conservation Act*, "A species is eligible to be listed as a vulnerable species if, in the opinion of the Scientific Committee, the species is likely to become endangered unless the circumstances and factors threatening its survival or evolutionary development cease to operate".

***Cercartetus nanus* (Desmarest, 1818).** In contrast to *N. brevis*, the Eastern Pygmy-possum *Cercartetus nanus* is well-known and has a very broad distribution that encompasses the coast and tableland habitats of New South Wales, as well as several localities on the far western slopes of the Great Dividing Range. Despite the great extent of the species' occurrence, however, *C. nanus* is small, cryptic, and apparently occurs at a small number of localities where it appears to be genuinely rare. In a detailed review of the literature, Bowen and Goldingay (2000) found that *C. nanus* had been reported in only 26 of 72 surveys carried out in ostensibly suitable habitat, and that most surveys (77%) turned up fewer than 10 individuals. Over several meetings, the Scientific Committee discussed whether such a widely-distributed species could be considered threatened. Much debate focused on whether standard sampling methods could adequately reveal the presence of *C. nanus*, whether the species had declined or was continuing to decline, and whether it was subject to ongoing threats.

With respect to sampling, the Committee gave weight to evidence that such disparate sampling techniques as live-trapping, spotlighting, and detection of hairs from sampling tubes and predator scats returned similarly low rates of detection. Further evidence from experts also suggested that *C. nanus* would usually be captured in standard Elliott traps if they were present at a site and sampling effort was sufficiently intense (A. Tulloch, University of Sydney, pers. comm. 2001), and that even the most efficient means of detecting the species (nest boxes or traps set in food trees) frequently returned negative results. Given that relatively few *C. nanus* have been recorded in massive sampling efforts using a variety of techniques throughout the known distribution of the species (Bowen and Goldingay 2000), these considerations confirmed that *C. nanus* is sparse over most of its range. Because intense survey effort over the last 25-30 years has produced most records of the Eastern Pygmy-possum, it is not possible to assess trends in population size or distribution. However, continuing fragmentation and loss of habitat across much of the species' range, inappropriate fire regimes and probable impacts from introduced predators and the introduced plant disease *Phytophthora cinnamomi* (Garkaklis *et al.* 2004) provide no confidence that *C. nanus* is secure. With this body of evidence, the Scientific Committee considered that *C. nanus* met the criteria specified in section 14 of the *Threatened Species Conservation Act*, and it was listed as a vulnerable species on 8 June 2001 (Appendix 2).

Endangered populations

Under section 11 of the *Threatened Species Conservation Act*, "A population is eligible to be listed as an endangered population if, in the opinion of the Scientific Committee, its numbers have been reduced to such a critical level, or its habitat has been so drastically reduced, that it is in immediate danger of extinction and it is not a population of a species already listed in Schedule 1, and:

(a) it is disjunct and at or near the limit of its geographic range, or

(b) it is or is likely to be genetically distinct, or

(c) it is otherwise of significant conservation value."

North Head population of the Long-nosed Bandicoot, *Perameles nasuta* Geoffroy, 1804. The Long-nosed Bandicoot is relatively common in forest habitats in eastern Australia and, as a species, is not considered to be at risk in New South Wales. However, the species has disappeared from much of its former range in the suburban and more developed areas of Sydney, and the population at North Head is probably the largest one remaining in the region. The population is confined to modified and less-disturbed habitats on the North Head peninsula and conforms to the requirement for a population to be "a group of organisms, all of the same species, occupying a particular area" (section 4, *Threatened Species Conservation Act*). In addition, the population can be considered of significant conservation value in that it is disjunct, has been much-studied (e.g., Scott 1995; Scott *et al.* 1999; Ravallion 2000; Chambers and Dickman 2002; Banks 2004), and is accorded considerable value by the local community (Reizes 1997). But, could the population be considered in *immediate* danger of extinction?

Previous research at North Head had suggested that about 100 *P. nasuta* may be present, but the number of individuals effectively contributing to subsequent generations was expected to be only 10-11% of the census population (Frankham 1995). Such a small population is likely to be at risk of chance disturbances, demographic stochasticity and genetic introgression. The population was known to suffer sporadic mortality due to the depredations of red foxes, stray (and probably) domestic cats and dogs, and continuous mortality due to collisions with cars. Future loss of habitat was expected due to proposed developments at North Head. Further research on a restricted suburban population of the Eastern Barred Bandicoot *P. gunnii* in Victoria had confirmed that a comparably small population of *Perameles* could decline precipitously within just 25 years (Lacy and Clark 1990; Minta *et al.* 1990). In light of all these observations, the Scientific Committee interpreted "immediate danger of extinction" to be in the order of 10-25 years for *P. nasuta*, and considered that the North Head population satisfied the criteria for listing. The final determination was gazetted on 28 February 1997 (Appendix 3). Recent simulation modelling by Banks (2004) has specified a range of extinction probabilities for the North Head bandicoots between 2010 and 2050, thus providing independent confirmation of the Committee's conclusions.

Endangered ecological communities

Under section 12 of the *Threatened Species Conservation Act*, "An ecological community is eligible to be listed as an endangered ecological community if, in the opinion of the Scientific Committee:

(a) it is likely to become extinct in nature in New South Wales unless the circumstances and factors threatening its survival or evolutionary development cease to operate, or

(b) it might already be extinct."

Under section 4 of the Act, an ecological community is defined as "an assemblage of species occupying a particular area."

Hygrocybeae community of Lane Cove Bushland Park. At the time of the nomination of this community, more than 20 species of fungi in the family Hygrophoraceae were known to occur in a small riparian area of the Lane Cove Bushland Park in north Sydney. Although the area occupied was prescribed and the species richness of Hygrocybeae apparently very high, the Scientific Committee debated several issues. Firstly, as few people are likely to be able to reliably recognize and identify Hygrophoraceae, it was not clear if the community might occur more extensively at other locations but had simply gone unnoticed. Secondly, the community at Lane Cove has been studied over many years (Young 1999; R. and E. Kearney, Sydney Fungal Studies Group, pers. comm. 2000) so that even species with ephemeral fruiting bodies had probably been detected. If dedicated, long-term surveys at other sites had not been carried out, could disjunct fragments of the community have been overlooked? Finally, what threats might extirpate a fungal community in a protected bushland park?

Several respondents suggested that sufficient surveys had been carried out in potentially suitable habitat to detect extensions of the community, if they occurred, both in the vicinity of the Lane Cove Bushland Park and in other parts of the Sydney Basin. Many species of Hygrocybeae are distinctive and strikingly colourful, suggesting further that surveys carried out at appropriate times should detect them. Despite reasonably extensive surveys having been made, the Committee also gave weight to the fact that the Lane Cove Bushland Park is the holotype site for six species and subspecies of Hygrophoraceae (Young 1999), with four of these not known to occur anywhere else. As several of the species have been recorded from tiny areas of only 1-2 m^2 in the Lane Cove Bushland Park, they are likely to be uniquely susceptible to trampling, dumping of rubbish and invasion by introduced weeds. The community as a whole is surrounded by housing and other developments, and is thus also susceptible to urban run-off, flash flooding and potentially toxic industrial pollutants (R. and E. Kearney, pers. comm. 2000). Although the Scientific Committee debated whether the loss of one or more species could compromise the community to the extent of rendering it extinct, the uniqueness of the assemblage and the array of potential threats to its survival left no doubt that it fulfilled the criteria of section 12 of the Act, and it was listed as an endangered ecological community on 3 March 2000 (Appendix 4). Subsequent research on this community has elevated the number of Hygrocybeae species known to occur, and further confirmed the significance of the Lane Cove site (Kearney and Kearney 2000; Young *et al.* 2001).

Key threatening processes

Under section 15 of the *Threatened Species Conservation Act*, "A threatening process is eligible to be listed as a key threatening process if, in the opinion of the Scientific Committee, it:

(a) adversely affects 2 or more threatened species, populations or ecological communities, or

(b) could cause species, populations or ecological communities that are not threatened to become threatened."

Anthropogenic climate change. During the 1980s and 1990s evidence accumulated that human activity was changing the world's climate at a rate considerably faster than had occurred previously. At the same time, developments in predictive modelling were beginning to confirm that the fine-scale distributions of many species of plants and animals are determined by climate, and that many would be at risk if either average climatic conditions or the frequency of extreme events such as floods and fires were to change (Busby 1988). "Greenhouse" warming, for example, could be expected to reduce the extent of cool refugia occupied by high-altitude mammalian specialists such as Leadbeater's Possum *Gymnobelideus leadbeateri* in Victoria, or the Mountain Pygmy-possum *Burramys parvus* and Broad-toothed Rat *Mastacomys fuscus* in New South Wales (Lindenmayer *et al.* 1991; Mansergh and Broome 1994; Green and Osborne 2003). High altitude plant communities would be similarly reduced in extent (Hughes and Westoby 1994). For many taxa, the adverse effects of climate change could be averted only if there were suitable areas for them to migrate to.

In view of these and other studies (e.g., Brereton *et al.* 1995), the Committee ultimately had little doubt that anthropogenic climate change met the criteria for listing as a key threatening process. While reviewing the evidence, however, the Committee debated the further issue of whether a listing of this process in schedule 3 of the Act would have any practical effect. Further, as listing of a key threatening process triggers preparation of a threat abatement plan, the Committee considered whether the listing of anthropogenic climate change might wastefully consume precious management resources. Climate change is occurring due to increased levels of CO_2 and other "greenhouse" gases in the atmosphere that are produced from the burning of fossil fuels, and to decreased plant biomass that could otherwise sequestrate much of the excess carbon. Clearly, it would be impossible for the NSW Government to halt the use of fossil fuel and, even if this could be done, it would have no effect on CO_2 production elsewhere in Australia and other parts of the world. Would the listing of anthropogenic climate change therefore be futile or even counter-productive?

Although little can be done to stem the production of greenhouse gases, the Committee took the view that adverse climatic effects on biota could be reduced if suitable habitats were available for species to migrate to. As the protected area network is currently too patchy to permit movement across the broader landscape (Pouliquen-Young 1999), the Committee considered that listing anthropogenic climate change would stimulate modelling of species movements in the future, and provide impetus to redesign the reserve network to accommodate them. After reviewing the evidence, the Committee was convinced that anthropogenic climate change met the criteria specified under section 15 of the *Threatened Species Conservation Act*, and it was listed as a key threatening process on 17 November 2000. The practicability of conservation planning for this process was noted in point 6 of the Committee's final determination (Appendix 5).

Discussion

Compilations of threatened taxa are of value for identifying taxonomic groups, biological attributes, geographical areas or other traits that may predispose species to endangerment, and provide trip wires for management intervention. Although the value of listing threatened species and other entities has been debated recently (Possingham *et al.* 2002; Brook *et al.* 2003; Lamoreux *et al.* 2003), lists underpin much conservation effort at state, national and international levels and help to direct the allocation of scarce resources (Burgess 2001; IUCN 2003). It is therefore imperative that the listing process is transparent, rigorous and effective.

The case studies provide an indication of the range of issues debated by the Scientific Committee in reviewing nominations. Other examples have been noted elsewhere (e.g., Dickman and Fleming 2002; Adam 2004). Additional issues that were canvassed by the Committee include how to delineate the geographical boundary of a population or community of mobile organisms; how to delineate the species composition of communities when not all organisms within a defined area could be distinguished or even sampled (e.g., many invertebrates, micro-organisms, fungal spores and plant seeds in the soil); how to deal with the problem of variation in species composition over time; whether communities could be considered endangered if they had been fragmented and degraded by invasion of introduced species but still occupied large areas; whether depauperate remnants of regrowth vegetation should still be considered part of communities; and whether certain taxa should be legally considered as 'fish' (e.g., dragonflies) or 'marine vegetation' (e.g., saltmarsh plants) because part of their life cycles take place in water, and hence be more appropriate for listing under the *Fisheries Management Act 1994*. There were, of course, many other issues of arcane, legal, philosophical or fundamental biological nature that enlivened debates of the Scientific Committee. However, the bottom line in all discussions was to ensure that decisions accorded with criteria set out in the *Threatened Species Conservation Act 1995* and were underpinned by sound science.

Despite the heavy workload, occasional invective and harangues from sections of the wider community, the Scientific Committee was able to maintain a clear focus during its first seven years of operation and handled some 600 nominations for listing or delisting of threatened species, populations, ecological communities and key threatening processes. Remarkably, virtually all decisions about listing were unanimous during this period. Among other factors, I suspect that this was due to the opportunities that were available for extensive debate. For example, except in the case of provisional listings that needed to be made quickly, members of the Committee had time to voice concerns about a nomination or the wording of a determination, to seek further information or expert opinion on specific issues, and to achieve broad consensus before a vote was taken. In addition, unanimity was assisted by adherence to scientific principles, and a common commitment to biodiversity conservation among members of the Scientific Committee. I hope the experiences gained and practices pioneered by the inaugural committee will be of use to members of both the present and future Scientific Committees.

Acknowledgements

I thank members of the Scientific Committee and its Executive for their support, Sue Chate for assistance with compiling information for this paper, and Paul Adam, Tony Auld, Sue Chate and Carol McKechnie for comments on the manuscript.

References

Akçakaya, H. R., Ferson, S., Burgman, M. A., Keith, D. A., Mace, G. M. and Todd, C. R. 2000. Making consistent IUCN classifications under uncertainty. *Conservation Biology* **14:** 1001-1013.

Adam, P. 2004. Opening a can of words. The importance of definitions in the NSW *Threatened Species Conservation Act*. Pp 59-67 in *Threatened Species Legislation: is it just an Act?* edited by P. Hutchings, D. Lunney and C. Dickman. Royal Zoological Society of New South Wales, Mosman, NSW.

Adam, P., Auld, T., Benson, D., Catling, P., Dickman, C., Fleming, M., Gunning, R., Hutchings, P., Kemp, D. and Shields, J. 1997*a*. The 10 Lords of the Universe respond to Lim. *Pacific Conservation Biology* **3:** 319-321.

Adam, P., Auld, T., Benson, D., Catling, P., Dickman, C., Fleming, M., Gunning, R., Hutchings, P., Kemp, D. and Shields, J. 1997b. The New South Wales Threatened Species Conservation Act. A response to Cardew. *Australian Planner* **34:** 204-207.

Anon. 1997. *Report of the Joint Select Committee upon the Threatened Species Conservation Act 1995*. Parliament of New South Wales, Sydney, NSW.

Banks, P. B. 2004. Population viability analysis in urban wildlife management: modelling management options for Sydney's quarantined bandicoots. Pp. 70-77 in *Urban Wildlife: More than Meets the Eye*, edited by D. Lunney and S. Burgin. Royal Zoological Society of New South Wales, Mosman, NSW.

Benson, S. 1997. Urban bush seized – conservation order threatens new homes. *The Daily Telegraph*, 31 May: 1-2.

Bowen, M. and Goldingay, R. 2000. Distribution and status of the Eastern Pygmy Possum (*Cercartetus nanus*) in New South Wales. *Australian Mammalogy* **21:** 153-164.

Brereton, R., Bennett, S. and Mansergh, I. 1995. Enhanced greenhouse climate change and its potential effect on selected fauna of south-eastern Australia: a trend analysis. *Biological Conservation* **72:** 339-354.

Brook, A., Zint, M. and de Young, R. 2003. Landowners' responses to an Endangered Species Act listing and implications for encouraging conservation. *Conservation Biology* **17:** 1638-1649.

Burgess, B. B. 2001. *Fate of the Wild: the Endangered Species Act and the Future of Biodiversity*. University of Georgia Press, Athens, USA.

Burgman, M. A. 2004. Expert frailties in conservation risk assessment and listing decisions. Pp. 20-29 in *Threatened Species Legislation: is it just an Act?* edited by P. Hutchings, D. Lunney and C. Dickman. Royal Zoological Society of New South Wales, Mosman, NSW.

Busby, J. R. 1988. Potential impacts of climate change on Australia's flora and fauna. Pp. 387-398 in *Greenhouse: Planning for Climate Change*, edited by G. I. Pearman. CSIRO Publishing, Melbourne.

Calver, M. C., Bradley, J. S. and Wright, I. W. 1999. Towards scientific contributions in applying the precautionary principle: an example from southwestern Australia. *Pacific Conservation Biology* **5:** 63-72.

Cardew, R. 1997. Editorial. *Australian Planner* **34:** 68.

Chambers, L. K. and Dickman, C. R. 2002. Habitat selection of the Long-nosed Bandicoot, *Perameles nasuta* (Mammalia, Peramelidae), in a patchy urban environment. *Austral Ecology* **27:** 334-342.

Dickman, C. R. 1997. The nominations process of the Threatened Species Conservation Act. Pp. 19-25 in *On the Brink: Your Bush, Their Habitat, Our Act*, edited by H. Webb. Nature Conservation Council of New South Wales, Sydney, NSW.

Dickman, C. R. and Fleming, M. R. 2002. Pest, or Passenger Pigeon? The New South Wales Scientific Committee's assessment of the status of the Grey-headed Flying-fox. Pp. 20-28 in *Managing the Grey-headed Flying-fox as a Threatened Species in New South Wales*, edited by P. Eby and D. Lunney. Royal Zoological Society of New South Wales, Mosman, NSW.

Frankham, R. 1995. Effective population size/adult population size ratios in wildlife: a review. *Genetical Research* **66:** 95-107.

Garkaklis, M. J., Calver, M. C., Wilson, B. A. and Hardy, G. E. St. J. 2004. Habitat alteration caused by an introduced plant disease, *Phytophthora cinnamomi*: a potential threat to the conservation of Australian forest fauna. Pp. 899-913 in *Conservation of Australia's Forest Fauna* (second edition), edited by D. Lunney. Royal Zoological Society of New South Wales, Mosman, NSW.

Green, K. and Osborne, W. S. 2003. The distribution and status of the Broad-toothed Rat *Mastacomys fuscus* (Rodentia: Muridae) in New South Wales and the Australian Capital Territory. *Australian Zoologist* **32:** 229-237.

Hughes, W. and Westoby, M. 1994. Climate change and conservation policies in Australia: coping with change that is far away and not yet certain. *Pacific Conservation Biology* **1:** 308-318.

IUCN. 2003. *2003 IUCN Red List of Threatened Species*. IUCN Species Survival Commission, Gland, Switzerland.

Jerogin, P. 1997. The developers perspective: working towards a more meaningful outcome. Pp. 136-141 in *On the Brink: Your Bush, Their Habitat, Our Act*, edited by H. Webb. Nature Conservation Council of New South Wales, Sydney, NSW.

Kearney, R. and Kearney, E. 2000. Significance of the Hygrocybeae community of Lane Cove Bushland Park in listings under the *NSW Threatened Species Conservation Act 1995* and under the *Australian Heritage Commission Act 1975*. *Australasian Mycologist* **19:** 64-70.

Keith, D. A. 1998. An evaluation and modification of World Conservation Union Red List criteria for classification of extinction risk in vascular plants. *Conservation Biology* **12:** 1076-1090.

Lacy, R. C. and Clark, T. W. 1990. Population viability assessment of the Eastern Barred Bandicoot in Victoria. Pp. 131-145 in *Management and Conservation of Small Populations*, edited by T. W. Clark and J. H. Seebeck. Chicago Zoological Society, Brookfield, Illinois, USA.

Lamoreux, J., Akçakaya, H. R., Bennun, L., Collar, N. J., Boitani, L., Brackett, L., Bräutigam, A., Brooks, T. M., da Fonseca, G. A. B., Mittermeier, R. A., Rylands, A. B., Gärdenfors, U., Hilton-Taylor, C., Mace, G., Stein, B. A. and Stuart, S. 2003. Value of the IUCN Red List. *Trends in Ecology and Evolution* **18:** 214-215.

Lim, L. 1997. The 10 Lords of the Universe – the New South Wales TSC Act's Scientific Committee. *Pacific Conservation Biology* **3:** 4-12.

Lindenmayer, D. B., Nix, H. A., McMahon, J. P., Hutchinson, M. F. and Tanton, M. T. 1991. The conservation of Leadbeater's Possum, *Gymnobelideus leadbeateri* (McCoy): a case study of the use of bioclimatic modelling. *Journal of Biogeography* **18:** 371-383.

Lunney, D., Curtin, A. L., Ayers, D., Cogger, H. G., Dickman, C. R., Maitz, W., Law, B. and Fisher, D. 2000. The threatened and non-threatened native vertebrate fauna of New South Wales: status and ecological attributes. *New South Wales National Parks and Wildlife Service Environmental and Heritage Monograph*, series no. **4:** 1-134.

Mansergh, I. and Broome, L. 1994. *The Mountain Pygmy-possum of the Australian Alps*. New South Wales University Press, Kensington, NSW.

Millsap, B. A., Gore, J. A., Runde, D. E. and Cerulean, S. I. 1990. Setting the priorities for the conservation of fish and wildlife species in Florida. *Wildlife Monographs* **54:** 5-57.

Minta, S. C., Clark, T. W. and Goldstraw, P. 1990. Population estimates and characteristics of the Eastern Barred Bandicoot in Victoria, with recommendations for population monitoring. Pp. 47-75 in *Management and Conservation of Small Populations*, edited by T. W. Clark and J. H. Seebeck. Chicago Zoological Society, Brookfield, Illinois, USA.

Possingham, H. P., Andelman, S. J., Burgman, M. A., Medellin, R. A., Master, L. L. and Keith, D. A. 2002. Limits to the use of threatened species lists. *Trends in Ecology and Evolution* **17:** 503-507.

Pouliquen-Young, O. 1999. *The Implications of Climate Change for Land-Based Nature Conservation Strategies*. Australian Greenhouse Office, Canberra.

Ravallion, J. 2000. Review of the monitoring component of the recovery plan for the endangered Long-nosed Bandicoot (*Perameles nasuta*) population at North Head, Manly, NSW. Unpublished report, Macquarie University, Sydney, NSW.

Reizes, J. 1997. Saving what's left … Manly's struggle achieves a world first for two common species. Pp. 46-52 in *On the Brink: Your Bush, Their Habitat, Our Act*, edited by H. Webb. Nature Conservation Council of New South Wales, Sydney, NSW.

Scott, L. K. 1995. *Preliminary report on the Long-nosed Bandicoot population at the Quarantine Station, North Head, Sydney Harbour National Park*. Institute of Wildlife Research, University of Sydney, Sydney, NSW.

Scott, L. K., Hume, I. D. and Dickman, C. R. 1999. Ecology and population biology of Long-nosed Bandicoots (*Perameles nasuta*) at North Head, Sydney Harbour National Park. *Wildlife Research* **26:** 805-821.

Smith, J. 1997. Skinning cats, putting tigers in tanks and bringing up baby: a critique of the Threatened Species Conservation Act 1995 (NSW). *Environmental and Planning Law Journal* **14:** 17-37.

Whelan, R., Brown, C. L and Farrier, D. 2004. The precautionary principle: what is it and how might it be applied in threatened species conservation? Pp. 49-58 in *Threatened Species Legislation: is it just an Act?* edited by P. Hutchings, D. Lunney and C. Dickman. Royal Zoological Society of New South Wales, Mosman, NSW.

Young, A. M. 1999. The Hygrocybeae (Fungi, Basidiomycota, Agaricales, Hygrophoraceae) of the Lane Cove Bushland Park, New South Wales. *Austrobaileya* **5:** 535-564.

Young, A. M., Kearney, R. and Kearney, E. 2001. Additions to the Hygrophoraceae of Lane Cove Bushland Park. *Australasian Mycologist* **20:** 79-86.

APPENDIX I

Appendix I.

Final Determination

The Scientific Committee, established by the Threatened Species Conservation Act, has made a Final Determination to list the beetle *Nurus brevis* Motschulsky, 1865 as an ENDANGERED SPECIES on Part 1 of Schedule 1 of the Act. Listing of endangered species is provided for by Part 2 of the Act.

The Scientific Committee has found that:

1. *Nurus brevis* (Coleoptera: Carabidae) is a large (up to 5cm in length), predatory ground beetle. It is flightless, heavily built and black with metallic green reflections.
2. *Nurus brevis* appears to have been confined to heavily timbered areas east of the Great Dividing Range on the north coast of NSW, including the "Big Scrub". Prior to the clearing of the "Big Scrub" rainforest, *Nurus brevis* is thought to have been relatively common. There are many collection records of *Nurus brevis* from the 19th and early 20th centuries.
3. The biology of *Nurus brevis* has not yet been studied in detail, but it is known that the beetles live in burrows and that the females nurture the young at the bottom of the burrow.
4. *Nurus brevis* had not been collected for many years and was thought to be extinct until 1972, when *Nurus brevis* was re-discovered by G. Monteith at Rotary Park, Lismore. However, by the early 1990s this population had declined and latest data indicate that *Nurus brevis* is extinct in Rotary Park (G. Williams, G. Carruthers, pers. comms). This population formed the basis of a nomination of Rotary Park to the Register of the National Estate (Carruthers, 1993, Greenslade, 1994). There are now only two known populations of *Nurus brevis*, both of which are near Mallanganee and are very isolated (G. Monteith, C. Reid pers. comms).
5. The survival of this species is currently threatened by an extremely restricted distribution, clearing of rainforest remnants, removal of fallen timber and ground cover, and beetle collecting activities.
6. In view of 1, 2, 3, 4 and 5 above, the Scientific Committee is of the opinion that the numbers of *Nurus brevis* Motschulsky, 1865 have been reduced to such a critical level and its habitats have been so drastically reduced, that it is in immediate danger of extinction.

Dr Chris Dickman
Chairperson
Scientific Committee

References

Carruthers, G. 1993. Nomination of Rotary Park, Lismore to the Register of the National Estate.

Greenslade, P. 1994. Heritage listing of invertebrate sites. *Memoirs of the Queensland Museum* 36 (1): 67 - 76.

Gazetted: 16/03/01

APPENDIX 2

Appendix 2.

Final Determination

The Scientific Committee, established by the Threatened Species Conservation Act, has made a Final Determination to list the Eastern Pygmy-possum *Cercartetus nanus* (Desmarest, 1818) as a VULNERABLE SPECIES on Schedule 2 of that Act. Listing of vulnerable species is provided for by Part 2 of the Act.

The Scientific Committee has found that:

1. The Eastern Pygmy-possum *Cercartetus nanus* (Desmarest, 1818) is a small arboreal marsupial that is distributed in the south-eastern corner of mainland Australia and in Tasmania. In New South Wales the species is found in coastal areas and at higher elevation in the south, but north of Newcastle at higher elevation only. Pygmy-possums are agile climbers that feed mostly on the pollen and nectar from banksias, eucalypts and understorey plants and will also eat insects, seeds and fruit.

2. Although the Eastern Pygmy-possum is broadly distributed, recent studies have shown that within this range the species appears to be patchily distributed and its overall abundance is low.

3. Despite a large number of intensive trapping programs undertaken in the eastern forests and woodlands of New South Wales in recent years, only a small number of captures (154) have resulted from a total trapping effort of 315,000 Elliott trap-nights and 57,000 pitfall trap-nights (Bowen and Goldingay 2000).

4. Other detection techniques such as spotlighting, predator scat analysis, hair tubes and trapping in trees have produced similar low rates of detection. Capture rates are highest for installed nest-boxes and traps set in flowering banksias. This may reflect a habitat preference or a more successful trapping method.

5. From these and more recent studies (A. Tulloch, pers. comm.), there were only six localities where more than 10 observations of Pygmy-possums have been made. These were the Pilliga area, New England Tablelands, Barren Grounds Nature Reserve-Budderoo National Park, Royal and Heathcote National Parks, Kioloa State Forest and the Eden area.

6. The factors threatening the survival of the Eastern Pygmy-possum include isolated sub-populations with little opportunity for dispersal which increases the risk of local extinction, clearing that results in habitat loss and fragmentation, inappropriate fire regimes that remove nectar-producing understorey plants, the loss of nest sites due to past intensive forestry and firewood collection, and predation by foxes and cats.

7. In view of 2, 3, 4, 5 and 6 above, the Scientific Committee is of the opinion that the Eastern Pygmy-possum *Cercartetus nanus* is likely to become endangered unless the circumstances and factors threatening its survival or evolutionary development cease to operate, and is therefore eligible for listing as a vulnerable species.

Associate Professor Paul Adam
Deputy Chairperson
Scientific Committee

References

Bowen, M. and Goldingay, R. 2000. Distribution and status of the Eastern Pygmy Possum (*Cercartetus nanus*) in New South Wales. *Australian Mammalogy* 21:153-65.

Turner, V. and Ward, S. 1995. Eastern Pygmy Possum *Cercartetus nanus*. In *The Mammals of Australia*, Ed. R. Strahan, pp 217-218. Reed Books: Sydney.

Gazetted: 08/06/01

APPENDIX 3

Appendix 3.

Final Determination

The Scientific Committee, established by the Threatened Species Conservation Act, has made a Final Determination to list the North Head population of the Long-nosed Bandicoot, *Perameles nasuta*, Geoffrey 1804, as an Endangered Population on Part 2 of Schedule 1 of the Act. The distribution of the population is defined on the map which forms part of this Final Determination.

Any submissions received following advertisement of the Preliminary Determination have been considered by the Scientific Committee.

Listing of Endangered Populations is allowed for by Section 11 of the Act.

The Scientific Committee has found that:

1. *Perameles nasuta* is not listed on Part 1 of Schedule 1 of the Act and thus populations of *P. nasuta* are eligible for consideration for inclusion on Part 2 of Schedule 1 of the Act.
2. *P. nasuta* has been reliably reported as occurring on North Head, within the area defined on the accompanying map.
3. The population of *P. nasuta* at North Head is estimated to be about one hundred animals, with fluctuations around this number over time.
4. The Scientific Committee is of the opinion that the effective population size is much less than the census population, and that, on the basis of the scientific literature the best available estimate of the effective population size would be approximately 10-11% of the census population.
5. There is evidence that one cause of mortality of *P. nasuta* at North Head is as a result of attacks by introduced predators (foxes, dogs, cats). Road kills have also been reported.
6. *P. nasuta* was once widespread in the Sydney region but many formerly recorded populations have become extinct. The North Head population is now isolated and disjunct. The nearest surviving population to that at North Head known to the Committee is in Pittwater Local Government Area.
7. Population genetic theory would suggest that a population as small as that of the North Head *P. nasuta* population would suffer effects of inbreeding depression and loss of genetic variation, affecting population viability.
8. In addition to genetic consequences of small population size, a small population occupying a limited, discrete area is at risk from stochastic disturbance events.
9. In light of the above, and taking into account discussion in relevant international scientific literature, the Scientific Committee is of the opinion that the North Head population of *P. nasuta* is in immediate danger of extinction.
10. While the North Head population of *P. nasuta* is disjunct it is not situated at or near the limit of the geographic range of the species.
11. The Committee is not aware of any genetic studies of the North Head *P. nasuta* population. However, the effects of fragmentation and isolation may result in local populations possessing distinct genetic features. There is thus a likelihood that the population may be genetically distinct.
12. The Scientific Committee is of the opinion that the North Head *P. nasuta* population is of significant conservation value on the grounds that it is:
    - a disjunct population
    - one of the few surviving populations within the Sydney region
    - a population which has been the subject of a number of scientific studies, and is thus an important reference population
    - accorded considerable value by the local community, and thus serves to promote conservation more generally.
13. As a consequence of points 1, 9, 11 and 12 above the Scientific Committee is of the opinion that the criteria specified in Section 11 of the Act are met and that the North Head Population of *P. nasuta* should be listed as an Endangered Population on Part 2 of Schedule 1 of the Threatened Species Conservation Act.

Dr Chris Dickman
Chairperson
Scientific Committee
Gazetted: 28/2/97

APPENDIX 3

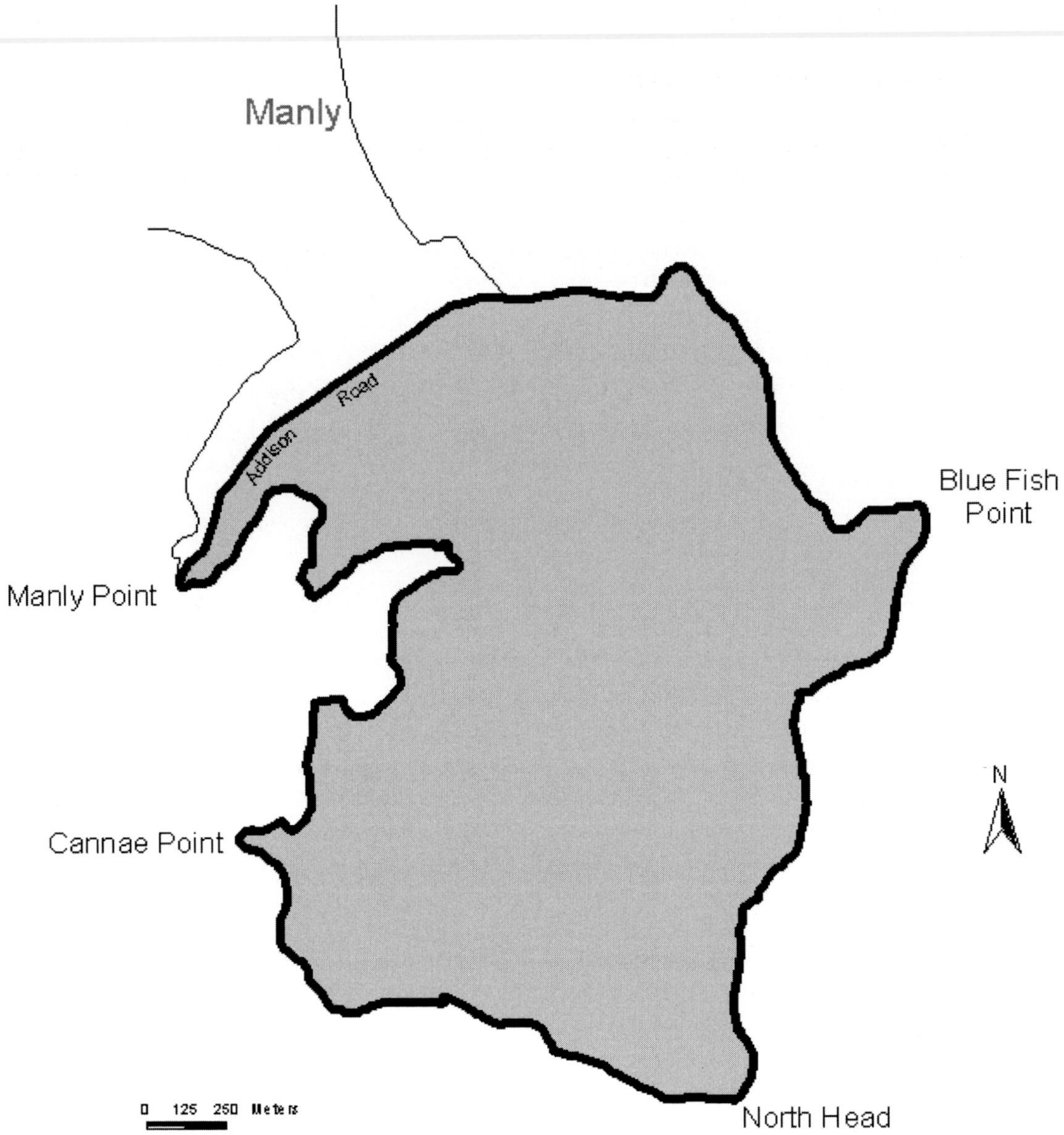

Map accompanying the Final Determination for the North Head population of the Long-nosed bandicoot

Appendix 4.

Final Determination

The Scientific Committee established by the Threatened Species Conservation Act, has made a Final Determination to list the Hygrocybeae Community of Lane Cove Bushland Park as an ENDANGERED ECOLOGICAL COMMUNITY under Part 3 of Schedule 1 of the Act. Listing is provided for under Part 2 of the Act.

The Scientific Committee has found that:

1. The Hygrocybeae Community of Lane Cove Bushland Park is an assemblage of more than 20 species of fungi in the family Hygrophoraceae (Fungi, Basidiomycota, Agaricales, Hygrophoraceae).
2. The Community is restricted to a core zone along the Gore Creek catchment in the Lane Cove Local Government Area in Sydney. The majority of species occur in the warm temperate gallery rainforest centred on the banks of the north-eastern arm of Gore Creek and its tributaries in Lane Cove Bushland Park. This core zone also extends to the wet sclerophyll catchment, north of the tributary junction with Gore Creek. A minority of the species in the assemblage is found in a buffer zone of dry sclerophyll between the perimeter of Lane Cove Bushland Park and outer edges of the gallery canopy and along Gore Creek in Osborne Park.
3. The following species have been recorded in the community:

Camarophyllopsis kearneyi	*Hygrocybe anomala* var. *ianthinomarginata*
Hygrocybe astatogala	*Hygrocybe aurantiopallens*
Hygrocybe aurantipes	*Hygrocybe austropratensis*
Hygrocybe cantharellus	*Hygrocybe cheelii*
Hygrocybe chromolimonea	*Hygrocybe erythrocala*
Hygrocybe graminicolor	*Hygrocybe helicoides*
Hygrocybe involutus	*Hygrocybe irrigata*
Hygrocybe kula	*Hygrocybe lanecovensis*
Hygrocybe lewellinae	*Hygrocybe mavis*
Hygrocybe miniata	*Hygrocybe reesiae*
Hygrocybe sanguinocrenulata	*Hygrocybe stevensoniae*
Hygrocybe taekeri	*Hygrocybe virginea*

   Other species in the Community have been collected but remain undescribed and unclassified, and other Hygrocybeae may be present.

4. Lane Cove Bushland Park is the holotype site for *Hygrocybe aurantipes, Hygrocybe austropratensis, Hygrocybe lanecovensis, Hygrocybe anomala* var. *ianthinomarginata, Camarophyllopsis kearneyi* and *Hygrocybe reesiae*. [Young, A.M., 1999, The Hygrocybeae (Fungi, Basidiomycota, Agaricales, Hygrophoraceae) of the Lane Cove Bushland Park, New South Wales. *Austrobaileya* **5**: 535 - 564].
5. The assemblage is not known to occur outside the Lane Cove Local Government Area. Furthermore, the number of species of *Hygrocybe* is very high compared with other known sites in Australia and overseas. Species will not have above-ground fruiting bodies at all times of the year. There may be differences depending on seasonal conditions and other factors.
6. Within Lane Cove Bushland Park, different species of Hygrocybeae have been reported from one to several specific locations. More species occur at the southern than at the northern end of the Park. The ecological requirements of most species are poorly known, but are likely to be associated with a dense tree canopy and sandstone rocks.
7. The Community is threatened by water-borne pollutants. Industrial pollutants occur particularly in the upper reaches of Gore Creek catchment and domestic contaminants arise from residential properties on the perimeter of Lane Cove Bushland Park. The Community is also at risk from encroachment by exotic weeds, dumping of rubbish and garden refuse, excess pedestrian traffic in areas sensitive to erosion, and inappropriate bush regeneration measures that disturb the forest canopy and native understorey plants.
8. In view of the small area occupied by the Community and the threats to its integrity identified in 7 above, the Scientific Committee is of the opinion that the community is likely to become extinct in New South Wales unless the circumstances threatening its survival cease to operate.

Dr Chris Dickman
Chairperson
Scientific Committee
Gazetted: 3/3/00

APPENDIX 5

Appendix 5.

Final Determination

The Scientific Committee, established by the Threatened Species Conservation Act, has made a Final Determination to list Anthropogenic Climate Change as a KEY THREATENING PROCESS on Schedule 3 of the Act. Listing of Key Threatening Processes is provided for by Part 2 of the Act.

The Scientific Committee has found that:

1. The distribution of most species, populations and communities is determined, at least at some spatial scale, by climate.
2. Climate change has occurred throughout geological history and has been a major driving force for evolution.
3. There is evidence that modification of the environment by humans may result in future climate change. Such anthropogenic change to climate may occur at a faster rate than has previously occurred naturally. Climate change may involve both changes in average conditions and changes to the frequency of occurrence of extreme events.
4. Response of organisms to future climate change (however caused) is likely to differ from that in the past because it will occur in a highly modified landscape in which the distribution of natural communities is highly modified. This may limit the ability of organisms to survive climate change through dispersal (Brasher & Pittock 1998; Australian Greenhouse Office 1998). Species at risk include those with long generations, poor mobility, narrow ranges, specific host relationships, isolate and specialised species and those with large home ranges (Hughes & Westoby 1994). Pest species may also be advantaged by climate change.
5. Modelling of the distribution of species under realistic climate change scenarios suggests that many species would be adversely affected unless populations were able to move across the landscape (for example, Brereton, Bennett and Mansergh 1995). Examples of species which would be at risk in New South Wales include:

Mammals	
Burramys parvus	Mountain Pygmy-possum
Potorous longipes	Long-footed Potoroo
Mastacomys fuscus	Broad-toothed Rat
Pseudomys fumeus	Smoky Mouse
Birds	
Leipoa ocellata	Malleefowl
Pedionomus torquatus	Plains-wanderer
Tyto tenebricosa	Sooty Owl
Calyptorhynchus banksii graptogyne	Red-tailed Black-cockatoo
Polytelis anthopeplus	Regent Parrot
Petroica rodinogaster	Pink Robin
Pachycephala rufogularis	Red-lored Whistler
Reptiles	
Delma impar	Striped Legless Lizard
Amphibians	
Litoria spenceri	Spotted Frog
Litoria raniformis	Southern Bell Frog
Pseudophryne pengilleyi	Northern Corroboree Frog
Pseudophryne corroboree	Southern Corroboree Frog
Flora	
Communities likely to become threatened include alpine vegetation communities (Busby 1988, Hughes & Westoby 1994).	

6. The present protected area network was not designed specifically to accommodate climate change, and the present biodiversity values of the protected area system may not all survive under different climatic conditions (see Pouliquen-Young, O. 1999). Conservation planning at the landscape scale could provide opportunities for species to respond to future climate change and the Threat Abatement Plan could address modifications to the present protected area network to account for climate change.
7. Fire is an integral part of the dynamics of many Australian ecosystems. Studies suggest that the risk of fire may increase in some areas as the climate changes and decrease in others with consequent changes to the species composition and structure of ecological communities (Brasher & Pittock 1998; NSW Scientific Committee 2000).
8. In view of the above, the Scientific Committee is of the opinion that Anthropogenic Climate Change adversely affects two or more threatened species or could cause species, populations or ecological communities that are not threatened to become threatened.

Dr Chris Dickman
Chairperson
Scientific Committee

References

Australian Greenhouse Office 1998. The National Greenhouse Strategy. Commonwealth of Australia, Canberra.

Brasher, R.E. and Pittock, A.B. 1998. Australasian Impacts of Climate Change. *An Assessment of Vulnerability*. In Watson, R.T., Zinyowera, M.C., Moss, R.H. and Dokken, D.J. "The Regional Impacts of Climate Change. An assessment of Vulnerability", IPCC Report. Australian Greenhouse Office, Canberra.

Brereton, R., Bennett, S. and Mansergh, I. 1995. Enhanced Greenhouse climate change and its potential effect on selected fauna of south-eastern Australia: a trend analysis. *Biological Conservation* **72**, 339-354.

Busby, J.R. 1988. Potential impacts of climate change on Australia's flora and fauna. In 'Greenhouse. Planning for climate change.'ed. G.I.Pearman, pp. 387-98, CSIRO.

Hughes, W. & Westoby, M. 1994. Climate change and conservation policies in Australia: coping with change that is far away and not yet certain. *Pacific Conservation Biology* **1**: 308-18

NSW Scientific Committee 2000. Final Determination to list High Frequency Fire as a Key Threatening Process in the Schedules of the Threatened Species Conservation Act 1995. NSW Scientific Committee, Sydney.

Pouliquen-Young, O. 1999. The implications of climate change for land-based nature conservation strategies. Australian Greenhouse Office, Canberra.

Gazetted: 17/11/00

Endangered populations: the concept in practice

Jack Baker

NSW Department of Environment and Conservation

PO Box 1967 Hurstville NSW 2220

ABSTRACT

The NSW *Threatened Species Conservation Act 1995* (TSC Act) is unique and effective in Australian jurisdictions in recognising threatened populations. Section 11 of the TSC Act provides for the listing of endangered populations. There were 28 endangered populations listed under the TSC Act (January 2003) for which there were two approved recovery plans, one publicly exhibited draft plan and four plans in preparation. Monitoring was being undertaken for only six populations. The TSC Act would be strengthened by: (i) adding an objective which states that recognising threatened entities is fundamental to the legislation; (ii) changing the definition of population to make it consistent with IUCN terminology; (iii) simplifying Section 11; (iv) better resourcing for faster preparation of plans and their implementation; and (v) including population monitoring in all recovery plans. The process of saving threatened species is occurring at the level of population; for example, through priority sites for threat abatement, translocation, replicates for research treatments and habitat protection. The objective of recognising and the process of listing threatened populations are fundamental to sound threatened species legislation.

Key words: threatened populations, TSC Act of New South Wales, definition of populations.

Introduction

Throughout the world, countries legislate in an effort to prevent the extinction of species and to conserve natural biodiversity. The topic of the forum asks the questions: Are the relevant transactions of deliberative assemblies (statutes, legislation etc), the **acts**, effective in preventing extinctions? Or are they a pretence, an **act**? Globally, the transactions are a pretence! My pessimistic belief, together with learned commentators (*e.g.* Suzuki 1990; Gould 1991; Flannery 1994; Wilson 2001), is that the ongoing catastrophe of worldwide anthropogenic threats to biodiversity and extinctions of species are symptomatic of the global industrialised society which is causing unsustainable resource extraction and despoliation of natural ecosystems.

In NSW, the *Threatened Species Conservation Act 1995* (TSC Act) deals with threatened plants and animals and Part 7A of the *Fisheries Management Act 1994* (FM Act) deals with threatened fish and marine vegetation as defined in these acts respectively. This paper focuses on the TSC Act. Is it effective or is it a pretence?

An effective threatened species act would have, as its first objective, the recognition of threatened entities (taxa, populations and ecological communities) within its jurisdiction. This important objective is missing from the TSC Act even though it is meant to be effected through the listing process which is an integral part of the Act. Regardless of the uncertainty about status of threatened species (Burgman 2004) and the limitations and potential misuses of threatened species lists (Possingham *et al.* 2002; Lamoreux *et al.* 2003), threatened species legislation cannot be effective unless it begins by recognising which entities are threatened. The TSC Act would be strengthened by being explicit that the recognition of threatened entities is the foundation of the Act and that (a) recognising and (b) saving threatened species are different objectives to be achieved through different processes. Failure to make this distinction in NSW has resulted in confusion about these separate roles and a demonstrable head-in-the-sand approach to amending and interpreting the legislation.

The TSC Act does have an objective to prevent extinction and promote recovery of threatened entities which is intended to be achieved through the process of preparing and implementing recovery plans and threat abatement plans. The TSC Act also achieves its objectives through protection and regulation but this is not assessed here. As at January 2003, there were 877 listings for which there were plans covering 335 listings (67 covered in approved plans, 15 in publicly exhibited drafts and 253 in plans in preparation) (NPWS 2003a). There were no plans for more than 60% of the listed threatened taxa, populations and ecological communities; most of the plans were only in preparation; many of the actions in the existing plans were unfunded, not being implemented and, for those being implemented, there is no reporting mechanism to assess the success or otherwise of those actions in promoting recovery; and of 16 key threatening processes listed, there was only one threat abatement plan approved and there was no plan for two of the most serious threats, land clearing and altered flow regimes. Judged by this alone, the TSC Act might be considered more pretence than it is effective in saving threatened species.

Endangered populations - the concept

Threatened species may be categorised as vulnerable, endangered or critically endangered depending on the increasing likelihood of them going extinct (IUCN 2001). A standard meaning of population is a group of individuals of a single species (Recher *et al.* 1986; Krebs 1994). Under the TSC Act and the FM Act, population means a group of organisms, all of the same species, **occupying a particular**

Pp 82 - 87 in *Threatened species legislation: is it just an Act?* 2004, edited by Pat Hutchings, Daniel Lunney and Chris Dickman. Royal Zoological Society of New South Wales, Mosman, NSW, Australia.

area. The focus of this paper is with this definition and Section 11 of the TSC Act, which explicitly provides for the listing of endangered populations. But first, three contextual notes.

Firstly, the concept of populations going extinct is simple and real; it is not a pretence. The Eastern Bristlebird *Dasyornis brachypterus* is a case in point. Within approximately a century of being named by Latham in 1801, the species was locally extinct in the Sydney area (Baker 1997). Over the last three decades, from Lake Tyers in eastern Victoria to Nadgee Nature Reserve in south-eastern NSW, 11 of 12 populations have become extinct (Baker 1997; Bramwell 1997; Clarke and Bramwell 1998) and, from the Conondale Ranges in south-eastern Queensland to Lismore in north-eastern NSW, 19 of 23 populations have also become extinct (Holmes 1989, 1997; D Stewart pers. comm. 2003).

Secondly, the legislation in NSW is unique among Australian jurisdictions in being the most explicit about, as well as being active in, listing threatened populations (Table 1). Under the TSC Act, 28 endangered populations had been listed as at January 2003 (Table 2) and a further two had been listed under the FM Act as at July 2003. Through adopting the IUCN (2001) categorisation (Queensland, South Australia, Western Australia and Northern Territory) or otherwise, the legislation in all other jurisdictions implies that threatened populations can be listed. For example, in Queensland, where threatened species are listed under the *Nature Conservation Act 1992*, a "species" means a species, subspecies, hybrid, variant, race, mutation or geographically separate population; in Victoria, where threatened species are listed under the *Flora and Fauna Guarantee Act 1988*, taxa which are below the level of sub-species and communities of flora or fauna narrowly defined because of their taxonomic composition, environmental conditions or geography are eligible for listing if there is a special need to conserve them. However, as at October 2003, none of the other states or territories had listed a threatened population. Under the Commonwealth *Environment Protection and Biodiversity Conservation Act 1999* (EPBC Act), threatened populations of species (and ecological communities) may be listed as sub-groups of species (and ecological communities) and two populations had been listed as at October 2003.

Thirdly, the *Threatened Species Conservation Amendment Act 2002* amended Section 11 of the TSC Act, most notably by adding the criterion that the population must be of conservation value at the State or regional level (regions as defined in Thackway and Cresswell 1995) and a footnote explaining that isolated populations of limited conservation value are intended to be excluded from listing. The meanings and means of assessment of "value" and "limited value" are not clear in the legislation. It seems to me, however, that this condones triage and it is possible that some isolated populations could go extinct because they are not protected by the legislation.

To understand the practice of listing endangered populations under the TSC Act in NSW, it is necessary to understand how populations may be eligible for listing. Since a population is a group of one species **in a particular area,** by legislation, the area must be delimited. In practice, the area has been delimited variously as a patch of habitat (*e.g.* habitat for Little Penguin *Eudyptula minor* in the Manly Point Area, along the shoreline where it is suitable for nesting and extending 100 m offshore), a geographic feature (*e.g.* the Squirrel Glider *Petaurus norfolcensis* on the Barrenjoey Peninsula), a polygon bounded by roads (*e.g. Dillwynia tenuifolia* at Kemps Creek), a local government area (*e.g. Persoonia hirsuta* and *Hibbertia incana* in the Baulkham Hills Shire), a bioregion (*e.g.* the Tusked Frog *Adelotus brevis* in the Nandewar and New England Tablelands Bioregions). These particular areas do not need to be biologically meaningful, for instance it is feasible to list a population delimited by a square.

The combination of the definition of population under the TSC Act and the criteria in Section 11 have been the cause of some angst. For instance, the Scientific Committee rejected the nomination to list the Scotts Head population of the Glossy Black-cockatoo *Calyptorhyncus lathami*. The final determination stated that the mobility of the species made definition of a population difficult and that the Scotts Head birds were part of a wider population which was not in immediate danger of extinction. In another case, the Scientific Committee listed the Gang-gang Cockatoo *Callocephalon fimbriatum* population in the Hornsby and Ku-ring-gai Local Government Areas because it was of significant conservation value as it is the last known breeding population of the species in the Sydney Metropolitan area, despite acknowledging that individual birds are likely to move across the population boundary.

Table 1. Threatened population listings for all Australian jurisdictions

Jurisdiction	Act	Population listings
Commonwealth	*Environment Protection and Biodiversity Conservation Act 1999*	2 (as at October 2003)
NSW	*Threatened Species Conservation Act 1995* *Fisheries Management Act 1994*	28 (as at January 2003) 2 (as at July 2003)
Victoria	*Flora and Fauna Guarantee Act 1988*	None (as at October 2003)
ACT	*Nature Conservation Act 1980*	None (as at October 2003)
Queensland	*Nature Conservation Act 1992*	None(as at October 2003)
Northern Territory	*Territory Parks and Wildlife Conservation Act 2000*	None (as at October 2003)
Western Australia	*Wildlife Conservation Act 1950*	None (as at October 2003)
South Australia	*National Parks and Wildlife Act 1972*	None (as at October 2003)
Tasmania	*Threatened Species Protection Act 1995*	None (as at October 2003)

Table 2. Summary of attributes of populations listed under the TSC Act (from NPWS 2003a)

Taxa	Number of populations	Number of plans	Number of populations for which there are recovery actions	Number of populations for which there is monitoring
Reptile	-	-	-	-
Frog	1	0	1	0
Bird	5	1 approved	4	2
Mammal	7	3 in prep. 1 publicly exhibited draft 1 approved	7	3
Invertebrate	1	0	1	1
Plant	14	1 in prep.	5	0
Totals	28	7	18	6

The TSC Act could be strengthened, in my opinion, by changing the definition of population and by simplifying Section 11 (see Box 1). Section 11(2) logically prevents populations of a listed endangered species from being listed as endangered. Section 11(1) gives the criteria for listing a population as endangered, depending on the opinion of the Scientific Committee as described in the final determination. This is consistent with the laudable practice of having an independent group of experts undertake the listing process on behalf of the government. The criteria for listing are that a population must: (i) be facing a high risk of becoming extinct in nature in NSW and (ii) be of conservation value at the State or regional level for at least one of the following reasons:

- It *is disjunct*.
- It *is or is likely to be genetically, morphologically or ecologically distinct*.

  These first two reasons can be summarised as geographically or otherwise distinct. Fragmentation of a species into isolated groups can lead to local extinctions as demonstrated by the example of the Eastern Bristlebird. Conversely, subgroups of a species which show distinct genetic, morphological or ecological characteristics are manifestly carrying the diversity which may lead to evolutionary development which the TSC Act seeks to conserve. Hence, these first two are sound criteria for recognising threatened populations and they are reasons about which the Scientific Committee has expertise in giving opinions.

- It *is near the limit of its geographic range*.

  This is a sound reason only if the population is also distinct. Otherwise, metapopulations of species (Levins 1970) are known and expected to wax and wane at the limit of their range without necessarily increasing the risk that the species will become extinct (Hastings and Harrison 1994).

- It *is otherwise of significant conservation value*.

  This is bothersome because "significant" and "value" are undefined and the Scientific Committee may be called upon to give opinions on non-scientific values. For instance, this criterion certainly should not be used to list populations on the basis of their cultural value because (i) this is not the intent of the TSC Act and (ii) it is neither within the charter nor the expertise of the Scientific Committee to give opinions on cultural values.

Box 1. TSC Act, Section 11

Populations eligible for listing as endangered populations

(1) A population is eligible to be listed as an endangered population if, in the opinion of the Scientific Committee, it is facing a high risk of becoming extinct in nature in New South Wales and it is of conservation value at the State or regional level for one or more of the following reasons:

(a) it is disjunct or near the limit of its geographic range,

(b) it is or is likely to be genetically, morphologically or ecologically distinct,

(c) it is otherwise of significant conservation value.

Note. The intention of the criteria is to exclude from listing isolated populations of limited conservation value.

(2) A population is not eligible to be listed as an endangered population if it is a population of a species already listed in Schedule 1.

In the Red List Criteria (IUCN 2001 Version 3.1), the term population is used in a specific sense that is different to its common biological usage. Population is defined as the total number of individuals of the taxon. **Subpopulations** are defined as geographically or otherwise **distinct groups** in the population between which there is little demographic or genetic exchange (typically one successful migrant individual or gamete per year or less). Much of the angst about listing threatened populations under the TSC Act would be removed if the IUCN definition (of subpopulation) was used and the only criterion under Section 11(1) was high risk of extinction. Furthermore, the footnote, which proposes to exclude isolated populations of limited conservation value from being listed, could be removed. Listing threatened populations under the FM Act (see Box 2) could also be improved with similar changes.

Box 2. FM Act, Section 220F(2)

Endangered populations

A population is eligible to be listed as an endangered population if it is a reproducing population but its numbers have been reduced to such a critical level, or its habitat has been so drastically reduced, that it is in immediate danger of extinction, and:

(a) it is a population of a vulnerable species listed in Schedule 5, or

(b) it is disjunct and at or near the limit of its geographic range and it is or is likely to be genetically distinct.

Endangered populations - in practice

Under the TSC Act, 28 populations were listed as endangered as at January 2003 (NPWS 2003a; Table 2). The list is dynamic. (i) New listing can be made. During January to October 2003 inclusive, two endangered populations were added. (ii) The definition of the particular area can change. This occurred for the Manly population of the Little Penguin (in October 1999) and the Tadgell's Bluebell *Wahlenbergia multicaulis* population in a group of Local Government Areas (LGAs) in Sydney (in November 2003). (iii) When a species is listed as endangered, populations of that species are not eligible to remain listed populations. This occurred with the Warrumbungles population of the Brush-tailed Rock-Wallaby *Petrogale penicillata* (in July 2003) and the Baulkham Hills Shire population of *Persoonia hirsuta* (in June 1998). (iv) The taxonomy of the species can be revised. For example, the population of *Hibbertia incana* in the Baulkham Hills LGA was removed from the list in October 2001 when the newly described *H. superans*, which included the population of *H. incana* (Toelken 2000), was listed as an endangered species.

In summary, the listings of endangered populations as at January 2003 (NPWS 2003a) were a Tusked Frog population, five separate bird species, seven mammal species, the beetle *Menippus fugitivus* population in the Sutherland Shire and 14 separate plant species (Table 2). The listings of endangered populations included eight populations of species listed as vulnerable: the Glossy Black-cockatoo, Koala (2 populations), Squirrel Glider (2), Broad-toothed Rat *Mastacomys fuscus*, Brush-tailed Rock-wallaby and the shrub *Dillwynia tenuifolia*.

Recovery planning for the populations listed as endangered has proceeded slowly. As at January 2003, plans existed for seven populations (NPWS 2003a; Table 2). There were approved plans for the Little Penguin in the Manly Point Area and the Brush-tailed Rock-wallaby in the Warrumbungles; the draft plan for the Koala population at Hawks Nest and Tea Gardens had been exhibited; and plans were in preparation for the Squirrel Glider in the Wagga Wagga LGA, the Long-nosed Bandicoot *Perameles nasuta* at North Head, the Broad-toothed Rat at Barrington Tops in the Gloucester, Scone and Dungog LGAs and the Tadgell's Bluebell in the Auburn, Bankstown, Canterbury and Strathfield LGAs. For 18 of the populations, one or more of the following recovery actions had been undertaken: surveys and mapping; research into causes of decline, ameliorative methods, biology and ecology; habitat management for pest animals, weeds, fire, site protection and site rehabilitation; and ex-situ husbandry and propagation. However, monitoring, which is fundamental to understanding the effect of management actions and the ongoing status of threatened entities, was occurring for only 6 of the 28 listed endangered populations. Judged by the above information, the TSC Act could be much more effective in the process of saving the endangered populations which it lists if more resources were provided for the preparation and implementation of recovery plans.

The spread of the listed endangered populations throughout NSW, based on the NPWS Directorate boundaries, was Central - the Sydney Basin (17), Northern - north-eastern NSW (7), Western (3) and Southern - south-eastern NSW (1) (NPWS 2003a). This is probably an indication that threats to biodiversity are more likely to be recognised in the more densely peopled areas.

If the changes to the definition and criteria for listing populations under the TSC Act, as proposed earlier in this paper were enacted, then only 1 of the 28 populations listed as at January 2003 (the Gang-gang Cockatoo population in the Hornsby and Ku-ring-gai LGAs) would have been ineligible for listing.

Widespread, vulnerable species are likely to have, and in the case of icons such as the Koala and the Glossy Black-cockatoo have had, multiple nominations for endangered populations. Presumably, every subset of a species in NSW is more likely to become extinct than the species. And, while it might be argued that populations of a vulnerable species are *ipso facto* of state or regional significance (Scientific Committee in lit.) and might thereby be eligible to be listed as endangered, the current *ad hoc* and often reactive practice of nominating and listing such populations can be inefficient and ineffective. For such species, it would be much more effective to be proactive through the recovery planning process by having as a first action: mapping the distinct populations, assessing their status and thereby ranking priorities for further recovery actions.

In practice, many of the actions in recovery plans and threat abatement plans work at the level of populations. The recovery plan for *Zieria prostrata* (NPWS 1998) has an action to identify, reserve and protect all populations of this endangered species. In the recovery plan for the vulnerable *Bertya* sp. Cobar-Coolabah, there is an action to develop management strategies for the Coolabah and Gibraltar Range populations which might include fencing and burning where these senescent populations occur (NPWS 2002). The recovery plan for the endangered Eastern Bristlebird (NPWS 2003b) recognised that there were few individuals (<2000) in a few disjunct populations, only two of which had more than 500 birds, and that the species is a poor disperser. Therefore, the plan included an action to translocate birds from one of the large populations in an attempt to re-establish another viable population. The threat abatement plan for the Red Fox *Vulpes vulpes* (NPWS 2001) identified 32 high-priority threatened species which are prey to the Fox and prioritised sites, some of which have distinct populations of a prey species, to concentrate the effort of reducing the impact of predation by the Fox.

For some species (*e. g.* Brush-tailed Rock-wallaby), distinct populations are used as replicates in experiments designed to test the efficacy of Fox removal in the recovery of that prey species. This model of focusing at the level of site/population will be used in all of the threat abatement plans which address pest species (Leys 2004).

Conclusions

The TSC Act would be more effective if it was strengthened in the following ways: (i) A new first objective should be added which states that recognising threatened entities is fundamental to the legislation (see Box 3). (ii) The definition of population could be made consistent with IUCN terminology (see Box 3). (iii) Section 11 could be greatly simplified (see Box 3). (iv) Recovery planning for endangered populations, as well as threatened species and ecological communities, could be better resourced for faster preparation and more thorough implementation. (v) All recovery plans for populations should include population monitoring. (vi) Recovery and threat abatement planning for species should continue to focus at the level of populations.

Recognising, listing and saving threatened populations is fundamental to sound threatened species legislation; without these, the TSC Act would just be an act.

Box 3. Suggested changes to TSC Act

Section 3 Objects of Act

(a) to recognise threatened species, populations, ecological communities and key threatening processes, and

(b) the other objects of the Act to be renumbered and to follow.

Section 4 Definitions

Populations mean geographically or otherwise distinct groups of a species between which there is little demographic or genetic exchange (typically one successful migrant individual or gamete per year or less).

Section 11 Populations eligible for listing as endangered populations

(1) A population is eligible to be listed as an endangered population if, in the opinion of the Scientific Committee, it is facing a high risk of becoming extinct in nature in New South Wales.

(2) A population is not eligible to be listed as an endangered population if it is a population of a species already listed in Schedule 1.

Acknowledgments

Stimulating discussions were provided by numerous colleagues including Jim Anderson, Peter Christie, Kate Dallimore, Paul Downey, Bill Faulkner, Tricia Hogbin, Dan Lunney, Paul Mahon, Alison Matthews, Keith McDougall, Michael Murphy, Suzanne O'Neil, Nick Sheppard, Kersten Tuckey and Graham Wilson. Suzanne O'Neil provided comments on earlier drafts of the manuscript. Graham Wilson, Michael Murphy, Pat Hutchings and Dan Lunney provided helpful reviews of the manuscript.

References

Baker, J. 1997. The decline, response to fire, status and management of the Eastern Bristlebird. *Pacific Conservation Biology* 3: 235-43.

Bramwell, M. 1997. *Action Statement - Eastern Bristlebird (Draft).* Department of Natural Resources and Environment, Victoria.

Burgman, M. 2004. Unacknowledged uncertainty in threatened species lists. Pp 20-29 , in *Threatened Species Legislation: is it just an Act?* edited by P. Hutchings, D. Lunney and C. Dickman. Royal Zoological Society of NSW, Mosman, NSW.

Clarke, R. and Bramwell, M. 1998. The eastern bristlebird *Dasyornis brachypterus* in East Gippsland, Victoria. *Australian Bird Watcher* 17: 245-53.

Flannery, T. F. 1994. *The Future Eaters.* Reed Books, Port Melbourne, Victoria.

Gould, S. J. 1991. *Bully for Brontosaurus.* Penguin Books, London.

Holmes, G. 1989. *Eastern bristlebird: species management plan for northern populations (draft).* Consultant's report to QLD NPWS & NSW NPWS.

Holmes, G. 1997. *Conservation status of eastern bristlebird in south-east Queensland and north-east New South Wales.* Consultant's report to Queensland Dept of Environment.

Hastings, A. and Harrison, S. 1994. Metapopulation dynamics and genetics. *Annual Review of Genetics and Systematics* 25: 167-88.

IUCN 2001. *IUCN Red List Categories: Version 3.1.* International Union for the Conservation of Nature and Natural Resources, Gland.

Krebs, C. J. 1994. *Ecology* (4th ed.). HarperCollins College, New York.

Lamoreux, J., Akcakaya, H. R., Bennun, L., Collar, N. J., Boitani, L., Brackett, D., Brautigam, A., Brooks, T. M., da Fonseca, G. A. B., Mittermeier, R. A., Rylands, A. B., Gardenfors, U., Hilton-Taylor, C., Mace, G., Stein, B. A. and Stuart, S. 2003. Value of the IUCN Red List. *Trends in Ecology and Evolution* 18: 214-5.

Levins, R. 1970. Extinction. *Lectures in Mathematics in Life Sciences* 2: 75-107.

Leys, A. 2004. Feral pests as key threatening process. Pp 102-106 in *Threatened Species Legislation: is it just an Act?* edited by P. Hutchings, D. Lunney and C. Dickman. Royal Zoological Society of NSW, Mosman, NSW.

NPWS, 1998. *Zieria prostrata* Recovery Plan. NSW National Parks and Wildlife Service, Sydney.

NPWS, 2001. *Threat Abatement Plan for Predation by the Red Fox* (Vulpes vulpes). NSW National Parks and Wildlife Service, Hurstville.

NPWS, 2002. Bertya *sp. Cobar-Coolabah Recovery Plan.* NSW National Parks and Wildlife Service, Hurstville.

NPWS, 2003a. *Saving our threatened native plants and animals. Recovery and threat abatement in action.* NSW National Parks and Wildlife Service, Hurstville.

NPWS, 2003b. *Draft Recovery Plan for the Eastern Bristlebird* (Dasyornis brachypterus). NSW National Parks and Wildlife Service, Queanbeyan, in preparation.

Possingham, H. P., Andelman, S. J., Burgman, M. A., Medellin, R. A., Master, L. L. and Keith, D. A. 2002. Limits to the use of threatened species lists. *Trends in Ecology and Evolution* **17**: 503-7.

Recher, H. F., Lunney, D. and Dunn, I. (Eds). 1986. *A Natural Legacy. Ecology in Australia.* (2nd ed.). Pergamon Press, Rushcutters Bay, NSW.

Suzuki, D. 1990. *Inventing the Future.* Allen and Unwin, North Sydney, NSW.

Thackway, R. and Cresswell, I. D. (Eds). 1995. *An Interim Biogeographic Regionalisation for Australia: a Framework for Establishing the National System of Reserves, Version 4.0.* Australian Nature Conservation Agency, Canberra.

Toelken, H. R. 2000. Notes on *Hibbertia* (Dilleniaceae) 3. *Hibbertia sericea* and associated species. *Journal of the Adelaide Botanic Gardens* **19**: 1-54.

Wilson, E. O. 2001. *The Diversity of Life* (2nd ed.). Penguin Books, London.

Invertebrates and threatened species legislation

Pat Hutchings

Member of Scientific Committee, 1996-2002

Australian Museum, 6 College Street, Sydney, NSW 2010

Email: path@austmus.gov.au

ABSTRACT

Invertebrates dominate both terrestrial and aquatic environments in terms of their diversity, yet few have been listed as threatened. Does this really mean that few species are threatened, or is it more likely that we just do not know enough about invertebrate populations? Realistically, given the slow acquisition of knowledge of invertebrates in Australia, it is argued that listing of invertebrates is suitable only for "iconic" species in order to raise community awareness. A more effective avenue is the conservation of habitats or communities for invertebrates as well as the implementation of legislation to identify and tackle key threatening processes. Until both the public and scientific community become aware of the importance of invertebrates in ecosystem functioning, there will be little pressure on conservation agencies to conserve invertebrates.

Key words: Invertebrates, threatened species legislation, conservation.

Introduction

Worldwide, invertebrates are represented by 38 phyla and it has been estimated that they contain over 30 million species of which most are arthropods, especially insects (Stork 1999). More than 315,000 Australian insects have been described, yet many more remain undescribed. This is certainly true for both the terrestrial and marine invertebrate fauna of Australia. For a review of the current status of our knowledge of the biodiversity of Australia's marine invertebrate fauna, see Ponder *et al.* (2002), and for terrestrial groups see Yen and Butcher (1997). It is difficult to see this situation changing in Australia given the current dearth of taxonomists in Australia and the minimal amount of funding provided for taxonomy. This lack of knowledge regarding Australia's invertebrate fauna contrasts strongly with our knowledge of the Australian vertebrate fauna which consists of a single phylum, the Chordata, and less than 2000 species (827 birds, 847 reptiles and amphibians, 224 mammals) and probably only a few species still remain to be described. As well as being well documented, most vertebrates have common names and there is a strong community awareness of their existence and in most cases it is much easier for people to relate to them than an earthworm or snail (Horwitz *et al.* 1999).

The requirements for listing under the NSW *Threatened Species Conservation Act 1995* (TSC ACT) are typically easier to fulfil for terrestrial vertebrates and vascular plants than for terrestrial invertebrates. The NSW *Fisheries Management Act 1994*, which covers aquatic plants and all animals except for marine mammals and reptiles, largely mirrors the TSC Act. While this paper discusses the NSW Acts, similar limitations are apparent with other State and Commonwealth Acts. All these Acts are based on the criteria developed by the World Conservation Union (IUCN) and in the Red List categories. They provide an explicit, objective, quantitative framework for classifying the risk of extinction for a species. The criteria provide quantitative thresholds for population size, trend, range size and modelled probability of extinction. They are used to assign species to categories of extinction risk that can be compared between different taxa (World Conservation Union 2001). These criteria have been widely adopted because of their wide applicability, objectivity and simplicity of use (Akçakaya *et al.* 2000). They include: documenting the size of the population, demonstrating that the numbers are declining and identifying the threatening processes which may be leading to these declines. Typically, these data are much easier to obtain for vertebrates than for invertebrates because usually there is some knowledge of the biology of the animal. A workshop, held in Sydney in 1997, looked at the usefulness of these criteria for invertebrates and agreed that they had limited applicability (Hutchings and Ponder 1999). However, no attempt to develop criteria more applicable to invertebrates was undertaken. Similar problems arise when considering non vascular plants. The reasons why the requirements for invertebrates are difficult to document include the fact that invertebrate populations fluctuate between seasons and between years. Hence estimates of population size and rates of change are extremely difficult to quantify. Marine species often have pelagic larvae and these may be transported considerable distances from the parent population and recruitment success is highly variable and often strongly influenced by external factors. Many invertebrates live in habitats occupying small areas, and if these are destroyed then the associated fauna also disappears. Examples of such specialised habitats include leaf litter, under logs or in rotting wood, or a particular plant with which the invertebrate is associated. Another complicating factor is that, in various stages of the life cycles, the invertebrate populations typically occupy different environments, for example butterflies have sedentary juvenile stages but their adult stages are highly mobile. In contrast, many marine invertebrates have free-swimming juveniles and sedentary adult stages.

Pp 88 - 93 in *Threatened species legislation: is it just an Act?* 2004, edited by Pat Hutchings, Daniel Lunney and Chris Dickman. Royal Zoological Society of New South Wales, Mosman, NSW, Australia.

An area of the IUCN criteria, which needs to be explored for listing of invertebrates, is the range criterion B. It has been used successfully for vascular plants (Auld personal communication). Distribution records from museum databases could be used to determine the extent of the occurrence of a species. The other components of the criterion are more qualitative and could take into account the fluctuations exhibited by the species over seasons and between years. Plants species, which exhibit massive fluctuations between years, are regarded as more vulnerable than species which exhibit little fluctuation, because threats operating during times of low populations may lead to extinction. The validity of using this range criterion needs to be tested on a range of invertebrates with varying reproductive strategies. In addition, the extent of museum records needs to be carefully assessed as to whether they do represent the sorts of habitats in which the species is likely to occur as well as the sampling techniques which were employed. This could be undertaken only by the relevant specialist.

An example of the complexity of invertebrate life cycles is the Bathurst Copper Butterfly *Paralucia spinifera*. This species lays its eggs on the native blackthorn *Bursaria spinosa* and the larvae feed on this plant. Eggs are laid either singly, or in groups up to five, on leaves towards the base of the host plant or on adjacent debris (Dexter and Kitching 1991) and they hatch within 14-17 days. The larvae of the butterfly graze on the host plant. The larval stages have a mutualistic relationship with the ant *Anonychomyrma itinerans,* which searches the host plant seeking newly-hatched larvae. During the first three larval instars, both larvae and ants are diurnal, but later instars become nocturnal. The ant protects the caterpillar from predation and hosts the pupae in its nest. In return, the ants receive nutritional secretions from the larvae (Dexter and Kitching 1991). This butterfly has been listed as an endangered species, and any factors which impact on either the ant or the host plant will also affect the butterfly. However, it is unclear if the butterfly has an impact on the host plant.

The Scientific Committee, which was established under the NSW *Threatened Species Conservation Act 1995*, must consider all the nominations it receives unless they are vexatious. Most nominations are unsolicited, they come primarily from the public, and often from urban areas. Of the faunal nominations, almost all are vertebrates, especially the larger species. The lack of invertebrate nominations is due to a combination of factors, including that they are small and largely hidden from view and the public's perception of invertebrates as having little value (Horwitz *et al.* 1999, Allen 1999).

Currently there are 167 vertebrates listed under the NSW *Threatened Species Conservation Act*. There are also 12 invertebrate species listed as endangered, one invertebrate population is listed as endangered, and one invertebrate species is listed as presumed extinct. Of these 12 endangered invertebrates, eight are insects, three are snails and one is an earthworm. Under the NSW *Fisheries Management Act* 1994, a single invertebrate is listed; a snail *Notopala sublineata,* known to occur only in the Murray-Darling system.

Similar patterns are found under the Commonwealth legislation, the *Environment Protection and Biodiversity Conservation Act* 1999 (EPBC Act) where 15 invertebrates are listed compared to 309 species of vertebrates. Among the states, the number of invertebrates to vertebrates listed varies slightly. The highest percentage is in Tasmania, of the 130 animal species listed, 56 (43%) are invertebrates. In Victoria, of 168 animal species listed, 21 (12.5%) are invertebrates. In Western Australia, 118 animal species are listed, 38 (32%) are invertebrates. In the Northern Territory, 86 animal species are listed, 35 (40%) are invertebrates. In Queensland, 270 animal species are listed, of which only 7 (2.6%) are invertebrates. In contrast, in South Australia none of the 68 listed animal species is an invertebrate.

Invertebrates are also protected under CITES (Convention on International Trade in Endangered Species of Wild Fauna and Flora, signed in 1973) and 2039 invertebrates are listed. This includes all scleractinian corals, some species of molluscs, insects and arachnids.

So the questions that must be asked are: why are so few invertebrates listed when they are numerically dominant in all ecosystems, and, more simply: how many invertebrates are threatened? Certainly those species confined to specialised but limited habitats, such as mound springs, are threatened (Ponder 1995, 1997, 2004) because that habitat is being degraded by reduced water flow and trampling by cattle. However, for most invertebrates, we just lack the information as to whether we should even be considering them for listing. For most invertebrates we lack basic information on the natural fluctuations of their populations. These fluctuations may change naturally by orders of magnitude between seasons or years. Also, as habitats become increasingly fragmented, knowing the gene flow between isolated populations of the species becomes critical in assessing the long-term viability of these populations (Clark and Richardson 2002, Clark 2004).

Another complicating factor is that the taxonomy of many groups is poorly known and often a species may be found to consist of a suite of sibling species, presumably with each having specific habitat requirements. For example, the polychaete *Marphysa* – known as the "bloodworm" - forms the basis of an important bait industry for recreational fishers. It was initially regarded as a widely distributed cosmopolitan species *M. sanguinea*. However, close examination revealed that the Australian material from Moreton Bay, Queensland, was an undescribed species. It was formally described by Hutchings and Karageorgopoulos (2003) as *M. mullawa*. Initially, it was thought that this species occurred south of Moreton Bay and along the New South Wales coast. Subsequent studies have shown that a suite of sibling species occurs along the NSW coast and that *M. mullawa* is currently only known from Moreton Bay. These species can be differentiated not only on morphological grounds, but also by their reproductive strategies (Hutchings unpublished). Some of these species appear to have distributions restricted to specific estuarine areas, often in seagrass beds. Hopefully, restrictions under the *Fisheries Management (General) Regulation* 2002 by NSW State Fisheries on bait digging will ensure their long term maintenance, and they still have to be formally described. The regulations prevent commercial and recreational fishers from digging in seagrass beds using either a spade or fork, although a yabbie pump is allowable. This example illustrates our lack of knowledge of even those species which have a commercial value. Such taxonomic impediments are widespread among marine invertebrate species (Ponder *et al.* 2002) and in many terrestrial groups.

There is also an apparent reluctance to list some of the colourful invertebrates which are collected and have a monetary value. Various interested groups perceive that listing will restrict their collecting ability. Certainly, listing of a species adds another layer of bureaucracy to obtaining permits to undertake research on the endangered species, and yet such research is critical to its long-term survival. This too has dampened the enthusiasm of some invertebrate specialists to nominate species in their groups for listing. There needs to be a closer working relationship between the regulation branch, which issues permits, with the scientists within the relevant department to ensure that permits for research, which will gather information necessary for managing that species, are given to the appropriate workers.

Another factor contributing to the lack of invertebrates being nominated for listing is the general ignorance within the community of the importance of invertebrates in the functioning of marine and terrestrial ecosystems. Most invertebrates are small, inconspicuous and very few are colourful or charismatic, although this latter characteristic relies upon the "human perception" of how a species acts or behaves. I contend that this is based upon ignorance, typically when one explains the reproductive behaviour of, say, a polychaete worm, then many people are amazed that such a "lowly" animal can do such things.

Conservation Outcomes of Listings

Despite these major hurdles to listing invertebrates, some species have been listed in most states, thus we also need to explore if listing has helped in its conservation? In New South Wales, once a species is listed there is a requirement that a recovery plan be developed within a reasonable time. However, of the 12 invertebrates listed, recovery plans have been developed for only three species. Once the recovery plan has been developed, its implementation depends on the necessary funds being allocated. The recovery plan for the Bathurst Copper Butterfly has certainly led to increasing public awareness of the species and encouraged plantings of its host plant, the native blackthorn, and also importantly the consideration of its habitat in hazard reduction burns. More details as to how the community, including school groups, has been involved in documenting the distribution and abundance of the species can be found in Nally (2004). More difficult to ascertain are the flow-on effects in raising the community awareness of other endangered species in the area. Certainly the local council has been highly supportive of the program. This is critical, because it is at the regional level where implementation occurs and largely determines if the species continues to survive.

Listing of the Cumberland Plain snail *Meridolum corneovirens* has raised the public awareness, although a recovery plan has not been developed. The snail occurs in the Cumberland Plain Woodland, Castlereagh Woodlands and the River Flat woodland, which are Endangered Ecological Communities found on the Cumberland Plain (Tozer 2003), for which recovery plans are being developed. The Cumberland Plain Woodland community - originally widespread across western Sydney - is now represented in many places by scattered fragments, which are often degraded or weed infested. Detailed mapping of these communities has also been undertaken as part of the development of recovery plans. However, even listing the snail and its associated ecological community has not prevented loss of snail habitat, which includes logs and debris under which the snail is found.

The reason for the loss of these communities in western Sydney is that these areas are regarded as prime sites for residential development. Nevertheless, one consequence of listing may have been the undertaking of a detailed study on the gene flow between these isolated snail populations by Clark and Richardson (2002). This study will be relevant to other sedentary terrestrial species, which live in isolated and fragmented habitats, and it will highlight the problems of conserving fragmented populations. Another consequence is a detailed mapping of the snail's distribution, undertaken by Clark (in prep), and additional sites have been found, but others have now been destroyed by development. This listing engendered complaints from consultants undertaking 8 part tests because of the difficulty in identifying these snails (Denny 2004).

The Lord Howe Island endemic earthworm *Pericryptodrilus nanus* was listed by the NSW Scientific Committee for several reasons. The species is restricted to the leaf litter on the summit of Mount Gower, Lord Howe Island, a World Heritage Area and this earthworm is an important food resource for the listed Lord Howe Island woodhen *Tricholimnas sylvestris*. Although it would have been preferable to have listed the entire leaf litter community, which contains a range of invertebrates, this was not feasible because all the other species are as yet undescribed. One of the desirable consequences of listing was the creation of an awareness of the value of the leaf litter community and for this to be taken into account when issuing permits for tourist activities and bird watching studies on the summit. To date, no recovery plan has been developed, although hopefully this listing has been taken into account when granting permission for various activities on Mount Gower. A Biodiversity Plan for all taxa on Lord Howe Island (excluding those under the NSW Fisheries Act) is currently being developed by the NSW Department of Environment and Conservation and this will include this fragile leaf litter community.

The way forward

In view of the large number of invertebrates present in both marine and terrestrial environments, and acknowledging that some are threatened, it is not an efficient process to list them apart from some iconic species, which can become flagships, that is the invertebrate equivalent to the koala. So what do we need to do?

Using the *Threatened Species Conservation Act* in NSW, two other options are available:- listing endangered ecological communities and listing key threatening processes. As the Scientific Committee in New South Wales came to terms with the complexities of listing ecological communities, and in response to comments received after advertising the preliminary determinations as required under the Act, we included in all the later listings two catch-all phrases, one which included the seed bank and some which included the invertebrates in the community without listing them. For those communities which tend to be defined largely by the vegetation, the listing will include, virtually by default, seeds in the soil which may or may not germinate given the right conditions and therefore the plants listed as defining that community may change over time. Similarly, the soil will contain a suite of invertebrates, many of which will be restricted to that particular community although in most cases they will still be undescribed. Loss of that community, or its degradation, will lead to local extinctions of some of the associated invertebrates, or even a complete loss of species. However, these losses have been poorly documented, although it is known that after land is cleared for agriculture it has a much poorer soil fauna, especially with regard to earthworms, than uncleared areas. However, this process can be reversed. Abandoned tropical pastures have been shown to be able to recover their native earthworm fauna (Sanchez-De Leon *et al.* 2003). Introduced European earthworm species are dominant in Sydney urban areas (Blakemore 1999) and have displaced native species, which may be more efficient at processing native Australian soils than introduced species although this has yet to be tested. There is also a dearth of information available on the impact of frequent fires on the invertebrate fauna, but the loss of this fauna, even on a temporary basis, would have major impacts on the large number of vertebrates that feed on them.

While several Key Threatening Processes (KTP) have been listed, it appears that the former NSW NPWS (as it was known until recently- now part of the Department of Environment and Conservation) had been concentrating its resources on developing recovery plans for species rather than on preparing Threat Abatement Plans, which are required under the legislation. No Threat Abatement Plans have been prepared for the two KTPs which have major impacts on terrestrial invertebrates, namely "Clearing of native vegetation" (gazetted September 2001) and "High frequency fire resulting in the disruption of life cycle processes in plants and animals and loss of vegetation structure and composition" (gazetted March 2000), both of which impact on soil and arboreal invertebrates (Baker 2004, Campbell and Tanton 1981).

In the marine environment, loss of particular habitats such as clearing of mangroves, salt marsh or seagrass beds, will lead to the local extinction of species restricted to these habitats. Changes to water flows may restrict freshwater habitats or allow salt water intrusions into the river, as well as allowing the accumulation of pesticides and excess fertilisers in soils leading to a change in soil structure. Drainage of wetlands to create agricultural land often exposes acid sulphate soils, which results in the leaching of acids into the river causing massive fish kills. This drainage, often as part of flood mitigation works, has been undertaken in northern NSW and Queensland. Other consequences of the loss of coastal wetlands, or clearing of vegetation on land, are increased soil erosion and the transport of excess fertilisers and pesticides down the rivers and into shallow coastal environments. Some pesticides, such as diuron, have been shown to have deleterious effect on the growth of seagrass beds (Haynes *et al.* 2000). This has been highlighted along the Queensland coast with the inshore waters of the Great Barrier Reef receiving excessive levels of sediments and pollutants leading to widespread degradation of inshore reefs (Baker 2003).

Instead of using Threatened Species legislation, there is a recent Commonwealth and Queensland Government initiative, the Reef Water Quality Protection Plan, to deal with the matter. One aspect of the plan is to minimise the loss of riparian habitats and to encourage their rehabilitation. Parts of the Commonwealth's *Environmental Protection and Biodiversity Conservation Act 1999* have been enacted as the Great Barrier Reef Marine Park is listed as a World Heritage Area, over which this Act has jurisdiction. The problems highlighted for the Queensland coast are not unique, and similar management strategies need to be put in place in NSW. The recently-established Catchment Management Authorities in NSW will become responsible for developing catchment management plans, which hopefully will consider river flows and the restoration of riparian vegetation. Previously, catchment and water management policies were developed by regionally-based committees with government and community representation (Fairfull and Williams 2003). These committees have now become statutory authorities, which presumably will improve the development and implementation of such policies. It also seems likely that threat abatement plans will be incorporated into these management plans (Leys 2004), but as to who will actually develop these plans is currently unclear.

Conclusions

So, to answer the question: "Is Threatened Species legislation appropriate or even useful for invertebrates?', one would have to conclude that it is of limited value and should be restricted to flagship species and to raise community awareness of the value of invertebrates. Of more concern is our lack of knowledge, and therefore our ability, to be able to identify species as being threatened. As we are unlikely to obtain that data in the near future, we should concentrate on conserving representatives of all our habitats and hopefully, by default, protect our invertebrate fauna.

This review of the legislation has aimed to highlight the need to raise the awareness of the numerous important roles, which invertebrates play both in the scientific community as well as the community at large. Invertebrates are critical for ecosystem functioning in all environments (New 1995, Snelgrove *et al.* 1997, Wall *et al.* 1997), such as in the breakdown of organic matter, recycling of organic material, and in the bioturbation and aeration of sediments. Sediments devoid of invertebrates are typically disturbed and polluted (Ward and Hutchings 1996). Most vertebrates (64%), including threatened vertebrate species, feed either entirely or largely depend on invertebrates (Lunney *et al.* 2000). Also, the loss of many insects will lead to changes in many key plant interactions, such as decreased pollination rates (Cane 2001, Cane and Tepedino 2001). All these examples draw attention to the real need to increase our awareness of the role that invertebrates play in all communities. Yet how to create that increased awareness is a major challenge. The majority of funds expended by various agencies on conservation, or even research, is geared towards vascular plants and vertebrates. There are far more vertebrate biologists than invertebrate biologists. This is reflected in the bias of the scientific community. There is thus a danger of ignoring the urgent need for more taxonomic and ecological work on invertebrates in both universities and government departments. Similarly, much more education is needed for senior management and bureaucrats on the key importance of the role and function of invertebrates in sustaining natural landscapes. So perhaps it is not surprising that the general community knows little about invertebrates, other than those which they wish to eradicate from their houses and gardens, such as cockroaches and snails and slugs. We invertebrate biologists have an uphill battle in front of us.

Acknowledgements

I would like thank my colleagues on the NSW Scientific Committee during my 7 year tenure for interesting discussions on the merits and problems of listing invertebrates, and especially to Tony Auld who give me insight into the IUCN criteria, as well as one of my reviewers and Dan Lunney for providing useful editorial comments.

References

Allen, J.T. 1999. Building pathways for marine invertebrate conservation. Pp 407-412 in *The Other 99%. Conservation and Biodiversity of Invertebrates,* edited by W. Ponder and D. Lunney. Transactions of the Royal Society of New South Wales, Mosman, NSW.

Akçakaya, H.R., Ferson, S., Burgman, M.A., Keith, D.A., Mace, G.M., and Todd, C.A. 2000. Making consistent classifications under uncertainty. *Conservation Biology* **14:** 1001-1013.

Baker, J. 2003. A report on the study of land sourced pollutants and their impacts on water quality in and adjacent to the Great Barrier Reef. A report prepared by an Intergovernmental Steering Committee to Premiers Department, Queensland Government www.premiers.qld.gov.au/about/reefwater.pdf

Baker, J. 2004. Endangered populations: the concept in practice. Pp 82-87 in *Threatened Species legislation: is it just an act?* edited by P. Hutchings, D. Lunney and C. Dickman. Royal Zoological Society of New South Wales, Mosman, NSW.

Blakemore, R.J. 1999. The diversity of exotic earthworms in Australia – a status report. Pp 182-187 in *The Other 99%" Conservation and Biodiversity of Invertebrates*, edited by W. Ponder and D. Lunney, Transactions of the Royal Zoological Society of New South Wales, Mosman NSW.

Campbell, A.J. and Tanton, M.T. 1981. Effects of fire on the invertebrate fauna of soil and litter of a eucalypt forest. Pp 215-241 in *Fire and the Australian Biota,* edited by A.M Gill, R.H. Groves and J.R Noble. Australian Academy of Science, Canberra.

Cane, J. H. 2001. Habitat fragmentation and native bees: a premature verdict? *Conservation Ecology* **5(1)**: 3 [online]. http://www.consecol.org/vol5/iss1/art3

Cane, J. H. and Tepedino, V.J. 2001. Causes and extent of declines among native North American invertebrate pollinators: detection, evidence, and consequences. *Conservation Ecology* **5(1)**: 1 [online]. http://www.consecol.org/vol5/iss1/art1

Clark, S.A. 2004. Native snails in an urban environment-conservation from the ground up. Pp 78-81 in *Urban Wildlife: more than meets the eye*, edited by D. Lunney and S. Burgin. Royal Zoological Society of New South Wales, Mosman, NSW.

Clark, S.A. in prep. Systematics, spatial analysis and conservation genetics of *Meridolum corneovirens* and related forms (Gastropoda:Cameridae) from the Sydney Region of Australia. PhD thesis, University of Western Sydney, Hawkesbury.

Clark, S.A, and Richardson, B.J. 2002. Spatial analysis of genetic variation as a rapid asssessment tool in the management of narrow range endemics. *Invertebrate Systematics* **16(4)**: 583-587.

Denny, M. 2004. Struggling with the Act- some observations by ecological consultants. Pp 164-171 in *Threatened Species legislation: is it just an act?* edited by P. Hutchings, D. Lunney and C. Dickman. Royal Zoological Society of New South Wales, Mosman, NSW.

Dexter E.M. and Kitching R.L. 1991. *Nomination for the Register of the National Estate.* Australian Heritage Commission (unpublished report).

Fairfull, S.J. and Williams, R.J. 2003. Community involvement in natural resource management: lessons for future water management in coastal catchments of New South Wales. Pp 55-61 in *Conserving Marine Environments. Out of sight out of mind*, edited by P. Hutchings and D. Lunney. Royal Zoological Society of New South Wales, Mosman, NSW.

Horwitz, P., Recher, H. and Majer, J. 1999. Putting invertebrates on the agenda: political and bureaucratic challenges. Pp 398-406 in *The Other 99%. Conservation and Biodiversity of Invertebrates* edited, by W. Ponder and D. Lunney. Transactions of the Royal Society of New South Wales, Mosman NSW.

Hutchings, P.A. and Karageorgopoulos, P. 2003. Designation of a neotype of *Marphysa sanguinea* (Montague, 1815) and a description of a new species of *Marphysa* from Eastern Australia. *Hydrobiologia* 496(1-3): 87-94.

Hutchings, P. and Ponder, W. 1999. Criteria for assessing and conserving threatened invertebrates. Pp 297-315. A summary of a workshop held in conjunction with the *Other 99% Conservation and Biodiversity of Invertebrates,* edited by W. Ponder and D. Lunney. Transactions of the Royal Zoological Society of New South Wales, Mosman, NSW.

Haynes, D., Ralph, P., Prange, J. and Dennison, B. 2000. The impact of the herbicide diuron on photosynthesis in three species of tropical seagrasses. *Marine Pollution Bulletin* **41**, 288-293.

Leys, A. 2004. Threat abatement plans: strategic pest management for biodiversity conservation. Pp 102-106 in *Threatened Species legislation : is it just an act?* edited by P. Hutchings, D. Lunney and C. Dickman. Royal Zoological Society of New South Wales, Mosman, NSW.

Lunney, D., Curtin, A.L., Ayers, D., Cogger, H.G., Dickman, C.R., Maitz, W., Law, B. and Fisher, D. 2000. The threatened and non-threatened native vertebrate fauna of New South Wales: status and ecological attributes. *Environmental and Heritage Monograph Series No. 4.* Pp 1-132. NSW National Parks and Wildlife Service, Hurstville.

Nally, S.C. 2004. Community involvement in the conservation of the endangered purple copper butterfly, *Paralucia spinifera* Edwards & Common (Lepidoptera: Lycaenidae). *Records of the South Australian Museum Monograph Series* No. 7: 39-49.

New, T.R. 1995. *An introduction to Invertebrate Conservation Biology.* Oxford University Press. Oxford UK 194pp.

Ponder, W.F. 1995. Mound spring snails of the Australian Great Artesian Basin. Pp 13-18. In *The conservation biology of molluscs*, edited by E.A. Kay. IUCN, Gland Switzerland.

Ponder, W.F. 1997. Conservation status, threats and habitat requirements of Australian terrestrial and freshwater Mollusca. *Memoirs of the Museum of Victoria* 56(2): 421-430.

Ponder, W. F. 2004. Endemic aquatic macroinvertebrates of artesian springs of the Great Artesian Basin – progress and future directions. *Records of the South Australian Museum Monograph Series* No. **7**: 101-110.

Ponder, W.F., Hutchings, P.A. and Chapman, R. 2002. Overview of the Conservation of Australia's marine invertebrates. A report for Environment Australia. http://www.amonline.net.au/invertebrates/marine_overview/index.html

Sanchez-De Leon, Y., Zou, Z., Borges, S. and Ruan, H. 2003. Recovery of Native Earthworms in Abandoned Tropical Pastures. *Conservation Biology* **17**: 999-1006.

Snelgrove, P., Blackburn, T., Hutchings, P.A., Alongi, D.M., Grassle, J.F., Hummel, H., King, G., Koike, I., Lambshead, P.J.D., Ramsing, N.B. and Solis–Weiss, V. 1997. The importance of marine sediment biodiversity in ecosystem processes. *Ambio* 26: 578–583.

Stork, N. E. 1999. Estimating the number of species on Earth. Pp 1-7 in *The Other 99%. The conservation and biodiversity of invertebrates*, edited by W. Ponder and D. Lunney, Transactions of the Royal Zoological Society of New South Wales, Mosman, NSW.

Tozer, M. 2003. The native vegetation of the Cumberland Plain, western Sydney: systematic classification and field identification of communities. *Cunninghamia* **8**: 1-75.

Tulau, M. 1999. Management of acid sulphate soils in New South Wales- Policy, organization, and regulation. *The Australasian Journal of Natural Resources Law and Policy* **6** (1), 1-32.

Wall, D.H., Blackburn, T., Brussaard, L., Hutchings, P.A., Palmer, M.A. and Snelgrove, P.V.R. 1997. Linking biodiversity and ecosystem functioning of soils and sediments. *Ambio* **26**: 556–562.

Ward, T. and Hutchings, P.A. 1996. Effects of zinc, lead and cadmium on the species composition of the infauna of polluted intertidal and subtidal marine sediments near a lead smelter, Spencer Gulf, SA. *Journal of Experimental Marine Ecology Progress Series* **135**: 123–135.

Yen, A. and Butcher, B. 1997. *An overview of the conservation of non-marine Invertebrates in Australia.* Environment Australia, Canberra.

World Conservation Union (IUCN) 2001. Red List categories and criteria, version 3.1. Species Survival Commission IUCN, Gland, Switzerland, and Cambridge UK.

Endangered ecological communities and landscape conservation in NSW: successes and failures in the Sydney Basin

Tony D. Auld and Mark Tozer

Biodiversity Research Group, Department of Environment and Conservation (NSW)

PO Box 1967 Hurstville NSW 2220

Email: tony.auld@environment.nsw.gov.au

ABSTRACT

We examine the nature and extent of listings of endangered ecological communities (EECs) under the *NSW Threatened Species Conservation Act 1995* (*TSC Act*), including the role of endangered ecological communities in conservation from the landscape to species level. We ask how well the implementation of the *TSC Act* is working for ecological communities in relation to the key ecological issues of habitat loss and fragmentation, conservation of remnants, and conservation of ecological processes and disturbance regimes. To do this we use examples of endangered ecological communities from the Sydney Basin.

Some 60 endangered ecological communities have been listed under the *TSC Act* since its inception and the rate of listing is likely to continue for some time yet. Most listings have been in coastal NSW, in particular in the Sydney Basin, with only a few in western NSW. This essentially reflects our current knowledge of vegetation associations in NSW and the fact that the majority of nominations have come from people who live in Sydney. In the Sydney Basin, for both the Duffys Forest EEC and the 12 EECs making up the Cumberland Plain, there has been extensive mapping of remnants and this has significantly increased both the awareness and acceptance of conservation issues relating to these communities. However, continued habitat loss and fragmentation of remnants is undermining their long-term viability. Any attempt to balance losses with reconstruction of ecological communities is not likely to contribute to their maintenance due to the time lags in reconstruction and importantly, our current lack of understanding of how to achieve restoration of complex ecological communities.

Greater success is occurring in ameliorating threats to remnants of EECs, in particular weed control and minimising disturbance, especially through support from local councils and volunteer groups. However, both the number of remnants and their proximity to urban areas will influence the ability to ameliorate threats in remnants due to resource constraints. The recognition that persistence of these EECs is also dependent upon allowing appropriate disturbance regimes is occurring in relation to fire management, with implementation success based on the degree of understanding of how fire affects flora and fauna. This consideration will need to be extended to maintenance of species interactions, ecological processes and other disturbance regimes if viable endangered ecological communities are to be maintained.

Based on this experience in the Sydney Basin, we would argue that the listing of threatened ecological communities in other parts of NSW is a useful conservation mechanism to raise awareness of conservation issues. Implementing recovery and the ongoing viability of such communities will be more dependent on developing an understanding of the processes that affect species persistence and interactions within communities and engaging support for conservation from local human communities.

Key words: clearing, fragmentation, disturbance regimes, threats, management, research

Introduction

The concept of recognising patterns in associations of species across the landscape has a long history, especially for vegetation-based associations. Such associations are often referred to as ecological communities. In NSW, the introduction of the *NSW Threatened Species Conservation Act* (*1995*) allowed for the protection of ecological communities that were considered to be endangered. This has been amended by the *NSW Threatened Species Amendment Act 2002*, which potentially allows for the listing of vulnerable ecological communities, although they are excluded from the licensing provisions of this Act. At the national level the *Endangered Species Protection Act 1992* (now replaced by the *Environmental Protection and Biodiversity Conservation Act 1999*) also allowed for identification and protection of endangered ecological communities. Other Australian states have also adopted threatened species legislation that includes the capacity to list ecological communities, e.g. Victoria and Tasmania.

In NSW, the *TSC Act* has several objectives, with the primary objective relating to biodiversity being 'To conserve biological diversity and promote ecologically sustainable development'. Most threatened species legislation is

Pp 94 - 101 in *Threatened species legislation: is it just an Act?* 2004, edited by Pat Hutchings, Daniel Lunney and Chris Dickman. Royal Zoological Society of New South Wales, Mosman, NSW, Australia.

focussed on listing of species and hence conservation at the species and within species (genetic) level of biodiversity. The listing and conservation of ecological communities allows for conservation at broader scales across the landscape including genetic, species and ecosystem diversity. Within threatened species legislation, only ecological communities and key threatening processes address conservation of biodiversity at a scale broader than the species scale. Other mechanisms that could address habitat conservation at this scale are not available in threatened species legislation. Endangered ecological communities have a number of roles in the conservation of:

i) ecological process, species interactions and ecological disturbance regimes;

ii) rare patterns in the landscape (e.g. mound springs);

iii) common patterns in the landscape that have been heavily impacted;

iv) common and rare species; and

v) variation in patterns of biota across the landscape.

In this paper, we briefly examine the nature and extent of listings of endangered ecological communities in NSW since the inception of the *TSC Act*. We then use two examples from the Sydney Basin (Duffys Forest and the suite of communities occurring on the Cumberland Plain) to examine the role that the *TSC Act* has played in the conservation of these communities. We specifically target the successes and failures of these listings in relation to: habitat loss and fragmentation; degradation of remnants, and maintenance of disturbance regimes that impact on species' life histories.

Listings of Endangered Ecological Communities (EECs) in NSW

In NSW, an independent Scientific Committee, established under the *TSC Act*, has been responsible for additions, changes or deletions to the existing schedules. The criteria for listing for EECs in the *TSC Amendment Act* are:

"An ecological community is eligible to be listed as an endangered ecological community if, in the opinion of the Scientific Committee:

(a) it is likely to become extinct in nature in New South Wales unless the circumstances and factors threatening its survival cease to operate, or

(b) it might already be extinct."

Hence, evidence of threat and/or decline is required for listing.

The definition of an ecological community is given in the *TSC Act* as 'an assemblage of species occupying a particular area' combining concepts of both species association and area of habitat, although ignoring ecological processes. Endangered ecological communities are listed on Schedule 1, Part 3 of the *TSC Act*. At the time the Act began in January 1996, no endangered or vulnerable communities were listed. Since that time, the Scientific Committee has listed some 60 EECs, up until October 2003. No vulnerable ecological communities have been listed, as this component of the amended *TSC Act* has not come into force. Fifty seven of the listed EECs have been largely based on plant floristic data (although some may specifically include other taxonomic groups in their definitions), one on wading birds, one on Hygrocybeae fungi and one on lichens. The rate of listings of EECs in NSW has risen to 8-15 per year from 1998 to 2002 (Fig. 1) after some 5 listings in the first 2 years of operation of the *TSC Act*. Independent of what taxonomic group the listings are based on, all taxa (including all life stages, such as seeds in a soil seed bank) within the EEC are considered to be a part of the community and are given some protection. Unlike native vertebrates (excluding fish), invertebrates, vascular and non vascular plants and fungi in NSW have no legislative protection unless they either occur within the National Parks and Wildlife Reserve system, or are listed as a threatened species or they occur within a listed EEC. A small number of vascular plants are also listed as protected if they are subject to harvesting or commercial utilisation. The distribution of the listed communities across NSW (Fig. 2) shows a strong bias towards the Sydney Basin and coastal NSW in general. The listings may also occur at different spatial scales (Fig. 2), partially in response to the scale of variation in floristic patterns, but also in relation to the scale of available data on ecological communities. In essence, both the preponderance of listings based on vegetation data and the concentrations of listings in the Sydney Basin, reflect the greater knowledge base and detailed vegetation survey effort in these areas, as well as the prevalence of nominations from people within Sydney. Also, given the large urban population in Sydney, there has been, and continues to be, significant loss of habitat in this area and hence, a high and continuing level of threat to existing ecological communities. There are some listings from arid and semi-arid NSW, including both EECs that were always very rare in the landscape and have been severely degraded (mound springs) and those that were once widespread, but have been heavily cleared and degraded (e.g. White Box Yellow Box Blakely's Red Gum Woodland).

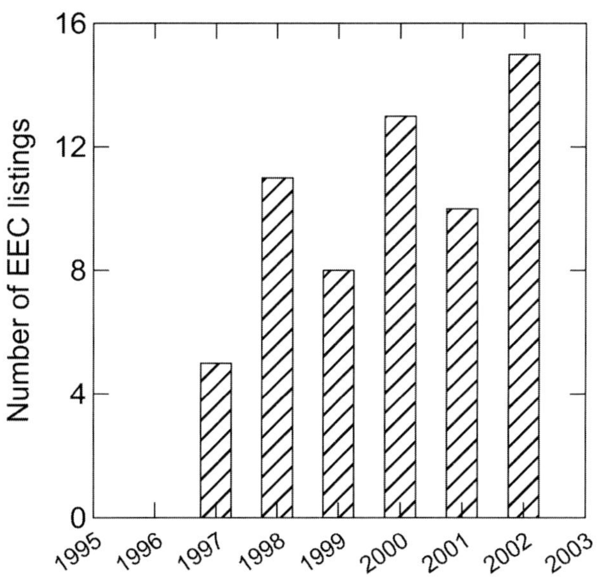

Figure 1. Annual frequency of listings of Endangered Ecological Communities by the NSW Scientific Committee under the *TSC Act*.

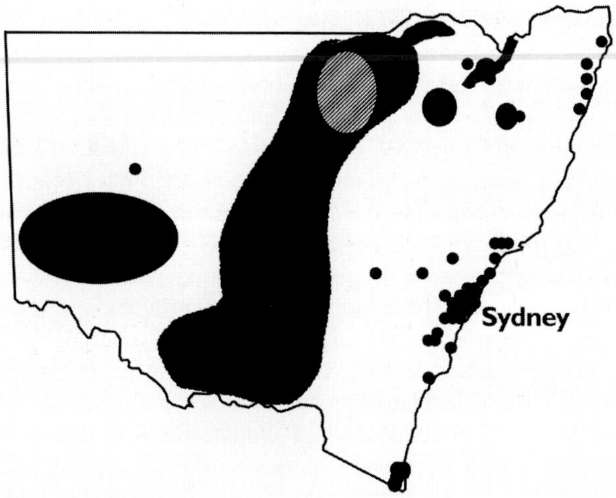

Figure 2. Approximate distribution of listed endangered ecological communities on Schedule 1, Part 3 of the NSW *TSC Act.* Large shaded or cross hatched areas represent boundaries of the distribution of a listed EEC. Remnants of the EEC may be very patchy and limited in occurrence across this distribution.

Key Issues for Conservation of Ecological Communities

Several factors currently influence the conservation of endangered ecological communities. Their listing highlights the degree of past habitat loss and continuing threats to these communities, but there is also a lack of quantitative data (and hence nominations) for ecological communities over much of NSW. We suggest that the factors that need urgent attention for managing listed EECs as viable entities are:

i) cessation of further loss and fragmentation of habitat. Clearing of vegetation has been the primary cause of decline for both species and ecological communities. The significance of this threat is recognised in its listing as a Key Threatening Process in NSW (NSW Scientific Committee 2001). Continued clearing of remnants of listed EECs will reduce their distribution and extent, their connectivity and the viability of the community. It will also increase edge effects and result in further degradation of remaining remnants.

ii) reversing degradation of habitat. Many remnants of existing EECs are small with large edges. They are prone to a range of human-caused disturbances and alterations such as roads, tracks, urban runoff and nutrification. This is frequently associated with invasion by a range of exotic species of plants and animals detrimental to the survival of native species in the remnants. In order to maintain existing remnants into the long term these threats must be mitigated.

iii) maintaining species life history processes and interactions. Species that make up ecological communities are not static through space and time. Progression of species through their life cycles and the maintenance of the ecological interactions between species is necessary for long-term viability. Disturbance regimes such as fire play a key role, and in many ecological communities they have been a key driving force in maintenance of species over evolutionary time scales. Consequently, to maintain EECs, even in degraded and fragmented landscapes, some attempt must be made to implement disturbance regimes that are appropriate for the whole suite of species within the community.

Success and Failures in Conserving Duffys Forest and Cumberland Plain Endangered Ecological Communities

We evaluate how well the listing of EECs has contributed to their conservation by considering the three key ecological factors listed above. We acknowledge that the implementation of existing listings on the *TSC Act* (as opposed to the listing process) also incorporates both social and economic considerations in making decisions that impact on the conservation of species and EECs. However, we believe that ultimately it is the ecological conservation outcomes that are critical in determining whether or not the *TSC Act* has been successful, irrespective of whether the social and economic consequences are palatable.

Location and descriptions of Duffys Forest EEC and EECs on Cumberland Plain

The Duffys Forest EEC (Smith and Smith 2000, NSW Scientific Committee 2002) is confined to a small area on the urban/bushland interface in northern Sydney. This community was always rare in the landscape and restricted to laterite rich soil on ridges and upper slopes (Benson and Howell 1994). It has been estimated that some 1450 ha existed prior to European settlement (Smith and Smith 2000). Today it exists in less than 50 small fragmented remnants comprising some 15% of the original area (Smith and Smith 2000). Most of the remnants are small (<10ha), have large edges and are internally fragmented by roads or other clearings. Just under half of the known remnants are contained within Ku-ring-gai Chase and Garigal National Parks. A full description of the floristics of Duffys Forest EEC can be found in Smith and Smith (2000).

The Cumberland Plain, located in western Sydney and stretching from Thirlmere in the south to Sackville in the north and to the edge of the Blue Mountains, contains 12 listed EECs (Tozer 2003) (Table 1). An estimated 13% of the vegetation that existed prior to European settlement remains today (Tozer 2003). There is a very large number of remnants and as the original area of

Table 1. Endangered ecological communities listed under *NSW TSC Act* occurring on the Cumberland Plain. Remnant areas derived from Tozer (2003).

Endangered Ecological Community	Approx. remnant area (ha)
Elderslie Banksia Scrub Forest	<1
Agnes Banks Woodland	98
Blue Gum High Forest	168
Western Sydney Dry Rainforest	338
Moist Shale Woodland	604
Castlereagh Swamp Woodland	616
Cooks River/Castlereagh Ironbark Forest	1,012
Sydney Turpentine Ironbark Forest	1,183
Shale/Gravel Transition Forest	1,721
Sydney Coastal River-Flat Forest	5,415
Shale/Sandstone Transition Forest	9,960
Cumberland Plain Woodland	11,054

vegetation was large (274,000ha), the area of remnants of EECs on the Cumberland Plain overall is much greater than for Duffys Forest (31,500 v 240 ha). The remnants vary in size, with most (97%) less than 80 ha and around half (51%) <3 ha (Tozer 2003). Some of the listed EECs on the Cumberland Plain are also very small, with few remnants, e.g. Agnes Banks Woodland, Elderslie Banksia Scrub, Blue Gum High Forest (see Table 1). The EECs with the most remnant habitat and number of remnants are Cumberland Plain Woodland and Shale/sandstone Transition Forest (Tozer 2003, Table 1). A full description of the floristics of EECs on the Cumberland Plain can be found in Tozer (2003), except for the Elderslie Banksia Scrub Forest, which is briefly described in NSW Scientific Committee (1999).

Loss and fragmentation of habitat

To achieve conservation of remnants of EECs across the landscape, and across the spatial and temporal ecological variation within each community, there should be no further loss of habitat or fragmentation of ecological communities once they are listed as endangered. Listing as endangered is a reflection of the high risk of extinction of EECs. Any further loss will compromise the viability of listed communities. Consequently, the major determinate of the success of the *TSC Act* for EECs should be a measure of the degree and rate of loss pre- versus post-listing.

Duffys Forest

One of the key aims of the recovery plan being developed for Duffys Forest is no further loss of remnants. For this community, there has been extensive targeted mapping of remnants (Smith and Smith 2000) as well as pre-listing mapping (Thomas and Benson 1985; Sheringham and Sanders 1993; Benson and Howell 1994). Further, there have been management actions under the *Grevillea caleyi* recovery plan (Scott *et al.* 1995), that cover a number of remnants of Duffys Forest. There has been some limited acquisition of remnants into reserves through the activities of the *Grevillea caleyi* recovery team (NPWS 2001). Some 48% of remnants are in national parks, while the remainder occur on a range of tenures from state and local government to private land (Smith and Smith 2000; NPWS recovery plan for Duffys Forest (in prep.)).

Further losses of remnant habitats are occurring. Such losses vary from the loss of transitional vegetation surrounding Duffys Forest EEC, loss of edges or parts of remnants to the loss of whole remnants. Such losses are mostly associated with continued clearing for housing, as well as infrastructure support such as roads and clearing for fire mitigation. There are currently no complete details on the extent of these losses relative to the amount and type of remnant habitat that remains. An assessment of the current rates of habitat loss is essential, but any further clearing will likely compromise the long-term viability of Duffys Forest EEC.

Cumberland Plain

The aim of the recovery plan being developed for all the EECs on the Cumberland Plain involves a net gain in vegetation cover. The specific goal is to achieve 30% of original landscape with protection and revegetation (NPWS 2002). To do this there has been extensive mapping of remnants and predictions of former habitat (Tozer 2003), combined with an assessment of core habitat areas and areas that may be a useful support for these core habitats. Such mapping has had a number of immediate benefits including: greater certainty in landuse planning decisions; greater acceptance of listed endangered ecological communities on the Cumberland Plain at federal (e.g. Cumberland Plain Woodlands, Shale/sandstone Transition Forest), state and local government levels and among local communities; increased awareness of EEC issues, including threats, ecological processes and disturbance regimes; and increased community awareness. This had led to some significant protection of remnants in conservation reserves. However, significant habitat loss is still occurring outside these reserves. There has been no attempt to assess the current rate of habitat loss, yet such data are important to gain insights into the effectiveness of the *TSC Act* for conservation of EECs on the Cumberland Plain.

Based on the available area of remnant habitat (Tozer 2003), significant areas will need to be revegetated to achieve the 15-30% landscape restoration goal for the remnant EECs on the Cumberland Plain (Fig. 3). Some of the areas identified as remnants in Fig. 3 may only contain intact tree overstorey, while the understorey may have been removed, while others may be cleared in the future (further clearing is not prevented by listing on the *TSC Act*). As well, some areas to be revegetated may have a partially intact understorey, even though the overstorey has been removed or thinned. Consequently, we believe the relative proportions illustrated in Fig. 3 are likely to be a best case scenario. Clearly more survey effort is needed to identify high quality understorey in areas with few or no trees to assist any planning for revegetation as these areas represent significant remnants for the conservation of biodiversity.

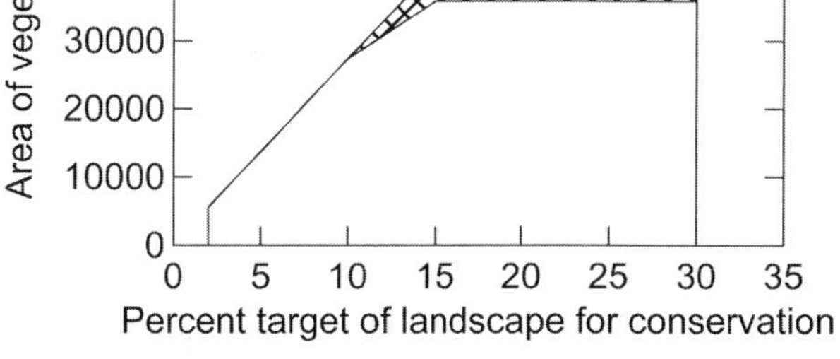

Figure 3. Estimated area of land needed for target restoration of landscape on the Cumberland Plain.

Filled area is essentially cleared and would need revegetation. Open area is remnant vegetation.

Although EECs on the Cumberland Plain are often considered collectively for conservation planning and restoration purposes, it must be emphasised that a 15-30% conservation target for the Cumberland Plain will not necessarily ensure adequate conservation of all EECs on the Cumberland Plain. First, the area of remnant bushland varies widely between individual EECs (Table 1) and the conservation of large remnant areas of vegetation may make no impact on the status of individual EECs unless they are specifically targeted. Second, for some EECs, a large proportion of the original habitat may be unavailable for restoration due to urban encroachment, therefore any restoration activities will have no impact for these communities. Third, ecological communities are not homogeneous units, rather they vary in structure and composition throughout their range. For example, Cumberland Plain Woodland encompasses floristic variation significant enough to warrant recognition of separate communities (Tozer 2003). Any conservation or restoration activity must encompass this variation to be effective in the conservation of the EECs. Fourth, the issue of just how much restoration or how many remnants are needed to maintain viable communities (i.e. communities that will persist into the future with the maintenance of species interactions, life cycles and ecological processes that drive species persistence in the community) is also currently unknown. As well the success, in term of restoration of species, species interactions and ecological processes, of revegetation on cleared or degraded habitat will be dependent upon the degree of soil disturbance in the past (Doherty 2000).

We suggest that areas that are currently essentially cleared habitat and are then subject to revegetation are not considered to be part of the listed EECs. The concept of offsets, whereby areas of remnant habitat are cleared and, as a trade off, some 1-20 times the area of cleared habitat are to be 'revegetated' has been suggested in some instances. This concept does not assist the conservation of EECs as remnant habitat is lost. Any 'revegetated' areas should not be considered to add to the area of remnant EECs as:

a) they take no consideration of the considerable time lag needed to potentially regrow communities. This is likely to be in the order of decades if not centuries.

b) based on current knowledge there is little chance of reconstruction of complex communities. Recent work on revegetation in the Cumberland Plain (Wilkins *et al.* 2003) suggests very little success.

Consequently, while it will be worthwhile to plan for potential revegetation, more data on where high quality remnants exist, and research into revegetation techniques are currently needed prior to any implementation of extensive revegetation. Additionally, we need an understanding of how to restore faunal communities within these habitats. Any revegetation should not be traded off against the key importance of retaining all existing remnants.

Maintenance of viability of remnants in the landscape

The viability of existing EEC remnants is frequently compromised by degradation of habitat. This may involve: further loss of habitat; loss of adjacent vegetation or transitional vegetation that may help buffer the EEC from adverse disturbance; invasion by exotic weeds; disturbance to habitat by road and track construction, rubbish dumping; firewood collecting; nutrient enrichment from adjacent urban runoff etc. Hence, control of these adverse impacts is needed to maintain these communities. Restoration of degraded areas may also be required to create large enough remnants that are viable or to reduce adverse effects of weedy incursions into remnants.

Duffys Forest

A program to ameliorate the threats to remnant patches of the EEC currently covers some 30% of the known area of remnant habitat. This is based on work carried out under the implementation of the recovery plan for the endangered plant *Grevillea caleyi* (Scott *et al.* 1995; NPWS 2001). This has involved weed management, disturbance control and development of plans of management for specific sites. Further work for the remainder of the known sites is being planned. The sandstone soils on which this EEC and surrounding vegetation occur appear to be relatively resistant to weed invasion, with invasion governed by physical disturbance, and in particular, nutrient enrichment of sites (Leishman 1990). Thus, the vegetation of even small remnant sites may be viable in the long term if adverse disturbance can be minimised. There is also an interaction between weed invasion and fire, with remnant sites that are long unburnt being susceptible to recruitment of weeds such as small-leaved privet *Ligustrum sinense*, lantana *Lantana camara* and the native *Pittosporum undulatum* (which is not native to the ridgetop sites of this EEC).

At the same time, further losses of habitat at the edges of remnant sites continue to occur. Some guidelines for buffer protection are proposed in the draft recovery plan for this EEC to minimise such losses. There is also some enforced habitat restoration occurring, with topsoil from remnant areas that are to be cleared being used to restore severely degraded nearby areas of former Duffys Forest habitat. The success of such restoration may take decades to judge.

Cumberland Plain

The magnitude of the number of small remnants, and their wide distribution across western Sydney, make amelioration of threats at remnant sites a major challenge. A number of significant protection/management measures are underway. These include remnant site management and threat amelioration by some local councils and volunteer bush regeneration groups. Techniques for regeneration of degraded remnants have been developed and are being implemented (Hawkesbury Nepean Catchment Management Trust 2000).

The richer clay soils on the Cumberland Plain may make remnants less resilient to weed invasion than sandstone-derived plant communities such as Duffys Forest. The key weeds of the Cumberland Plain (olives *Olea eurpoea* subsp. *africana* and bridal creeper *Myrsiphyllum asparagoides*) are bird dispersed (Tozer 2003) and appear to have the ability to out-compete natives. Control of these weeds is a priority (Tozer 2003) and will be influenced by the extent of connectivity of the landscape in relation to the avian seed dispersers of these weeds. Hence, great care should be taken in any attempts to 'reconstruct' the landscape and create corridors to ensure such reconstructed areas do not simply serve as a pathway for dispersal of weed seeds. As most local native species have only local seed dispersal, the concepts of linkages and corridors being necessary for remnant viability may not apply to many vascular plants and more active management may be needed to restore species that have been lost. The role of such restored areas or corridors for the movement of vertebrates and invertebrates also requires further study.

Management of species' life histories and interactions via appropriate ecological process/disturbance regimes

Essential for the long-term viability of ecological communities is an appreciation that the species composition will to some extent vary both spatially and temporally. The component species have evolved in response to a number of disturbance regimes that impact on species persistence, species interactions and the movement of individuals through the life cycle of a species. Here we use fire as an example of a key disturbance regime that impacts on a number of EECs. We ask how effective management of the Duffys Forest and the EECs in Cumberland Plain has been in relation to fire.

Duffys Forest

Research on the ecology of the endangered plant *Grevillea caleyi* has informed management in relation to the potential impacts of fire in Duffys Forest. *G. caleyi* is a characteristic species of Duffys Forest and acts as an indicator of the impacts of fire frequency of the community. It is a slow growing species that has low levels of seed production and much of the seed produced is consumed by the native bush rats *Rattus fuscipes* and swamp wallabies *Wallabia bicolor* (Scott *et al.* 1995; Auld and Denham 2001; Regan *et al.* 2003). It is likely to be one of the slowest plant species in the EEC to replenish its seedbank after a fire. Consequently high frequency fire may impact on survival of this and other species in the community (see NSW Scientific Committee 2000; Regan *et al.* 2003; Regan and Auld in press). Other key indicator species, such as fire-sensitive *Persoonia*, appear to have a similar requirement to *G. caleyi*.

If fire is excluded for a long interval (>20-25 years) then a number of key structural components of the understorey of Duffys Forest will be absent or reduced (Auld and Scott unpubl). Long unburnt remnants are also prone to invasion by bird-dispersed weeds such as privet and lantana. Hence, infrequent fire is also likely to be a threat to the community. However, the impact of other components of the fire regime on species persistence within this community is less well understood. Guidelines for fire management for Duffys Forest (Table 2) have been developed (Scott *et al.* 1995; NPWS 2001; NPWS in prep.) as a guide to land managers and fire management authorities. These guidelines have been implemented through Rural Fire Service, NPWS and local councils. However, with the recent implementation of the Rural Fire Service Bushfire Environmental Assessment Code, there may be increased pressure on remnants of Duffys Forest that occur in asset protection or strategic management zones, e.g. including habitat clearing or the application of high fire frequency.

Cumberland Plain

Less is known about the impacts of fire on plants from shale soils on the Cumberland Plain. Hence the development of fire management guidelines is less advanced than for Duffys Forest. However, the planning for EECs on the Cumberland Plain does encourage incorporation of fire management guidelines. The impact of asset protection and strategic management zones on EECs on the Cumberland Plain remains to be assessed. Fire frequency requirements for different vegetation types as part of the Bushfire Environmental Assessment Code will be applied to EECs on the Cumberland Plain until further research informs management.

Table 2. Guidelines for fire management for Duffys Forest EEC. The overall objective is to ensure conservation of all sites by appropriate fire management practices. Guidelines based on Scott *et al.* (1995) and Auld unpubl.

Fire regime component	Suggested guideline
Fire Frequency	1) prevent fire from burning sites which are <8-12 years old;
	2) prevent any site from being burnt by three consecutive fires at <5 year intervals;
	3) allow fire to burn a site if it is >15-20 years old.
Fire Intensity	1) fire intensity effect little known. Some species require significant soil heating effect for germination. Fires of varying intensities recommended.
Fire Season	1) no evidence of seasonal effects: late summer/autumn burns more likely to be favourable for regeneration than winter/spring burns. More research is needed.
Fire size/patchiness	1) if fires are to be implemented specifically for management of a Duffys Forest site, the area to be burnt should be large to reduce the impact of seed predators post-fire, and to reduce the potential for increasing the fire frequency at the site;
	2) if planned fires are required for bush in the vicinity of, but not in a Duffys Forest site, give the site a large buffer zone of surrounding unburnt vegetation. This applies for wildfires as well, if such controls are possible.

Future Directions

Based on the three key ecological issues described above, we recognise several key future management and research issues. Some of these apply specifically to the Cumberland Plain, but most are applicable to the conservation of all EECs.

Research should be focussed on:

i) understanding of key processes that allow species persistence, e.g. disturbance regimes (e.g. fire on Cumberland plain), pollination, dispersal, predation/grazing, in relation to remnant size and degree of fragmentation.

ii) understanding how threats acting on the community affect population persistence of component species;

iii) estimating current rates of habitat loss within EECs; and

iv) developing techniques for increasing the effectiveness of restoration programs.

For management, key issues are:

i) development and release of recovery plans to guide management;

ii) on-ground threat mitigation activities;

iii) development of site based plans of management for remnants;

iv) involvement of community groups in conservation activities; and

v) survey of areas with little tree cover for remnants with good understorey (e.g. Cumberland Plain).

As well, there will be issues relating to resource planning and allocation. With the increasing number of listings of endangered ecological communities (Fig. 1), as well as species and key threatening processes, there will not be sufficient available resources for all aspects of management or research that are needed for conservation. Priorities and development of key issues across EECs will be required. At the same time, there will be other ecological communities that are in decline and may be considered common, but declining, near threatened or even vulnerable. Addressing such landscape-scale decline needs to be a priority. One way will be to address key threatening process at the landscape scale (rather than simply in relation to high priority threatened species). We suggest that this will also prevent communities and species from becoming threatened.

Summary

Listing of EECs under the NSW *TSC Act* has lead to significant improvements in awareness and acceptance of conservation needs for protection of these communities across the landscape. For that reason alone, listing has been beneficial for conservation of these communities. Combined with this increased awareness, there is a number of sound on-site management solutions to threats, often based at local government and community levels. This has enhanced the viability of remnants of EECs, although both the size and number of remnants will influence management success as will proximity to urban areas.

In contrast to these successes, there has been continued loss and fragmentation of habitat of listed EECs. This continues to undermine the on-site threat mitigation and habitat restoration activities. In a scenario of continued habitat loss, the long-term viability of remnants remains in doubt and will depend on the resilience of the species making up each community to threats that impact on them. The resilience is likely to vary among species across EECs and hence across the landscape. At present our understanding of the dynamics of ecological communities and the species that comprise them is limited and predictions of the degree of resilience of particular communities remain equally limited. Research and management initiatives should be directed at understanding the role of ecological process, disturbance regimes and species interactions in maintaining species within ecological communities.

Acknowledgements

We thank the *Grevillea caleyi* recovery team and the Duffys Forest Working group for conservation initiatives relating to Duffys Forest. M. Brock, P. Hutchings and D. Lunney made helpful comments on the manuscript.

References

Auld, T.D. and Denham, A.J. 2001. The impact of seed predation by mammals on post-fire seed accumulation in the endangered shrub *Grevillea caleyi* (Proteaceae). *Biological Conservation* 97; 377-385.

Benson, D. and Howell, J. 1994. The natural vegetation of the Sydney 1:100 000 map sheet. *Cunninghamia* 3(4); 677-787.

Doherty, M. 2000. The conservation value of regrowth native plant communities: a review. Report to NSW Scientific Committee. http://www.nationalparks.nsw.gov.au/PDFs/Regrowth_of_native_plants.pdf. Accessed 8th April 2004.

Hawkesbury Nepean Catchment Management Trust. 2000. Bringing the bush back to western Sydney. Best practice guidelines for bush regeneration of the Cumberland Plain.

Leishman, M. 1990. Suburban development and resultant changes in the phosphorus status of soils in the area of Ku-ring-gai, Sydney. *Proceedings of the Linnean Society of NSW* 112; 15-25.

NPWS. 2001. *Grevillea caleyi* R.Br. (Proteaceae) Draft Recovery Plan for public comment. NSW National Parks and Wildlife Service, Hurstville.

NPWS. 2002. Biodiversity Strategy Case Study. Cumberland Plain Subregion, Sydney Basin Bioregion NSW SB8: Cumberland. A report to the National Land and Water Resources Audit.

NSW Scientific Committee. 1999. Elderslie Banksia Scrub Forest ecological community. NSW Scientific Committee Endangered Ecological Community Final Determination. http://www.nationalparks.nsw.gov.au/npws.nsf/Content/Elderslie+b

anksia+scrub+forest+community+endangered+ecological+community+listing. Accessed 8th April 2004.

NSW Scientific Committee. 2000. High frequency fire resulting in the disruption of life cycle processes in plants and animals and loss of vegetation structure and composition.

NSW Scientific Committee. Key Threatening Process Final Determination. http://www.nationalparks.nsw.gov.au/npws.nsf/Content/Ecological+consequences+of+high+frequency+fires+key+threatening+process+declaration. Accessed 8th April 2004.

NSW Scientific Committee. 2001. Clearing of native vegetation. NSW Scientific Committee Key Threatening Process Final Determination. http://www.nationalparks.nsw.gov.au/npws.nsf/Content/Clearing+of+native+vegetation+key+threatening+process+declaration. Accessed 8th April 2004.

NSW Scientific Committee. 2002. Duffys Forest ecological community in the Sydney Basin Bioregion. NSW Scientific Committee Endangered Ecological Community Final Determination. http://www.nationalparks.nsw.gov.au/npws.nsf/Content/Duffys+Forest+vegetation+community+in+the+Sydney+Basin+Bioregion+endangered+ecological+community+listing. Accessed 8th April 2004.

Regan, H.M., Auld, T.D., Keith, D.A. and Burgman, M.A. 2003. The effects of fire and predators on the long-term persistence of an endangered shrub, *Grevillea caleyi*. *Biological Conservation* 109; 73-83.

Regan, H.M. and Auld, T.D. 2004. Using population viability analysis for management of an endangered Australian shrub, *Grevillea caleyi*. In *Species Conservation and Management: Case Studies*, edited by H.R. Akçakaya, M.A. Burgman, O. Kindvall, C. Wood, P. Sjogren-Gulve, J. Hatfield, and M. McCarthy. Oxford University Press.

Scott, J., Marshall, A. and Auld, T.D. 1995. Conservation research statement and recovery plan for *Grevillea caleyi*. ANCA Endangered Species Project No. 456.

Sheringham, P.R. and Sanders, J.M. 1993. Vegetation survey of Garigal National Park and surrounding Crown Lands. A report for the NSW National Parks and Wildlife Service.

Smith, P. and Smith, J. 2000. Survey of the Duffys Forest Vegetation Community. Unpublished Report to NSW National Parks and Wildlife Service and Warringah Council.

Thomas, J. and Benson, D.H. 1985. Vegetation survey of Ku-ring-gai Chase National Park. National Herbarium of New South Wales, Royal Botanic Gardens, Sydney.

Tozer, M. 2003. The native vegetation of the Cumberland Plain, western Sydney: systematic classification and field identification of communities. *Cunninghamia* **8**; 1-75.

Wilkins, S. Keith, D.A. and Adam, P. 2003. Measuring success: evaluating the restoration of a terrestrial plant community on the Cumberland Plain, Sydney, Australia. *Restoration Ecology* 11; 489-503.

Threat abatement plans: strategic pest management for biodiversity conservation

Andrew Leys

Department of Environment and Conservation (NSW), PO Box 1967, Hurstville, NSW 2220.

Email: andrew.leys@npws.nsw.gov.au

ABSTRACT

In Australia, invasive terrestrial pest species have been implicated in the decline of a number of threatened species. Under the NSW *Threatened Species Conservation Act 1995*, pest threats can be listed as Key Threatening Processes which require the preparation of Threat Abatement Plans to ameliorate the threats. This paper presents an overview of how the threat abatement planning process is being used to plan initiatives to reduce the biodiversity impacts of invasive species. The approach taken in four threat abatement plans (foxes *Vulpes vulpes*, feral cats *Felis catus*, bitou bush *Chrysanthemoides monilifera* and gambusia *Gambusia holbrooki*) is used to illustrate this process.

Key words: key threatening process, threat abatement plan, invasive pest species, foxes, feral cats, bitou bush, gambusia.

Introduction

Throughout the world, invasive terrestrial pest species are acknowledged as a major cause of biodiversity loss (IUCN 2000). In Australia, it has been suggested that introduced terrestrial pest species are second only to land clearing as the greatest threat to native biodiversity (Roush 2002; Martin 2003). The most recent estimate of the annual cost of weeds to Australian agriculture is $4 billion (Martin 2003). This report suggests that weeds appear to be implicated in the extinction of at least four native plant species and are known to be adding pressure to a further 57 now under threat. As outlined later in this chapter, this is a gross underestimation of the species threatened by weeds.

Recently, the Cooperative Research Centre for Pest Animal Control has estimated that the national cost of invasive animal pests is conservatively, $720 million per year (McLeod 2004). However, with the exception of foxes, feral cats and carp, this estimate does not include the cost of any impacts on biodiversity.

Threat abatement planning process

The development of threat abatement plans is a statutory requirement under the *Threatened Species Conservation Act 1995* (TSC Act). Although there has been a history in Australia of plans to control pest species, the focus has been primarily to protect agricultural enterprises e.g. wild dog control to protect the sheep industry (Fleming and Harden 2003). The threat abatement planning process is being used to plan and implement statewide, strategic pest control initiatives to protect biodiversity, independent of land tenure. This process has been recognised as the accepted model by the NSW Legislative Council Inquiry into Feral Animals (New South Wales Legislative Council General Purpose Standing Committee No. 5. 2002) and by an independent review of the pest management programs being undertaken by the then NSW National Parks and Wildlife Service (English and Chapple 2002). The NSW National Parks and Wildlife Service, now part of the Department of Environment and Conservation, is required to prepare a threat abatement plan to abate or ameliorate the adverse effects on threatened biodiversity of any key threatening process listed under the TSC Act.

The threat abatement planning process provides a framework to minimise the impacts of key threatening processes on threatened biodiversity or on entities that may become threatened. This also allows for priorities to be established for the allocation of limited resources. The planning process is based on the assumption that the eradication of pests over large areas, is rarely if ever possible. The main objectives of threat abatement plans are to:

- target pest control across all land tenures at the species/sites where the impacts of pests are likely to be greatest;
- develop best practice guidelines that maximise the effectiveness of control programs while minimising non-target impacts;
- establish monitoring programs to demonstrate these impacts and to measure the effectiveness of the resultant control programs;
- identify knowledge gaps and develop research proposals where information is lacking; and
- increase community education and involvement.

Pp 102 - 106 in *Threatened species legislation: is it just an Act?* 2004, edited by Pat Hutchings, Daniel Lunney and Chris Dickman. Royal Zoological Society of New South Wales, Mosman, NSW, Australia.

A number of pest threats has been listed as key threatening processes under the TSC Act:

- Predation by the European red fox *Vulpes vulpes*.
- Competition by the feral European rabbit *Oryctolagus cuniculus*.
- Predation by the feral cat *Felis catus*.
- Predation, habitat degradation, competition and disease transmission by feral pigs *Sus scrofa*.
- Predation by the plague minnow *Gambusia holbrooki*.
- Predation by the ship rat *Rattus rattus* on Lord Howe Island.
- Competition from feral honeybees *Apis mellifera*.
- Importation of red imported fire ants *Solenopsis invicta*.
- Introduction of the large earth bumblebee *Bombus terrestris*.
- Invasion of native plant communities by bitou bush and boneseed *Chrysanthemoides monilifera*.
- Invasion of native plant communities by exotic perennial grasses.

Also, the Scientific Committee (the independent body that considers listings under the TSC Act) has made Preliminary Determinations to list the following pest threats as key threatening processes:

- Herbivory and environmental degradation by feral deer (various species)
- Competition and habitat degradation by feral goats *Capra hircus*.

Approach taken in current NSW threat abatement plans

The approach being taken in four approved, or near completed, threat abatement plans involving invasive pest species (foxes, feral cats, bitou bush and gambusia).

1. Foxes

A *Threat Abatement Plan for Predation by the Red Fox* (Fox TAP) was completed in 2001 and is now being implemented across New South Wales by the Department of Environment and Conservation, the Department of Primary Industries (State Forests) and the Department of Lands (NSW National Parks and Wildlife Service 2001). It was the first threat abatement plan to be completed under the TSC Act and it takes a new approach to fox management by identifying priorities for fox management across all land tenures. The threatened species considered most at risk from fox predation are identified along with 81 priority sites for fox control. Collaborative fox control programs have been established at most of these sites with the respective land managers.

At many of these sites, monitoring programs have been incorporated to assess the effectiveness of the fox control programs and the recovery of the target threatened species. In most cases the control programs have not been running for sufficient time for a measurable increase in the recovery of the threatened species. However, some preliminary results are encouraging. For example, a program to protect the yellow-footed rock wallaby *Petrogale xanthopus* in the Mutawintji Nature Reserve and the Mutawintji National Park, has resulted in a significant increase in the wallaby population since fox control began in 1995 (DEC 2004a). Positive responses to fox control have also been reported for shore birds such as little terns and pied oystercatchers (DEC 2004a).

In addition, the Fox TAP identifies *Best Practice Guidelines for Fox Control* which seek to maximise the effectiveness of control programs while minimising the potential non-target impacts.

2. Feral cats

In some areas of New South Wales, feral cats have significant impacts on ground birds and small native mammals and have caused the failure of attempts to re-introduce some threatened species (Dickman 1996). Hence, *Predation by the Feral Cat* is listed as a key threatening process under the TSC Act. A draft threat abatement plan has been prepared and should be available for public comment in late 2004 or early 2005.

Feral cats are difficult to study because they are wary of traps, shy and elusive. Little is known about their impacts on threatened species and the best ways to reduce these impacts. The draft *Threat Abatement Plan for the Feral Cat* (Feral Cat TAP) will address this lack of information by using a modelling approach, similar to that used in the Fox TAP to identify the threatened species most at risk from feral cat predation. Currently, available control options (mainly trapping) are ineffective for broad-scale cat control and there are no poison baits registered for control of feral cats in NSW. Hence, the Feral Cat TAP does not propose implementing these measures on a significant scale although they may be used in localised areas to protect threatened species. Best-practice guidelines for trapping will be developed to improve and evaluate its effectiveness as a control technique.

Research is being conducted in other states to develop a toxin and baiting system which is humane and cat-specific (Fisher *et al.* 2001). The Feral Cat TAP will focus on trials to evaluate this control method when it becomes available. In the meantime, the Feral Cat TAP will identify priority species most at risk of cat predation and the locations for priority control. Also, the Feral Cat TAP will list actions to educate and inform the community about the impacts of feral cats on wildlife and to gain community support for implementing feral cat management programs once effective control options become available.

3. Bitou bush

Bitou bush is a highly invasive coastal shrub which now occurs on more than 80% of the New South Wales coastline (Thomas and Leys 2000). In September 2004, a draft *Threat Abatement Plan for Invasion of Native Plant Communities by Bitou Bush* (Bitou TAP) was released for public comment (DEC 2004b). The approach taken with this plan is similar to that which was adopted in the Fox TAP in that it focuses on:

- developing a strategic framework for delivering control of bitou bush to areas of high conservation value (in terms of threatened biodiversity);
- monitoring the effectiveness of control programs in terms of recovery of threatened biodiversity;
- developing and promoting best practice management;
- fostering community education, involvement and awareness; and
- identifying and filling knowledge gaps.

The Bitou TAP identifies 63 plant species, two endangered plant populations and nine endangered ecological communities (EECs) as being at greatest risk from bitou bush (DEC 2004b). Of these, 11 plant species, 2 endangered plant populations and 4 EECs are at serious risk of extinction if action is not undertaken to combat the problem. The plan also identifies the sites at which these threatened species occur and ranks them for control. From a list of over 600 sites along the coast, 60 priority sites have been identified at which control action is essential for the conservation of the species at greatest risk (DEC 2004b).

4. Gambusia

Gambusia is a contributing factor to the decline of frogs (threatened or otherwise) as well as other aquatic life such as native fish and macro-invertebrates (NSW National Parks and Wildlife Service 2003). An approved *Threat Abatement Plan for Predation by Gambusia – The Plague Minnow* (Gambusia TAP) was published in 2003 (NSW National Parks and Wildlife Service 2003). Management options for this pest are limited to localised control through pond draining and re-stocking. In the long-term, genetic research such as that being undertaken for carp, may be applicable to gambusia. The concept of this technology is to target genes that control sexual development of fish to exclusively produce males. The aim is to release these modified fish to mate with carp in the wild. Carp populations would become biased towards males and gradually decline over many generations. The work is currently in the very early stages.

The Gambusia TAP focuses on the following actions:

- research to determine the impacts of gambusia on threatened frogs species;
- research to develop more effective control options;
- a survey of the distribution of gambusia in areas occupied by key frog species in order to determine priority sites which can be maintained free of gambusia; and
- public education and awareness programs aimed at minimising human dispersal of gambusia to water bodies that are not already infested.

Discussion

The threat abatement planning process is a science-based mechanism for delivering strategic initiatives which tailor pest management programs to specific biodiversity outcomes e.g. recovery of specific threatened species, endangered populations of ecological communities. A key objective is the setting of performance measures to assess the effectiveness of each threat abatement plan. This is considered to be an essential part of effective pest management (Braysher 1993; Braysher and Saunders 2003).

In this chapter, I have outlined the approach being taken with four species-specific threat abatement plans with very different objectives. Threat abatement plans for foxes and bitou bush have identified priority threatened species, populations and ecological communities and the locations where protection of these species is most critical. In this way, resources can be allocated to protect threatened biodiversity at the localities where the benefits of control will be greatest.

For feral cats and gambusia, where effective options are not available for broad-scale control, or where information on the impacts on threatened biodiversity is lacking, research and other actions are identified in the threat abatement plans. For the Feral Cat TAP, a reference group was formed to allow key stakeholders to have input into the plan. Because there are sections of the community opposed to any cat control, this is an important step in gaining community support.

Where critical populations of a threatened species have been determined to be at risk on private lands, the Department of Environment and Conservation discusses with landowners what options they may wish to consider to protect the species. The threat abatement planning process does not use coercive measures on privately owned lands to protect threatened species at priority sites identified in the TAPs. Instead, a cooperative approach is fostered where the landowner/manager is encouraged to work with the Department of Environment and Conservation and other government agencies to control the pest in a way that will protect the threatened species. For example, support has been provided to private landowners for fox control to protect Malleefowl *Leipoa ocellata*, Plains-wanderer *Pedionomus torquatus* and Bush Stone-curlew *Burhinus grallarius* in western New South Wales. Also, in developing the draft Bitou TAP, the Department of Environment and Conservation has been encouraged by the positive response shown by private landowners when they were approached to discuss the protection of priority sites on their lands e.g. Telstra, St. Michael's Golf Club.

The draft Bitou TAP is the first threat abatement plan for any weed species in Australia. It has taken considerable time to prepare, partly because of the need to develop an effective model to rank plant species threatened by bitou bush that can be followed in the future. Unfortunately, information on the impact of invasive weed species on biodiversity is lacking and the use of a modelling approach is essential. Grice *et al.* (2004) have recently reviewed all available information published on this subject and they found that the ecological impacts of very few invasive plant species have been quantified (16 species). This is in marked contrast to the large number of exotic species that has naturalised in Australia (2,732 species according to Groves *et al.* 2003).

Because there are more than 200 environmental weeds that impact on natural ecosystems, there is potentially a large number that could be listed as key threatening processes, each of which would require the preparation of a threat abatement plan. To examine solutions to this problem, the Department of Environment and Conservation held a workshop on listing weeds as Key Threatening Processes (KTPs) under the TSC Act. Participants at this workshop represented all relevant NSW government agencies, the NSW Scientific Committee, the Commonwealth Department of Environment and Heritage, the Cooperative Research Centre for Australian Weed Management and local government. The main outcome was that the listing of weeds as KTPs should be undertaken through the listing of individual species, and wherever possible meaningful groups of weed species (e.g. exotic perennial grasses) (Downey and Leys 2004). The concept of a generic listing of environmental weeds as occurs under the Victoria *Flora and Fauna Guarantee Act 1988* is not possible under the TSC Act (Downey and Leys 2004). In addition, the workshop participants considered such a generic listing as an unacceptable outcome, for it removed the emphasis from the worst species or groups to all weeds. The other main outcome of the workshop was to establish a small working group to prioritise the weed species that impact upon biodiversity, especially to threatened species or ecological communities in NSW. The recommendations of this working group will be used to develop nominations for listing weeds as KTPs under the TSC Act, either by DEC or by other stakeholders (Downey and Leys 2004).

Threat abatement plans utilise an adaptive management approach. Pest control at priority sites identified in threat abatement plans has to be effective in reducing the pest threat to levels where the recovery of threatened biodiversity can occur. Hence, the monitoring component of threat abatement plans is essential to measure the reduction in pest density from pest control and, more importantly, the resultant recovery of threatened biodiversity.

Experience gained during the first four years of implementing the Fox TAP has shown that field staff depend on support from the Fox TAP Coordinator (Dr Paul Mahon) to implement the plan. This has been particularly important for the monitoring process where many field staff have not had the necessary skills to undertake wildlife monitoring programs or the expertise to design and conduct appropriate experiments. In addition, a uniform approach to monitoring is required to allow collation and analysis of the data. Hence, sufficient support for the coordination of threat abatement plans is essential for the threat abatement planning process to be able to deliver effective outcomes.

In 2003, the NSW Government announced major changes to the way natural resource management will be delivered. A key component of the changes is the establishment of Catchment Management Authorities (CMAs) which have replaced the 19 Catchment Management Boards and 20 Regional Vegetation Committees that existed previously. The CMAs are responsible for managing natural resource matters including responsibility for developing catchment action plans which will integrate existing regional vegetation management plans, catchment blueprints and investment strategies. It is anticipated that CMAs will recognise the impacts of invasive species on natural resources and that relevant actions identified in national and NSW threat abatement plans will be included in catchment action plans.

Invasive pest species are having significant adverse impacts on national and state biodiversity values. The threat abatement planning process is a framework that enables priorities to be established for the allocation of limited resources to reduce the adverse impacts of listed key threatening processes. The Fox TAP is a good example of how this is being implemented in the field. The listing of invasive pest species as key threatening processes and the preparation and implementation of threat abatement plans is a task that will require considerable input in the future. To be successful, this process will need the support and involvement of government agencies, CMAs and the community.

Acknowledgments

The contributions of Dr Paul Mahon (Fox TAP), Dr Paul Downey and Jeff Thomas (Draft Bitou TAP), Ron Haering (Gambusia TAP) and Rebecca Larkin and Catherine Price (Draft Feral Cat TAP) in preparing individual threat abatement plans are gratefully acknowledged.

References

Braysher M. 1993. *Managing Vertebrate Pests: Principles and Strategies*. Australian Government Publishing Service, Canberra.

Braysher M. and Saunders 2003. *PESTPLAN. A Guide to Setting Priorities and Developing a Management Plan for Pest Animals*. Bureau of Rural Sciences, Canberra.

DEC 2004a. *Controlling Animal Pests in NSW National Parks*. NSW Department of Environment and Conservation, Hurstville.

DEC 2004a. *Draft Threat Abatement Plan for Invasion of Native Plant Communities by Bitou Bush/Boneseed* (Chrysanthemoides monilifera). NSW Department of Environment and Conservation, Hurstville.

Dickman, C.R. 1996. *Overview of the Impacts of Feral Cats on Australian Native Fauna.* Australian Nature Conservation Agency, Canberra.

Downey, P.O. and Leys A.R. 2004. Weeds as key threatening processes: Implications for managing environmental weeds. Papers and Proceedings of the 14th Australian Weeds Conference, Wagga Wagga. Eds B.M. Sindel and S.Johnson, pp. 454-57. (R.G. and F.J. Richardson, Melbourne).

English, A.W. and Chapple, R.S. 2002. *A Report on the Management of Feral Animals by the New South Wales National Parks and Wildlife Service*. Report to the NSW Minister for the Environment, July 2002.

Fisher, P., Marks, C. and Johnston, M. 2001. New techniques for feral cat management: Biological, chemical or fertility control? Proceedings of the 12th Australasian Vertebrate Pest Conference, Melbourne. pp. 150-55. (Department of Natural Resources and Environment, Melbourne).

Fleming, P. and Harden, R. 2003. *Guidelines for Preparing a Working Plan to Manage Wild Dogs*. NSW Agriculture, Orange.

Grice, A.C., Field, A.R. and McFadyen, R.E.C. 2004. Quantifying the effects of weeds on biodiversity: Beyond Blind Freddie's test. Papers and Proceedings of the 14th Australian Weeds Conference, Wagga Wagga. Eds B.M. Sindel and S.Johnson, pp. 464-68. (R.G. and F.J. Richardson, Melbourne).

Groves, R.H., Hosking, J.R., Batianoff, G.N., Cooke, D.A., Cowie, I.D., Johnson, R.W., Keighery, G.J., Lepschi, B.J., Mitchell, A.A., Moerkerk, M, Randall, R.P., Rozefelds, A.C., Walsh, N.G. and Waterhouse B.M. 2003. *Weed Categories for Natural and Agricultural Ecosystem Management*. Bureau of Rural Sciences, Canberra.

IUCN (The World Conservation Union) 2000. *IUCN Guidelines for the Prevention of Biodiversity Loss Caused by Alien Invasive Species.* Website: http://www.iucn.org/themes/ssc/pubs/policy/invasivesEng.htn (Species Survival Commission, Invasive Species Specialist Group, IUCN, Switzerland).

Martin, P. 2003. *Killing us softly – Australia's Green Stalkers: A Call to Action on Invasive Plants, and a Way Forward.* CRC for Australian Weed Management, Adelaide.

McLeod, R. 2004. *Counting the Cost: Impact of Invasive Animals in Australia.* Cooperative Research Centre for Pest Animal Control, Canberra.

New South Wales Legislative Council General Purpose Standing Committee No. 5. 2002. *Feral Animals.* Legislative Council Parliamentary Paper No. 158. New South Wales Parliament, Sydney.

NSW National Parks and Wildlife Service. 2001. *Threat Abatement Plan for Predation by the Red Fox (Vulpes vulpes).* NSW National Parks and Wildlife Service. Hurstville.

NSW National Parks and Wildlife Service. 2003. *Threat Abatement Plan for Predation by Gambusia holbrooki – The Plague Minnow.* NSW National Parks and Wildlife Service. Hurstville.

Roush, R. 2002. CEO's Report. In: *Annual Report 2001-2002, Cooperative Research Centre for Australian Weed Management.* Cooperative Research Centre for Australian Weed Management, Adelaide. pp. 3-4.

Thomas, J. and Leys, A. 2002. Strategic management of bitou bush (*Chrysanthemoides monilifera* ssp. *rotundata*). In: Spafford Jacob, H. Dodd, J. and Moore, J.H. (eds) 13th Australian Weeds Conference: Papers and Proceedings, 8-13 September 2002, Perth. Shannon Books, Melbourne.

Case studies in implementing the NSW *Threatened Species Conservation Act* 1995 for invertebrate conservation

Michael J. Murphy[1] and Simon Nally[2]

[1]Threatened Species Unit, Conservation Programs and Planning-Northern
Locked Bag 914 Coffs Harbour NSW 2450

[2]Threatened Species Unit, Conservation Programs and Planning-Central
PO Box 1967 Hurstville NSW 2220

Environment Protection and Regulation Division, NSW Department of Environment and Conservation
Email: michael.murphy@environment.nsw.gov.au and simon.nally@environment.nsw.gov.au

ABSTRACT

The difficulties in applying threatened species legislation, which has a general vertebrate and vascular plant focus, to invertebrates, have been widely recognised. Nevertheless, such legislation provides valuable opportunities to promote invertebrate conservation that were previously available only for high profile, charismatic vertebrate species. This paper presents three case studies, two land snails Lord Howe Placostylus *Placostylus bivaricosus* and Mitchell's Rainforest Snail *Thersites mitchellae*, and the lepidopteran Purple Copper Butterfly *Paralucia spinifera*, to demonstrate some of the achievements of the NSW *Threatened Species Conservation Act* 1995 in this regard, including significant increases in the known range of the species, reduction of threatening processes at key sites, improved understanding of the species' ecology, consideration of the species in land use planning decisions, involvement of local communities in recovery programs and an increase in community awareness of broader invertebrate conservation issues. The Act can be effective as one of a range of tools to positively promote and build support for invertebrate conservation in the public arena.

Key words: invertebrate; conservation; recovery; legislation; NSW *Threatened Species Conservation Act* 1995; biodiversity; land snail; *Placostylus bivaricosus; Thersites mitchellae;* lepidoptera; *Paralucia spinifera;* habitat; community.

Introduction

The New South Wales *Threatened Species Conservation Act* 1995 (TSC Act), including amendments introduced by the *Threatened Species Conservation Amendment Act* 2002, is the key legislation with respect to threatened terrestrial species in NSW. The objects of the TSC Act are to conserve biodiversity, prevent extinction and promote recovery, protect critical habitat, eliminate or manage threatening processes, ensure impacts of proposed actions are properly assessed, and encourage conservation through cooperative management. The Act allows for the recognition of threatened biodiversity at three different levels: that of the population, species and ecological community. Species and ecological communities can be listed as *endangered* (higher risk category) or *vulnerable* (lower risk category), while populations are eligible for listing only in the *endangered* category. Species may also be listed as *presumed extinct* in NSW. The TSC Act allows for the identification of *critical habitat* for endangered species and ecological communities. The Act also enables a strategic, direct approach to common threats to biodiversity through the listing of *key threatening processes*. The TSC Act aims to conserve threatened species, populations and ecological communities and their habitats through the preparation and implementation of recovery plans and threat abatement plans, the identification and protection of critical habitat, consideration of threatened taxa in environmental planning decision making, and through other provisions including licensing controls, stop work orders and joint management agreements.

In common with similar legislation around the world, the listing of fauna under the TSC Act is heavily weighted towards vertebrates, with approximately 95% of the fauna species listed coming from 1% of the actual fauna. This is partly a reflection of broader community values (the appeal of the 'cute and cuddly'), partly due to data availability and partly legal and historical precedence. Threatened species legislation has generally been established originally with vertebrates in mind, and the legislative criteria and thresholds developed for listing vertebrates do not apply equally as well to invertebrates (Hutchings and Ponder 1999; Kitching 1999; Mawson and Majer 1999).

The difficulties faced in applying legislation, such as the TSC Act, to invertebrate conservation include:

1. the sheer number of invertebrate species. (The number of terrestrial invertebrates in Australia could be 10 million or more species (Horwitz, Recher and Majer 1999), compared to approximately 2000 terrestrial vertebrate species);
2. the poor taxonomic understanding of many invertebrate groups, with a significant proportion of taxa undescribed or predicted to be awaiting discovery (New 1995; Stork 1999);
3. the even poorer understanding of the distribution, ecology and conservation status of species in many invertebrate groups (Yen *et al.* 1990; New 1995).

Pp 107 - 124 in *Threatened species legislation: is it just an Act?* 2004, edited by Pat Hutchings, Daniel Lunney and Chris Dickman. Royal Zoological Society of New South Wales, Mosman, NSW, Australia.

(While the biology and ecology of some invertebrate groups such as butterflies are relatively well-known, others such as Australian land snails are very poorly known, with basic biological and ecological information lacking for all but a few taxa, and most species never studied as living animals (Bishop 1981; Smith 1992));

4. the limited funding support and specialist expertise to address this data deficit (Yen *et al.* 1990; Hutchings and Ponder 1999; New 1999);
5. the lack of knowledge amongst some sections of the invertebrate science community (professional and amateur) concerning the functioning of threatened species legislation and associated reservation concerning the potential value thereof (Gunning 1999; Greenslade 1999); and
6. the generally poor awareness and valuation by the broader community of invertebrate conservation.

Notwithstanding, the TSC Act enables the listing and protection of threatened species, populations and communities of terrestrial invertebrates, and is thus an advance on its predecessor, the NSW *Endangered Fauna (Interim Protection) Act* 1991, which was restricted to the consideration of four classes of vertebrates (birds, mammals, amphibians and reptiles). Sixteen invertebrate species or populations had been listed under the TSC Act by April 2004: 1 earthworm, 3 land snails and 12 insects (Table 1).

This paper discusses, by way of presenting three separate case studies (two land snails and one butterfly), some of the challenges and successes experienced by the authors in implementing the TSC Act for the conservation of invertebrates in NSW. The aim is to demonstrate that the listing of invertebrate species under the TSC Act can result in significant conservation benefits to individual species and, through their promotion as flagship species, make a valuable contribution to broader conservation goals. The approved recovery plans for all three species can be seen on the NSW Department of Environment and Conservation (DEC) website www.nationalparks.nsw.gov.au.

A brief background to land snail conservation

Land snails are currently undergoing a global mass extinction of which few people outside of malacological circles are aware. Molluscs as a group make up approximately 6% of known global fauna species diversity, yet they comprise 18% of threatened fauna species and 43% of recorded modern fauna extinctions (IUCN Species Survival Commission 2003). This is despite the fact that only about 3% of the world's mollusc species have been assessed for conservation status, compared to better known groups such as birds and mammals where virtually 100% have been assessed (IUCN Species Survival Commission 2003). Australia has the second highest number of IUCN-listed threatened mollusc species (after the USA), with up to 25% of mollusc species potentially at risk (Ponder 2000; IUCN Species Survival Commission 2003). The land snails and freshwater snails are the two mollusc groups under greatest threat (Ponder 2000; IUCN Species Survival Commission 2003). Major threats to land snails include loss of habitat, frequent fire, and predation and habitat degradation by feral species (Ponder 1997). Three land snail species are currently listed under the TSC Act (Table 1). Positive outcomes achieved through the recovery programs developed following the listing of two of these species are briefly described below.

Table 1. Invertebrates listed under the TSC Act (as at April 2004)

	Species	Date of listing under TSC Act	Status under TSC Act
butterfly	*Paralucia spinifera* Edwards & Common, 1978 (Purple Copper Butterfly)	Nov 1996	Endangered
moth	*Synemon plana* Walker, 1854 (Golden Sun Moth)	Nov 1996	Endangered
phasmid	*Dryococelus australis* Montrouzier, 1855 (Lord Howe Island Phasmid)	Dec 1996	Endangered
land snail	*Thersites mitchellae* (Cox, 1864) (Mitchell's Rainforest Snail)	March 1997	Endangered
land snail	*Placostylus bivaricosus* (Gaskoin, 1855) (Lord Howe Placostylus)	April 1997	Endangered
land snail	*Meridolum corneovirens* (Pfeiffer, 1851) (Cumberland Land Snail)	Aug 1997	Endangered
dragonfly	*Petalura gigantea* (Leach, 1815) (Giant Dragonfly)	Jan 1998	Endangered
beetle	*Menippus fugitivus* (Lea)	March 2000	Endangered Population (Sutherland Shire)
beetle	*Nurus atlas* Castelnau, 1867	March 2001	Endangered
beetle	*Nurus brevis* Motschulsky, 1865	March 2001	Endangered
earthworm	*Pericryptodrilus nanus* Jamieson 1977	March 2001	Endangered
beetle	*Hybomorphus melanosomus* (Saunders & Jekel, 1885) (Lord Howe Island Ground Weevil)	Dec 2002	Presumed Extinct
butterfly	*Argyreus hyperbius* (Linnaeus, 1763) (Australian Fritillary)	Dec 2002	Endangered
butterfly	*Ocybadistes knightorum* Lambkin & Donaldson, 1994 (Black Grass-dart Butterfly)	Dec 2002	Endangered
moth	*Phyllodes imperialis* Druce (ANIC 3333) (southern subspecies) (Pink Underwing Moth)	Dec 2003	Endangered
cockroach	*Panesthia lata* Walker, 1868 (Lord Howe Island wood-feeding Cockroach)	Feb 2004	Endangered

Case Study 1: Lord Howe Placostylus *Placostylus bivaricosus*

The *Placostylus* (Family Bulimulidae) is a genus of large, ground-dwelling land snails found in the south-west Pacific from the Solomon Islands, Vanuatu, Fiji and New Caledonia to Lord Howe Island and the northern extremity of New Zealand (Abbott 1989; Parrish *et al.* 1995; Ponder *et al.* 2003). It has been suggested that this distribution reflects an ancestral distribution on a once extensive but now largely submerged Gondwanan continental fragment on the eastern side of the Tasman rift (Hedley 1893). Many *Placostylus* species are considered to be at risk of extinction (IUCN Species Survival Commission 2003). Threats include clearing of habitat, predation by introduced vertebrate species such as rodents, pigs and Eurasian thrushes (Parrish *et al.* 1995; Sherley *et al.* 1998) and, in some areas, over-harvesting of wild snails for human consumption (Salas *et al.* 1997). New Caledonian *Placostylus* species were recently described in the popular print media as rare but much-loved when simmered in white wine and garlic (Guy 2004).

Placostylus bivaricosus (Gaskoin, 1855) (Fig. 1) is endemic to Lord Howe Island, and is most closely related to the New Zealand species *Placostylus ambagiosus* and *Placostylus hongii* (Ponder *et al.* 2003). It has a conical shell up to 80 mm long, medium to dark brown in colour (weathering to white in older animals) and with a thickened lip in mature adults. The body colour is black. *Placostylus bivaricosus* was previously abundant on sandy and calcarenite-derived soils of the central lowlands of Lord Howe Island, and also occurred more sparingly on the volcanic soils of the northern hills and southern mountains (Brazier 1889). Three Recent subspecies are currently recognised on the basis of shell morphology: *P. bivaricosus bivaricosus* (Gaskoin, 1855) from the central lowlands and northern hills, *P. bivaricosus cuniculinsulae* (Cox, 1872) from a small island (Blackburn Island) in the Lord Howe lagoon, and *P. bivaricosus etheridgei* (Hedley, 1891) from the southern mountains. Ponder *et al.* (2003) proposed that the population from the summit of Mount Gower (in the southern mountains) may also have been distinct. One fossil taxon, *P. bivaricosus solidus* (Brazier, 1889), is known from Pleistocene calcarenite deposits of the central lowlands of Lord Howe Island.

Historical accounts and fossil evidence indicate that *P. bivaricosus* was formerly widespread and common on Lord Howe Island (Brazier 1889; Hedley 1891; Smithers *et al.* 1977; Sutherland and Ritchie 1977). The decline of the species was first noted in the 1940s (Iredale 1944), and it is now recognised internationally as critically endangered through listing on the *2003 IUCN Red List of Threatened Species* (IUCN Species Survival Commission 2003), in the category CR B1+2abcde. *Placostylus bivaricosus cuniculinsulae* is presumed to be extinct. The original forest cover on Blackburn Island has been largely cleared and no specimens have been collected there for over a century (Ponder and Chapman 1999). *Placostylus bivaricosus etheridgei* has declined in range and abundance and was considered likely to be extinct (NSW Scientific Committee 1997a), but it may survive as a number of small, isolated local populations (Ponder and Chapman 1999). *Placostylus bivaricosus bivaricosus* remains extant but is endangered and has declined in range and abundance (NPWS 2001a).

Figure 1: Lord Howe Placostylus *Placostylus bivaricosus*. This large, ground-dwelling land snail is endemic to Lord Howe Island, where it is threatened by loss of habitat and predation by introduced species, particularly the Black Rat *Rattus rattus*. Preparation of the recovery plan for the species involved assessment of current status through field survey, identification of past and current threats, review of scientific literature detailing research into related *Placostylus* species and consultation with the Lord Howe Island community. Photograph by Michael Murphy.

Little information is available on the ecology of *P. bivaricosus*, and behavioural and life history studies are a priority for future research to assist conservation efforts (NPWS 2001a). Live animals are found under well-developed, moisture-retaining leaf litter in evergreen closed forest, often, but not exclusively, in the vicinity of Banyan Figs *Ficus macrophylla columnaris* (NPWS 2001a; Murphy pers. obs.). Recent records of *P. bivaricosus* are from *Drypetes australasica-Cryptocarya triplinervis* evergreen closed forest or *Howea fosteriana* evergreen closed forest, or in ecotones between these two associations, using the vegetation associations described and mapped by Pickard (1983). Ponder and Chapman (1999) noted that *P. bivaricosus* was generally sparse or absent in areas developed as palm plantations, but that mixed palm and broad-leaf forest provided good habitat. *Placostylus bivaricosus* is generally active at night. Nocturnal behaviour in the related New Caledonian species *Placostylus fibratus* was found to be related to a preference for lower temperatures and higher humidity levels (Salas *et al.* 1997). *Placostylus* species are unable to seal the shell aperture during dry conditions and are therefore prone to desiccation (Sherley 1994).

In areas of suitable habitat, *P. bivaricosus* occurs at an average density of 0.24 live adults and 0.33 live juveniles per square metre (Ponder and Chapman 1999), with localised densities of up to two live animals per square metre in appropriate microhabitat patches within these areas. These densities are comparable to those observed in New Zealand *Placostylus* species (Parrish *et al.* 1995). Accumulated shells of dead *P. bivaricosus* can remain in the soil and leaf litter for many years and, in some places, can occur at densities of up to 30 shells per square metre (Ponder and Chapman 1999; NPWS 2001a).

The diet of *P. bivaricosus* is thought to be the fallen dead leaves of broadleaf trees (Ponder and Chapman 1999). New Zealand *Placostylus* species feed on the fallen leaves of a range of native broadleaf tree and shrub species (Parrish *et al.* 1995; Sherley *et al.* 1998). *Placostylus bivaricosus* lays small clutches of eggs in the soil beneath leaf litter, probably during the warmer months, and the eggs hatch into small-shelled snails about 6.7 mm in length and 5 mm in width (Ponder and Chapman 1999). Hatchling and juvenile mortality is high (Ponder and Chapman 1999). Growth rate and lifespan of *P. bivaricosus* are unknown, but related *Placostylus* species in New Zealand reach maturity at 3 to 5 years and may live for 20 years or more (Parrish *et al.* 1995).

Recovery Program for *Placostylus bivaricosus*

Placostylus bivaricosus was listed as an endangered species under the TSC Act in April 1997 (NSW Scientific Committee 1997a). A recovery plan for the species was developed by the NSW National Parks and Wildlife Service (NPWS) (now part of the DEC) between 1998 and 2001. A draft recovery plan, prepared in accordance with the TSC Act, was publicly exhibited in December 1998 (the first recovery plan for a fauna species in NSW) and a final recovery plan was approved in October 2001 (NPWS 2001a). The recovery plan identifies habitat clearing and modification for development, and predation and habitat disturbance by exotic vertebrate fauna (predominantly the Black Rat *Rattus rattus* and possibly the Common Blackbird *Turdus merula* and Song Thrush *Turdus philomelos*), as the major causes of the species' decline. Herbicide and pesticide use, invasion of habitat by exotic weeds, trampling and habitat disturbance by domestic stock, and predation and habitat disturbance by domestic poultry are listed as additional current threats. Predation and habitat disturbance by feral pigs *Sus scrofa* and habitat disturbance by feral goats *Capra hircus* are listed as previous threats, with pigs successfully eradicated from the island and eradication of feral goats almost completed. The objectives of the recovery plan are to identify and protect remaining habitat and extant populations of the species (with a priority given to conserving maximum surviving genetic diversity), identify and ameliorate current threats, support and coordinate relevant research and encourage community involvement in the conservation and recovery of the species (NPWS 2001a). Some of the outcomes achieved through the recovery program are outlined below.

Increased knowledge of extant distribution

At the time of listing under the TSC Act in 1997, the only recent records of live *P. bivaricosus* were three site records from a general invertebrate survey on Lord Howe Island by the Australian Museum in 1971 (at Windy Point, the east end of North Bay and the east end of Neds Beach) (Smithers *et al.* 1977; NSW Scientific Committee 1997a). One of these sites (Windy Point) had since been destroyed through development of the Lord Howe Island airfield in the mid 1970s. Later identification of hatchling *P. bivaricosus* from leaf litter samples collected from Big Slope and Little Slope in the 1970s indicated that populations may also have been extant in these remote coastline localities in the southern mountains area at that time (Ponder and Chapman 1999).

The Australian Museum was engaged by the NPWS in 1999 to undertake a targeted survey for *P. bivaricosus* as part of the development of the recovery plan. The survey recorded live or freshly dead *P. bivaricosus* at 24 sites out of a total of 59 sites examined (Ponder and Chapman 1999). Additional records have also been obtained through increased community awareness. *P. bivaricosus* is now known from approximately 5 localities comprising 26 sites in the central lowlands and northern hills (Ponder and Chapman 1999; NPWS 2001a). Investigation of the Big Slope and Little Slope localities in the south of Lord Howe Island is a priority for proposed future surveys (NPWS 2001a).

Statistical modelling of predicted habitat for *P. bivaricosus* was completed by the NPWS in 2002 (NPWS 2002). Maps of predicted habitat were derived for pre-settlement (pre-1833) and the present time (Fig. 2) through a process of overlaying and statistical analysis of environmental variable spatial data and *P. bivaricosus* site data using an ArcView Geographic Information System and S-plus statistical software. Environmental variables which proved to be significant in driving the statistical model were calcium availability and elevation. Correlation with calcium-rich soils reflects the requirement of land snails for calcium for shell production (Stanisic 1994, 1997). Correlation with elevation probably in part also reflects differences in calcium availability between the calcareous sands and calcarenite of the central lowlands and the volcanic geology of the northern hills and southern mountains, and is consistent with early descriptions of the species' distribution (Brazier 1889).

The resulting maps showed three classes of habitat quality: category A and category B for high quality habitat (78-100% and 55-77% likelihood of occurrence respectively); a single category of lower quality habitat (17-54% likelihood of occurrence); with the remaining areas (0-16% likelihood of occurrence) classified as unsuitable habitat (NPWS 2002). These maps provide the best guide to the historical and current distribution of *P. bivaricosus* on Lord Howe Island. Analysis of this mapping indicated significant spatial overlap between human settlement of Lord Howe Island and high quality snail habitat, with 50% of the highest quality snail habitat (category A) cleared since 1833, compared to only 3% of the lower quality snail habitat (NPWS 2002), supporting the contention that loss of habitat has been a major factor in the decline of the species.

Development control and land use planning

Listing under the TSC Act triggers an obligation for approval authorities under the *NSW Environmental Planning and Assessment Act* 1979 (EP&A Act) to consider the impacts of proposed developments and activities on a threatened species, population or

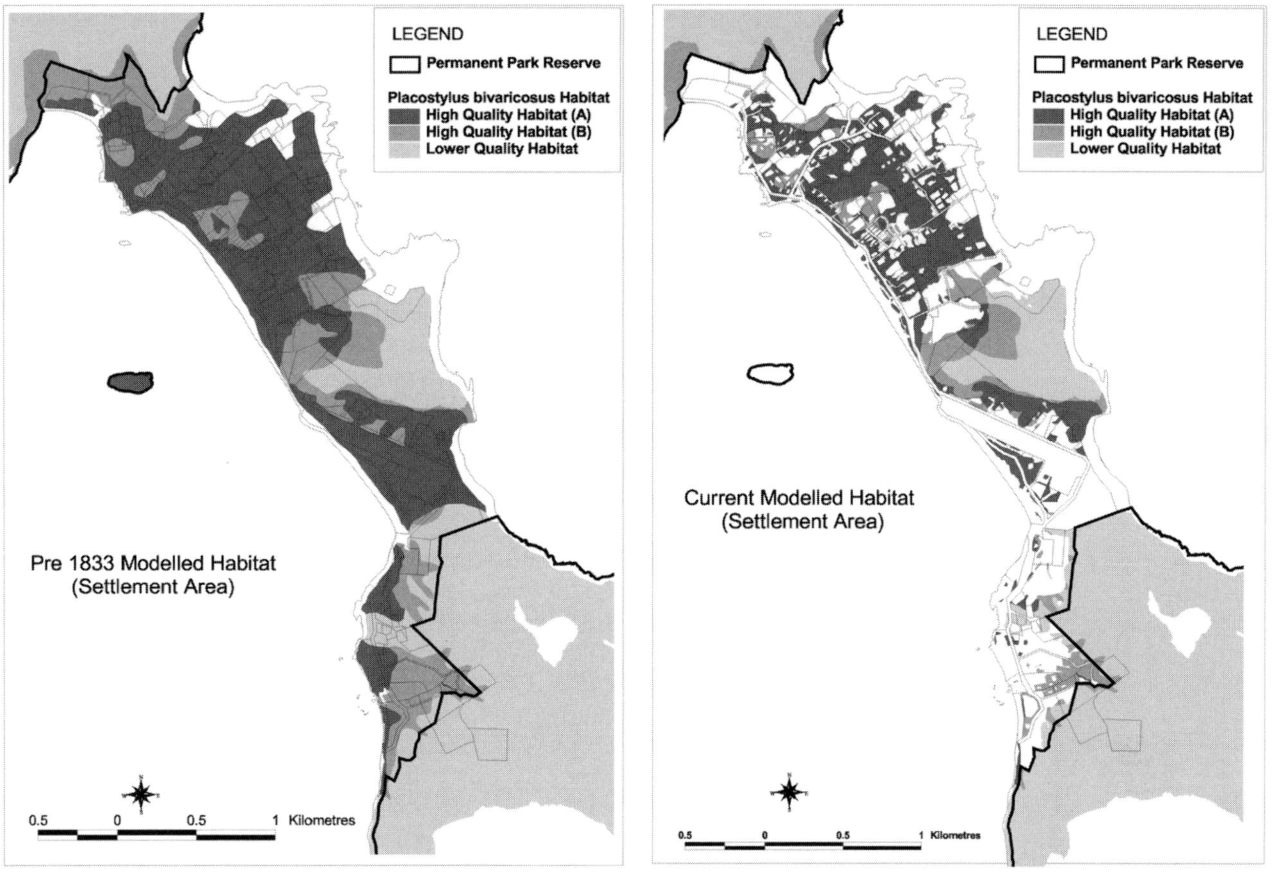

Figure 2: Modelled habitat of *Placostylus bivaricosus* in the Lord Howe Island Settlement area. Left: pre-1833 modelled habitat; Right: current modelled habitat (From NPWS 2002). Statistical modelling of habitat extended recognition and consideration of the species' habitat beyond simple point locality records. Past land clearing on Lord Howe Island has been most extensive in the highest quality habitat areas.

endangered ecological community. Where a significant effect on threatened species cannot be avoided, or where critical habitat will be affected, the approval authority must seek the concurrence of the Director-General of DEC (or, where a Minister is the approval authority, must consult with the Minister for the Environment). The recovery plan for *P. bivaricosus* includes, as an appendix, guidelines for environmental impact assessment to assist those required to prepare or review such assessments in relation to *P. bivaricosus*. These guidelines include information concerning field survey methods, habitat assessment and significance of likely impacts on the species (per section 5A of the EP&A Act), and were designed to ensure that consideration of the species in environmental impact assessment and land use planning is robust, with a precautionary approach taken where necessary information is lacking.

The main approval authority on Lord Howe Island is the Lord Howe Island Board (LHIB) under the *Lord Howe Island Act* 1953. The map of current predicted snail habitat (NPWS 2002) is being used by the LHIB as a tool to assess the likely effect of individual development proposals on *P. bivaricosus* (T. Wilson LHIB pers. comm), thereby extending recognition and consideration of the species' habitat beyond simple point locality records. On a strategic level, this mapping has also been included as a data layer in the development of a revised *Regional Environmental Plan* for Lord Howe Island under the EP&A Act, which is currently in preparation by the NSW Department of Infrastructure Planning and Natural Resources in consultation with the LHIB and DEC. The mapping will be used in identifying significant areas to be protected from development.

Rodent control

The Black Rat *Rattus rattus* was accidentally introduced to Lord Howe Island in June 1918 following the grounding of the ship *Makambo* (Hutton 1990), and was in prolific numbers across the Island by 1930 (Billing 1999). Within 25 years of their arrival, *P. bivaricosus* was in noticeable decline (Iredale 1944). Rats prey extensively on *P. bivaricosus*, particularly on juvenile snails, and are considered to be a major predator of this species and a significant threat to its survival (Ponder and Chapman 1999; NPWS 2001a). Evidence of rat predation on *P. bivaricosus* is readily observable in the form of distinctively damaged empty shells (Fig. 3) which can be easily found on the forest floor (Murphy pers. obs.). Predation by the Black Rat on Lord Howe Island was listed as a key threatening process under the TSC Act in May 2000, partly on the basis of their impact on *P. bivaricosus* and other endemic land snail species including *Epiglypta howinsulae* and *Gudeoconcha sophiae* (NSW Scientific Committee 2000).

Figure 3: Shells of *Placostylus bivaricosus* showing spiral pattern of shell damage typical of rat predation (compare with Fig. 2 of Sherley *et al.* 1998). Photograph by Michael Murphy.

Rat predation has similarly been identified as a significant threat to *P. ambagiosus* populations in New Zealand, restricting the recruitment of adult snails by removing medium to large-sized juveniles and resulting in progressively ageing snail populations which could ultimately die out (Sherley *et al.* 1998). At one site studied, 73% of empty *P. ambagiosus* shells collected prior to commencement of rodent control were the result of rodent predation, and there was a significant increase in snail numbers following the commencement of rodent baiting, due mostly to a relative increase in the number of juvenile snails (Sherley *et al.* 1998).

The LHIB has been running a regular, warfarin-based rodent control program on Lord Howe Island since 1986. The program has a primary focus on the protection of the palm seed industry, but it also includes a number of significant sites for the conservation of *P. bivaricosus* (NPWS 2001a). The LHIB is currently considering the feasibility of a proposal for the total eradication of introduced rodents on Lord Howe Island.

Genetics research

Conservation of the maximum surviving genetic diversity within *P. bivaricosus* is identified as an objective in the recovery plan (NPWS 2001a). A preliminary assessment of the genetics of extant populations of *P. bivaricosus bivaricosus* was undertaken in 2001 with joint funding provided by the NPWS and Australian Museum (Colgan and Ponder 2001). The results suggested two genetic clades with distinctive shell morphology within the extant populations, with a more elongated-shelled form found predominantly in the northern part of the Island (and north-eastern part of the Settlement area) and a broader-shelled form from the central and western parts of the Island (Colgan and Ponder 2001). The results of this study were included in the identification of priority sites for protection in the rodent control program and further genetics research is proposed (NPWS 2001a).

Habitat restoration

The original evergreen closed forest on Blackburn Island (in the Lord Howe lagoon) has been predominantly cleared since European settlement (Pickard 1983), with only a few individuals of the original tree species remaining, and the forest has been replaced by a grassland of Rhodes Grass *Chloris gayana*. *Placostylus bivaricosus* formerly occurred on the island but has not been recorded there for over a century (Ponder and Chapman 1999). Rats are currently not present on offshore islands in the Lord Howe Island group, including Blackburn Island. An action proposed in the recovery plan for *P. bivaricosus* is the restoration of an area of evergreen closed forest habitat on Blackburn Island to create a rat-free area of habitat to be used as a release site in the event that the translocation of *P. bivaricosus* is considered necessary at a future date (NPWS 2001a). Establishment of translocated populations on offshore islands or in areas of regenerated habitat has been employed as an *ex situ* conservation measure for *P. ambagiosus* and *P. hongii* in New Zealand (Sherley 1994). The LHIB has recently commenced the initial work on a habitat restoration project on Blackburn Island (T. Wilson LHIB pers. comm.). A broader outcome of this work will be the provision of a rat-free, closed forest habitat for a range of additional threatened and significant species, including the endangered cockroach *Panesthia lata* (NSW Scientific Committee 2004).

Future actions

High to medium priority actions in the recovery plan for *P. bivaricosus* which remain to be implemented include scientific survey of areas of high potential habitat for *P. bivaricosus etheridgei* in the southern mountains area (including Big Slope and Little Slope), establishment of long-term monitoring sites in both rodent baiting and non-rodent baiting areas, establishment of a captive breeding program, research into the impact of the Common Blackbird and Song Thrush, research into the ecology of *P. bivaricosus*, preparation of community awareness material and nomination of the species for listing under the Commonwealth EPBC Act. At the time of writing this paper, planning was underway for a pilot captive breeding project for *P. bivaricosus* on Lord Howe Island which will incorporate an ecological research component and local community involvement. *Placostylus bivaricosus* has also been included in a series of six bookmark-style brochures currently being produced by the DEC for distribution to Lord Howe Island residents and visitors to highlight threatened species and the actions which can be taken to assist their conservation and recovery.

Case Study 2: Mitchell's Rainforest Snail *Thersites mitchellae*

The *Thersites* (Family Camaenidae) is a genus of large land snails found in the coastal lowlands and adjacent ranges of north-east New South Wales and south-east Queensland (Abbott 1989; Bishop 1978; Smith 1992). *Thersites mitchellae* (Cox, 1864) (Fig. 4) has a shell up to 55 mm wide and 50 mm high with a strongly elevated spire, deep reddish chestnut to black in colour with two prominent yellow bands. The body colour is dark grey to black with a thin lighter line on the dorsal midline. The species was formerly common in lowland rainforest and paperbark *Melaleuca quinquenervia* swamp forest on the coastal plain of north-east NSW between the Richmond and Tweed rivers, typically on basaltic alluvium (Bishop 1978; Smith 1992; NPWS 2001b). Coastal north-east NSW has experienced major development over the last century, initially for agriculture and now increasingly for urban development, and the coastal lowland habitat for *T. mitchellae* has been reduced in area by at least 90%, with the remaining habitat remnants generally small and fragmented (NSW Scientific Committee 1997b; NPWS 2001b). The largest known surviving population and largest single area of remaining coastal lowland habitat is on Stotts Island, in the Tweed River near Murwillumbah (NPWS 2001b; NPWS 2001c). *Thersites mitchellae* is recognised internationally as endangered through listing on the *2003 IUCN Red List of Threatened Species* (IUCN Species Survival Commission 2003), in the category EN C2a. Available ecological information was summarised by Murphy (2002a). Most of the sites where the species is known to survive are located on slightly elevated ground on the margins of coastal wetlands. Live animals have been recorded from within and under leaf litter on the ground and under bark on Fig (*Ficus* spp.) trees. The species is considered to be herbivorous, feeding on leaf litter, fungi and lichen, and lays eggs beneath leaf litter.

Figure 4: Mitchell's Rainforest Snail *Thersites mitchellae*. This large land snail is endemic to the far north coast of NSW. It is threatened by loss of habitat, with over 90% of its coastal lowland habitat already cleared, and many of the remaining habitat remnants at risk of clearing. Recovery efforts have included increasing protection of key habitat areas, ensuring the species is appropriately considered in land use planning and encouraging local community support and involvement. Photograph by Michael Murphy.

Recovery Program for *Thersites mitchellae*

Thersites mitchellae was listed as an endangered species under the TSC Act in March 1997 (NSW Scientific Committee 1997b). A draft recovery plan was publicly exhibited in July 2000 and the final recovery plan was approved in June 2001 (NPWS 2001b). The recovery plan identifies loss of habitat as the major threat to *T. mitchellae*, with fire, weeds and feral predators listed as additional threats. The objectives of the plan are to identify and protect remaining habitat and extant populations of the species, maximise the protection of the population on Stotts Island, improve the protection and management of other populations and encourage community involvement in the conservation and recovery of the species (NPWS 2001b). The recovery program was summarised by Murphy (2002a) and the following is an update on some of the outcomes.

Critical Habitat declaration and listing under Commonwealth legislation

Stotts Island in the Tweed River has approximately 120 ha of habitat for *T. mitchellae* and an estimated population of several hundred snails, and the Island is considered pivotal to the continued survival of the species (Stanisic 2000). The entire Island was declared to be *critical habitat* under the TSC Act in November 2001 (NPWS 2001c), the first such declaration for any species in NSW, affording it the highest level of protection under the Act. *Thersites mitchellae* was listed as *critically endangered* under the Commonwealth *Environment Protection and Biodiversity Conservation Act* 1999 (EPBC Act) in July 2002, being the first land snail species to be listed and one of only two land snails currently listed (April 2004). Both the critical habitat declaration and EPBC Act listing have increased the public profile of *T. mitchellae*, and land snails in general, as well as increasing the level of legislative protection afforded to the species. The NSW recovery plan for *T. mitchellae* (NPWS 2001b) was adopted as the Commonwealth recovery plan under the EPBC Act in February 2004.

Increased knowledge of extant distribution

Only two site records of *T. mitchellae* were known in the 70 years prior to listing under the TSC Act: from Stotts Island near Murwillumbah and Cumbebin Wetland near Byron Bay (NSW Scientific Committee 1997b). It is now known from eight localities in the coastal lowlands comprising approximately 30 sites (Fig. 5). The increase in sites since listing is due to a combination of targeted surveys undertaken by the Queensland Museum with funding from the NPWS (Stanisic 1998; 2000), records from the public as a result of increased community awareness, and inclusion of the species in standard development-related environmental impact assessment surveys. The known current range of *T. mitchellae*, from Banora Point near Tweed Heads south to Lennox Head, now approximates its historical distribution on the coastal lowlands, although the actual area of occupancy within this range has been substantially reduced (Murphy 2002a).

A 1980 record of *T. mitchellae* by a local resident (supported with voucher specimens), from a site (elevation 230 m AHD) at Wilsons River near Mullumbimby (Fig. 5) suggested that, contrary to accepted wisdom, the species' distribution might extend beyond the coastal lowlands into adjacent upland areas. The specific location of the Wilsons River site was only determined in 1999, and a targeted search of the site at that time found no sign of *T. mitchellae* (Stanisic 1999). Noting that the snail fauna there comprised upland species (including *Ngairea corticicola*, *Pedinogyra rotabilis* and *Thersites richmondiana*) not usually associated with *T. mitchellae*, it was suggested that the 1980 record might have been the result of accidental and unsuccessful translocation (Stanisic 1999). Nevertheless, further investigation of the Wilsons River area and other upland areas was recommended (Stanisic 1999; NPWS 2001b). At the time of writing this paper, additional upland records of *T. mitchellae* (from sites at elevations between 400 and 515 m AHD near Mount Chowan in Mount Jerusalem National Park) (Fig. 5) were received from a local resident. The identification was based on a brochure prepared by the NPWS for that purpose. These records (supported with voucher specimens) indicate that *T. mitchellae* may also occur in some upland areas adjacent to the coastal lowlands, and that the distribution and habitat preference parameters of the species may need to be reconsidered, with implications for recovery efforts.

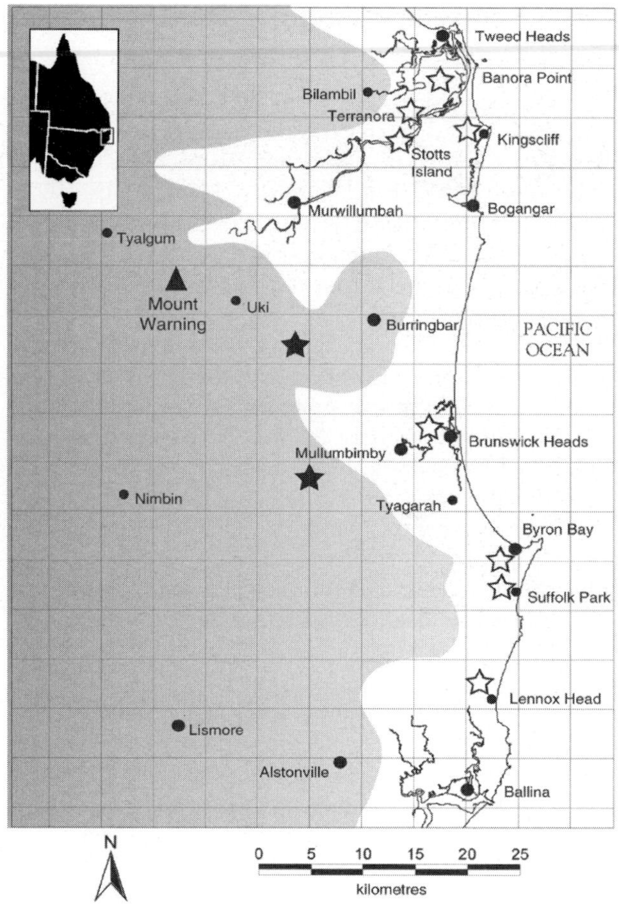

Figure 5: Recent records of *Thersites mitchellae*. The only known records in the 70 years prior to listing under the TSC Act in 1997 were from Stotts Island and Byron Bay. Open stars represent coastal lowland localities where the species has been recorded since 1997. Closed stars represent upland localities where the species has been found by local residents (the lower one represents Wilsons River and the upper represents Mount Chowan – see text). Shaded area represents land over 100 m elevation.

Development control and land use planning

Loss of habitat for development continues to be a major threat to *T. mitchellae*, with many of the remaining areas of coastal lowland habitat under threat of development (Murphy 2002a). Prior to listing as a threatened species, *T. mitchellae* received no consideration in environmental impact assessment and land use planning decisions. The recovery plan for *T. mitchellae* includes, as an appendix, environmental impact assessment guidelines. They are similar to those prepared for *P. bivaricosus*, and were designed to ensure that consideration of the species in environmental impact assessment and land use planning was robust and precautionary. *Thersites mitchellae* has been a significant consideration in a number of major development proposals on the NSW far north coast since listing under the TSC Act, requiring habitat protection measures to be negotiated. The listing of the species under the EPBC Act affords an additional safeguard, in that the Commonwealth Government now also an approval role in relation to development proposals likely to result in a significant impact on *T. mitchellae*.

Community awareness

The DEC (and previously the NPWS) has raised the public profile of *T. mitchellae* as a threatened species and as a flagship for land snail conservation since TSC Act listing through numerous local, regional, state and national media stories, including radio, television, newspaper and popular magazines (Murphy 2002a). The NPWS produced and distributed a brochure detailing how *T. mitchellae* can be identified, describing its habitat, outlining the threats to the species and listing actions members of the community can undertake to assist its conservation and recovery. The NPWS has also supported community-based educational and awareness efforts, including the funding of interpretive display material prepared by the Byron Environment Centre to highlight a significant area of habitat in Cumbebin Wetland, Byron Bay.

School education can be a powerful strategy for invertebrate conservation through the promotion of flagship taxa (New 1995; Sands 1999). A Teacher Resource Book titled *Mitch the Rainforest Snail* (NPWS 2003) (Fig. 6) was recently prepared and released by the DEC, with funding provided by the Foundation for National Parks and Wildlife. The book provides a range of classroom resource material suitable for school students in stages 2 and 3 (Year 3 to year 6), promoting *T. mitchellae* as a flagship for student awareness of molluscan diversity and land snail conservation issues on the NSW north coast, and has been positively received. Over 350 published copies have been distributed to primary school teachers, predominantly within the Tweed/Byron/Ballina target area where the species occurs, but also more widely in NSW. An electronic version of the book can be seen online on the DEC website www.nationalparks.nsw.gov.au.

Friend or Feral?

Native land snails are an important part of our natural environment. There are over 2000 different types of native land snail in Australia, with over 90% of these living in rainforests. Others live in bushland and woodlands, and a few are even found in deserts.

A rainforest has many different types of soft-leaved trees that grow closely together, allowing only a small amount of sunlight to filter through to the forest floor. A rainforest is shady and damp with plenty of food. Most native land snails only feast on fungus and decaying plants, while some eat other snails.

1. How many different words can you make from the word: RAINFOREST

If you can find:
30 words you're good
60 words you're exceptional
90 words you're a word genius!

2. Research rainforests and write a report on what you have found out. Find out about the different types of rainforest in Australia.
   On the internet go to www.bigvolcano.com.au/natural/rftypes.htm to read more about the different types of rainforest.

The most widespread type of snail in Australia is the feral European Snail, whose relatives first arrived in Australia around 200 years ago. It is usually these snails who are found in large numbers in gardens, eating vegetables and other plants. European snails like to eat leaves of introduced plants.

Some native snails look a bit like the European Snail, but others come in lots of different shapes and sizes. If you find a funny looking snail, it's probably a local! Native snails are usually found in the bush. Natives can also be found visiting gardens, especially if there are small areas of bushland nearby.

3. What is a feral animal?
4. List 3 feral animals that are a threat to the Australian environment. Why are they a threat? Give one reason for each.
5. Where do you think you are most likely to find a European Snail?

Mitch the rainforest snail NSW National Parks and Wildlife Service, 2003 3

Figure 6: Cover and sample page from *Mitch the Rainforest Snail Teacher Resource Book* (NPWS 2003). This book uses the endangered *Thersites mitchellae* to promote awareness of mollusc diversity and land snail conservation in primary schools. The book received a Whitley Award from the Royal Zoological Society of New South Wales in 2004, judged by the Society to be the best book in the teachers' resource category.

Future actions

High to medium priority actions in the recovery plan for *T. mitchellae* which remain to be implemented include scientific survey of areas of high potential habitat on the coastal plain and adjacent foothills (including the Wilsons River area), statistical modelling of predicted habitat, research into the ecology of *T. mitchellae*, establishment of long-term monitoring sites and assessment of additional areas for identification as critical habitat. At the time of writing, records of *T. mitchellae* from sources including development-related environmental impact assessment surveys, community reports and opportunistic field surveys were being collated and validated in preparation for a habitat modelling project.

A brief background to Lepidoptera conservation

Despite their prominence in community-based conservation actions internationally, Lepidoptera in Australia remain poorly known, with only 10 500 of the estimated 21 870 species having been described (Yen and Butcher 1997). Sands and New (2001) assessed the conservation status of 624 Australian Lepidoptera taxa, and recommended that 15 taxa were threatened according to criteria modified from those used in the 1994 IUCN Red List. Such a small proportion may result from the combination of the lack of data available for most species, and that many entomologists prefer that species are not included on statutory lists as this often triggers the requirement to obtain permission to collect or undertake research on the listed species (Greenslade 1999; Gunning 1999). Conversely, the broader community has embraced the opportunity to participate in many species-based management programs, such as those for the Eltham Copper Butterfly *Paralucia pyrodiscus*, the Purple Copper Butterfly *Paralucia spinifera*, and the Richmond Birdwing Butterfly *Ornithoptera richmondia* (Sands and New 2001). Five Lepidopteran species (three butterflies and two moths) are listed under the TSC Act (as at April 2004, Table 1). The recovery program of one of these species is detailed below.

Case Study 3: Purple Copper Butterfly *Paralucia spinifera*

The Purple Copper Butterfly *Paralucia spinifera* (Edwards and Common 1978) (Family: Lycaenidae) (Fig. 7) has been described as one of Australia's rarest butterflies (Dunn *et al.* 1994). It is known from about seven localities comprising 37 sites on the NSW Central Tablelands, all above 900 metres elevation and generally bounded by Bathurst, Lithgow and Oberon. The larvae of *P. spinifera* graze on native Blackthorn *Bursaria spinosa* ssp. *lasiophylla* (Braby 2000; NPWS 2001d). *Paralucia spinifera* is generally found within, or on the edges of, eucalypt woodland with a grassy understorey and patches of the larval food plant, where the aspect and local vegetation structure allow direct sunlight to reach the food plants for a proportion of the day (NPWS 2001d).

Figure 7: Purple Copper Butterfly *Paralucia spinifera* adult. This species was only described in 1978 and is known only from a small area on the NSW Central Tablelands. Photograph by Simon Nally.

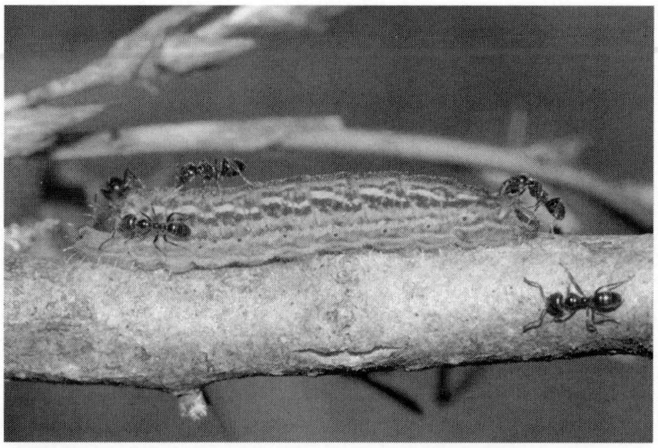

Figure 8: *Anonychomyrma itinerans* ants attending larva of Purple Copper Butterfly *Paralucia spinifera.* The larvae of *P. spinifera* have a mutualistic association with *A. itinerans,* with the ants protecting them from potential predators and feeding on 'honeydew' exuded by the larvae. Photograph by Simon Nally.

The larvae of *P. spinifera* have a mutualistic association with the ant species *Anonychomyrma itinerans* that attends its larvae and pupae (Braby 2000; NPWS 2001d) (Fig. 8). Although relatively common amongst Lepidoptera, the larvae – ant association engenders fascination both within researchers and the general community. After an apparent autumn and winter absence, the *A. itinerans* return to patrol the *B. spinosa* subsp. *lasiophylla* during the spring larval emergence of *P. spinifera*. Although attentive to the early larval instars and defensive of the *B. spinosa* subsp. *lasiophylla*, *A. itinerans* do not alter the behaviour of the early instar larvae. However, the activation of the larvae's 'honeydew' glands at the third or fourth instar is coincident with a marked change in larval behaviour. Whilst previously remaining on *the B. spinosa* subsp. *lasiophylla* both day and night, the larvae become nocturnal, being shepherded by *A. itinerans* to and from chambers within the subterranean *A. itinerans*' nest. *Anonychomyrma itinerans* constantly check the availability of 'honeydew' from the larvae's glands, and activity that stimulates the retraction of the glands. *Anonychomyrma itinerans* continue their behaviours of attacking potential predators that come into contact with the *B. spinosa* subsp. *lasiophylla*, and of causing the larvae to retreat underground when such a threat is detected.

Recovery Program for *Paralucia spinifera*

Paralucia spinifera was listed as endangered under the TSC Act in November 1996, on the basis that; there were three known sites, it was threatened by habitat disturbance (grazing of the host plant by sheep, goats and cattle, clearing for the establishment of pine plantations, invasion of exotic weeds and activities of feral pigs) and overcollecting, and only one site was secured within a conservation reserve (NSW Scientific Committee 1996). The recovery program, developed between 1997 and 2001, included surveys to identify unrecorded locations and the preparation of a recovery plan. A draft recovery plan, prepared both in accordance with the TSC Act and Commonwealth EPBC Act, was publicly exhibited in 2000. A final recovery plan was approved in June 2001 (NPWS 2001d). Since 1997, the DEC (and previously the NPWS) has implemented a program to acquire an understanding of the ecology of the species, and to prevent the operation of processes that threaten its survival. The recovery plan identified a range of habitat modification and disturbance factors as the key processes threatening the species. Accordingly, the recovery plan proposed habitat identification and survey, site management and protection, research, and community involvement as key program areas (NPWS 2001d). Some of the successes are outlined below.

Potential habitat identification and survey

At the time of listing under the TSC Act, *P. spinifera* was known from three locations (NSW Scientific Committee 1996). Further sites had been recorded, but these details were not made available to the Scientific Committee. It was expected to be found at further sites. In 1997, the NPWS and volunteers undertook helicopter-based aerial surveys to find potential habit. Topographic maps in the Bathurst-Lithgow-Oberon areas were used to record areas of *B. spinosa* ssp. *lasiophylla*, the larval host plant, at sites with minimal canopy cover. Due to time and land tenure constraints, ground-truthing of potential habitat sites was generally confined to accessible areas of public land. By the end of the flying season in the spring of 1998, further sites of *P. spinifera* had been recorded at Lithgow, and at new sites in the Lidsdale and Cheetham's Flat areas.

Surveys for flying adults of this butterfly are made difficult due to a range of factors:

- peak flying activity occurs over only a three-week period;
- preliminary data suggest that flying activity is related primarily to incident solar radiation, even at temperatures as low as 5°C. Accordingly survey work can be confounded by localised or unpredicted cloud conditions;
- time of flying varies between sites because aspect and tree canopy cover affect solar radiation reaching the *B. spinosa* ssp. *lasiophylla* patch;
- within a patch of *B. spinosa* ssp *lasiophylla*, apparent numbers of flying adults may be consistently recorded over a number of years, fluctuate over several years (Sands and New 2001, Nally pers. obs.), or annually decline to a level where adults are unrecordable (Nally 2003), possibly indicating an underlying metapopulation structure;
- there are both extensive stands and numerous small and discrete areas of potential habitat in this locality.

In response to these difficulties, the DEC has encouraged amateur entomologists and the broader community to be involved in the survey to identify new sites. All these activities have resulted in records of 37 sites where flying adults have been observed, and a doubling of the range (Fig. 9). Notably, this range increase was recorded by a seven year-old Oberon resident, Heather Wilson, (Fig. 10), who had identified the butterfly from a brochure prepared for this purpose.

In response to the helpful advice of numerous amateur and professional entomologists, combined with the results of the potential habitat modelling, DEC plans to facilitate further community-based survey in predicted potential localities in NSW and the ACT.

Management of threatening processes at recorded sites

Seven sites in the vicinity of Lithgow have been the subject of community-based weed control programs coordinated by Lithgow Tidy Towns, Lithgow-Oberon Landcare Association, Conservation Volunteers Australia and the DEC. This has involved herbicide treatment to large stands of Broom *Cytisus scoparius*, and intensive hand weed control of Hawthorn *Crataegus monogyna*, Radiata Pine *Pinus radiata*, Blackberry *Rubus fructicosus* and Cotoneaster *Cotoneaster* sp.. This has reduced the weed load over about three hectares of habitat. Four sites have been the subject of weed control by the landholder

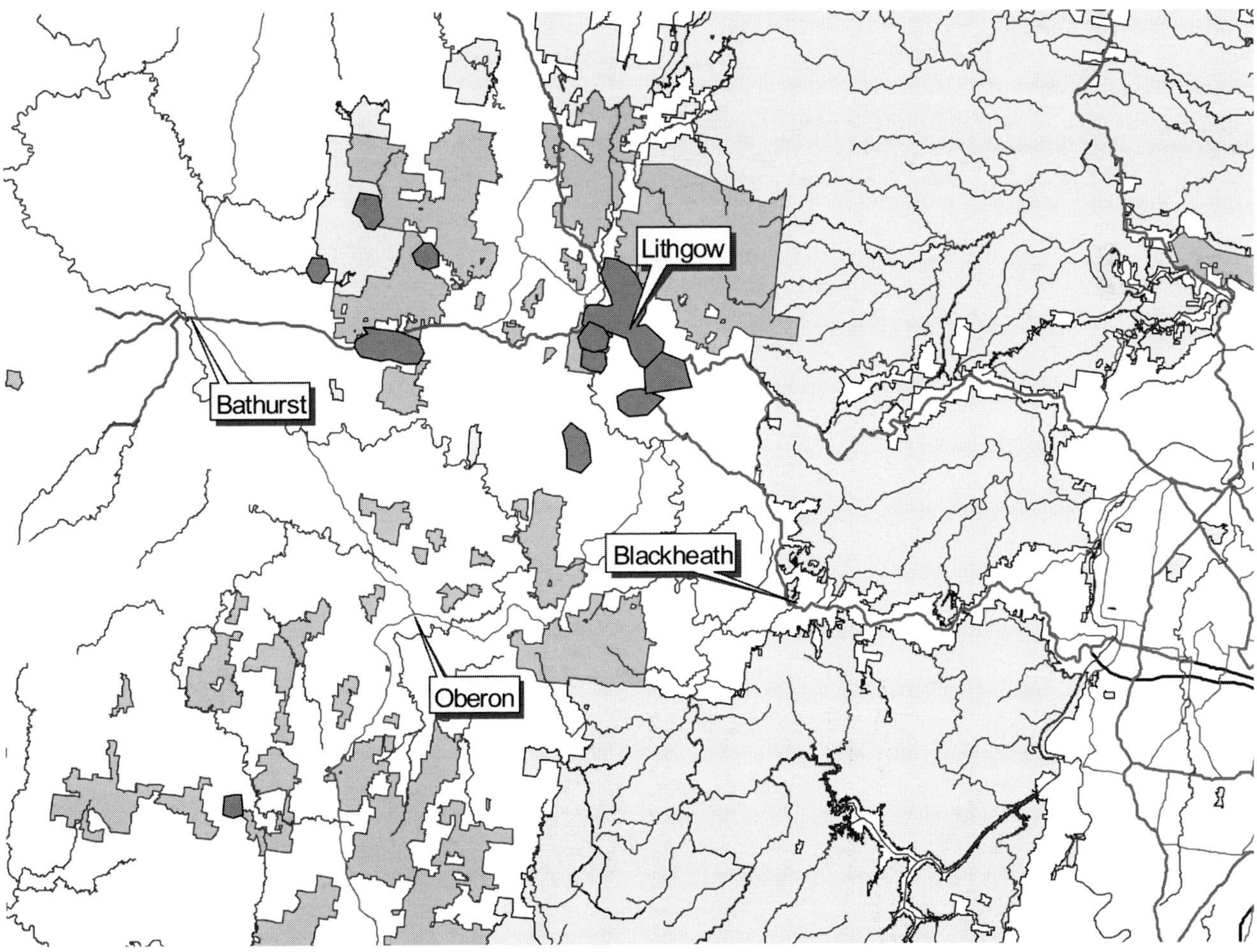

Figure 9: *Paralucia spinifera* distribution and range extension (bright green areas represent a one kilometre buffer around each site, pink hatched areas are State Forest, green hatched areas is the Winburndale Nature Reserve, red lines represent main roads, and blue lines represent drainage lines).

Girl finds rare butterflies

Nature lover: Heather Willson

By DESIREE D'COSTA

A SEVEN-YEAR-OLD girl with a love of insects has discovered a new population of one of Australia's most endangered butterflies.

Heather Willson's discovery of a colony of purple coppers at her grandparents' NSW property had increased their known habitat by 50km, Simon Nally of the National Parks and Wildlife Service said.

"We're very happy with the find," Mr Nally said.

"At the moment, there are 35 sites located in two main areas east of Bathurst, at Yetholme and Lithgow, covering 48km. The site Heather found last year is south of Oberon."

Last year, the NPWS told Heather's grandparents a colony of the butterflies might be on their property, sparking Heather's investigation.

"I'm not surprised that Heather found them, because a butterfly flies at eye height for a child and children are very good at focusing on small things," Mr Nally said.

"These butterflies aren't found anywhere else in the country, so we're very pleased with Heather's efforts.

"The CSIRO has established that the population Heather found is genetically distinct from all other populations, making the find even more valuable."

Heather's mother, Kate, says she has always been interested in nature.

"Her room is like a compost heap," Ms Willson said.

"She's always collecting colourful leaves and bark; she's very intrigued with the outdoors.

"I'm amazed that she found the butterfly, and really proud of her."

Found: The endangered butterfly

Figure 10: Extract from the *Sunday Telegraph* 12 October 2003 in response to DEC media release (Courtesy, News Limited, 12 October 2003, Desiree D'Costa)

with assistance from the DEC in the Lithgow and Yetholme areas. A further five sites on public land have been subject to similar weed control activities undertaken by their respective management agencies.

Augmentation and habitat linkage plantings have commenced on three sites, achieved through a combination of community-based and management agency activities. Both NSW State Forests and Commonwealth Department of Defence have installed fences around recorded habitat to prevent its loss through vehicular damage. The DEC is implementing feral animal and weed control programs at two sites within Winburndale Nature Reserve, and Evans Shire Council has commenced the process to close an unsealed road, where traffic-generated dust is settling on habitat.

All the above activities are ongoing, with a recovery team monitoring progress and priorities. Further discussion of community participation in these recovery actions is in Nally (2003).

Research

In order to identify potential habitat areas within the broad distribution of *B. spinosa* ssp. *lasiophylla*, predictive habitat modelling was undertaken. Due to inconsistent availability of vegetation data layers, and the recent review of *Bursaria* by Cayzer *et al.* (1999), vegetation could not be included as a variable in the predictive modelling. Accordingly, the modelling analysed the statistical relationship between the known locations and climate, terrain and substrate variables. The modelling showed the taxon has a strong relationship with temperature range and rainfall seasonality combined with elevation, which served as another climatic surrogate, with these factors delineating a 'climatic envelope' in which this species may occur (Eco Logical Australia Pty Ltd 2003). The model predicts that the following general areas represent areas offering similar conditions to those of the known habitat: western edge of Tweed Caldera rim; Barrington Tops; between Mudgee to Goulburn; south-east ACT to Bombala along the eastern slopes of the Great Dividing Range; Tumut to Mt Kosciusko along the western slopes of the Great Dividing Range; and small areas east of Mittagong and Moss Vale, and south east of Braidwood. The initial survey effort will focus on the area between Oberon and Goulburn.

In 2003 the CSIRO, in collaboration with the DEC and the Natural Heritage Trust Endangered Species Program, completed a genetic analysis of *P. spinifera*. The results showed a minor genetic differentiation between the main site clusters at Lithgow and Yetholme, with increased differentiation for the disjunct population located at Mount David south west of Oberon, and that there were generally high levels of genetic diversity within populations. Clarke and Grosse (2002) concluded:

> ...that there are few, if any, genetic concerns for the species at this stage. If habitat loss and fragmentation are impacting this species, the effects are not reflected in the genetic architecture of the species at this point in time, suggesting that population sizes and/or levels of gene flow, are sufficiently large to maintain genetic diversity and variation.

The genetic analysis indicates that the management effort should be directed to as many representative populations from the two groups centred on Yetholme and Lithgow as feasible in order to conserve genetic diversity within the species, and to specific sites that display unique genetic variation (Clarke and Grosse 2002).

Increased knowledge of the mutualistic relationship between *P. spinifera* and attendant ants has the potential to guide effective management of known sites, as well as provide insights into the creation of effective habitat linkages between sites. During 2003, the University of Wollongong established a partnership with the DEC to study the ant species assemblages in areas of *B. spinosa* ssp *lasiophylla* occupied and unoccupied by *P. spinifera*. Preliminary results show that top-dominant ant species assemblages varied between occupied and unoccupied habitat (Sam Luccitti University of Wollongong, pers. comm.). The significance of this finding is that ant surveys may be an effective tool in identifying potential or occupied habitat within the extensive areas of *B. spinosa* subsp. *lasiophylla*. The study has also provided indications of both seasonal and cyclical variation in ant assemblages that may reflect the temporary nature of habitat suitability.

Reservation of habitat

The DEC has commenced processes, pursuant to the provisions of the *National Parks and Wildlife Act* 1974, to dedicate a Nature Reserve west of Lithgow to conserve potential habitat for *P. spinifera*. A survey undertaken by 20 volunteers from the NSW Society for Insect Studies recorded two sites occupied by *P. spinifera* within the proposed Nature Reserve. Notably, one of these sites appears to be unaffected by human activities, possibly due to its remote location. All other sites, except for one within Winburndale Nature Reserve, have been highly disturbed by direct human intervention. This survey has effectively increased the number of sites within conservation reserves from two to four.

Future actions

High to medium priority actions in the recovery plan for *P. spinifera* which remain to be implemented include; ground-truthing and refining the habitat modelling, identification and survey of areas of potential habitat, research into the ecology of *P. spinifera* and *A. itinerans*, and extending community involvement activities. At the time of writing, the ground-truthing of modelled habitat was being planned for the coming spring emergence of adult butterflies.

Conclusions and future directions

We do not intend to delve into the issue of the relevance or applicability to invertebrates of state, national, or international criteria and processes for the assessment of conservation status. Hutchings and Ponder (1999), Sands and New (2002) and others have adequately canvassed these matters. The point we do make is that, as imperfect as these criteria and processes may be to invertebrates, it is better to apply and utilise threatened species legislation as best as able as one tool for invertebrate conservation than to ignore the opportunities it presents.

Several authors have questioned the value of utilising the provisions of the TSC Act towards invertebrate conservation (e.g. Gunning 1999; Hutchings 2004). The three case studies demonstrate that the TSC Act can make a valuable contribution to the conservation of individual threatened invertebrate species. The recovery programs for *Placostylus bivaricosus*, *Thersites mitchellae* and *Paralucia spinifera* have resulted in both the reduction of threatening processes at key sites for these species and significant increases in the known range of the species over the period of the recovery program. The combination of targeted research and fieldwork, and the involvement of amateurs and community groups, has improved the understanding of the ecology of these species, which is essential if they are to be properly considered in conservation and land use planning decisions (Murphy 2002b). There has also been an increase in the level of awareness in the broader community concerning these species and a commensurate increase in the involvement of local communities in their conservation and recovery. It is important that all tiers of Government maintain their commitment to the implementation of recovery plans to sustain this momentum.

A common theme in the three invertebrate recovery programs has been the promotion of these species as flagships for greater community awareness of broader invertebrate issues, including the unique nature of many native Australian invertebrates, the pivotal role of invertebrates in ecosystems, the critical conservation status of many taxa and the need for protection of habitats. With the scant handful of invertebrate species listed under the TSC Act (16 as at April 2004, Table 1) to represent 99% of the state's fauna, this is an important component of recovery programs for all listed invertebrates. Numerous authors have identified the critical role of flagship taxa in stimulating wider community awareness of, and support for, invertebrate conservation (e.g. Yen *et al.* 1990; New 1995; Van Praagh 1997; Horwitz *et al.* 1999; Kitching 1999; Lunney 1999; Mawson and Majer 1999; Sands 1999). It is imperative that the promotion of broader community awareness of the serious issues facing invertebrate conservation be continued. The media profile established by threatened species legislation, such as the TSC Act, can be used as an effective avenue for promoting invertebrates. Cassis (1999) noted with optimism that the popular media and the general public were becoming more responsive to positive messages about the relevance of invertebrates. The media coverage regarding the proposed listing of the native cockroach *Panesthia lata* under the TSC Act (Fig. 11) further demonstrates the value of the TSC Act in promoting invertebrate conservation issues. It is unlikely that such a story would have featured as prominently (or at all) in the absence of the proposed listing under the TSC Act.

The provisions of the TSC Act for multi-species conservation initiatives, such as the listing of endangered ecological communities and key threatening processes, have not yet been fully utilised for invertebrate conservation (Murphy 2002a). No invertebrate assemblages have yet been listed as endangered ecological communities under the TSC Act (April 2004). Invertebrate assemblages with high endemism, restricted distributions and which are under threat, for example some land snail assemblages occurring on restricted limestone outcrops threatened by land clearing or mining, would be worth considering for listing under the TSC Act. A number of key threatening processes which have been listed under the TSC Act are of relevance to invertebrate conservation. These include predation by the Ship Rat on Lord Howe Island (NSW Scientific Committee 2000), clearing of native vegetation (NSW Scientific Committee 2001a) and loss or degradation of sites used for hill-topping (i.e. congregation for mating) by butterflies (NSW Scientific Committee 2001b).

Kitching (1999) stressed the need for members of the invertebrate science community to work together to positively promote invertebrate conservation and build sympathy and support in the public arena. We suggest that the lack of sufficient collaboration and mutual support within some sections of the invertebrate science community is a significant factor in limiting the broader acceptance of the need to actively conserve invertebrates and reducing the efficacy of extant conservation programs.

Salvation at hand for cockroach that's more than a pretty face

Stephanie Peatling
Environment Reporter

Not all creatures worth saving are cute like koalas, or awe-inspiring like whales. Others that you would rather stomp on or hit with a newspaper can need a hand too. Even cockroaches.

The native-wood feeding cockroach, now found only on two tiny islands off Lord Howe Island, is about to have an image overhaul in a last-ditch attempt to save it from extinction.

The NSW Scientific Committee will today recommend the listing of the cockroach, *Panesthia lata*, as an endangered species.

Michael Murphy, threatened species officer with the NSW National Parks and Wildlife Service, said many people did not appreciate the need to protect the less telegenic members of the foodchain.

"Biodiversity is not just made up of lovely birds and mammals," Mr Murphy said.

"Ninety-nine per cent of creatures out there are the forgotten invertebrates that keep the wheels turning on the whole ecosystem."

And, he notes, this particular endangered roach is not as ugly as

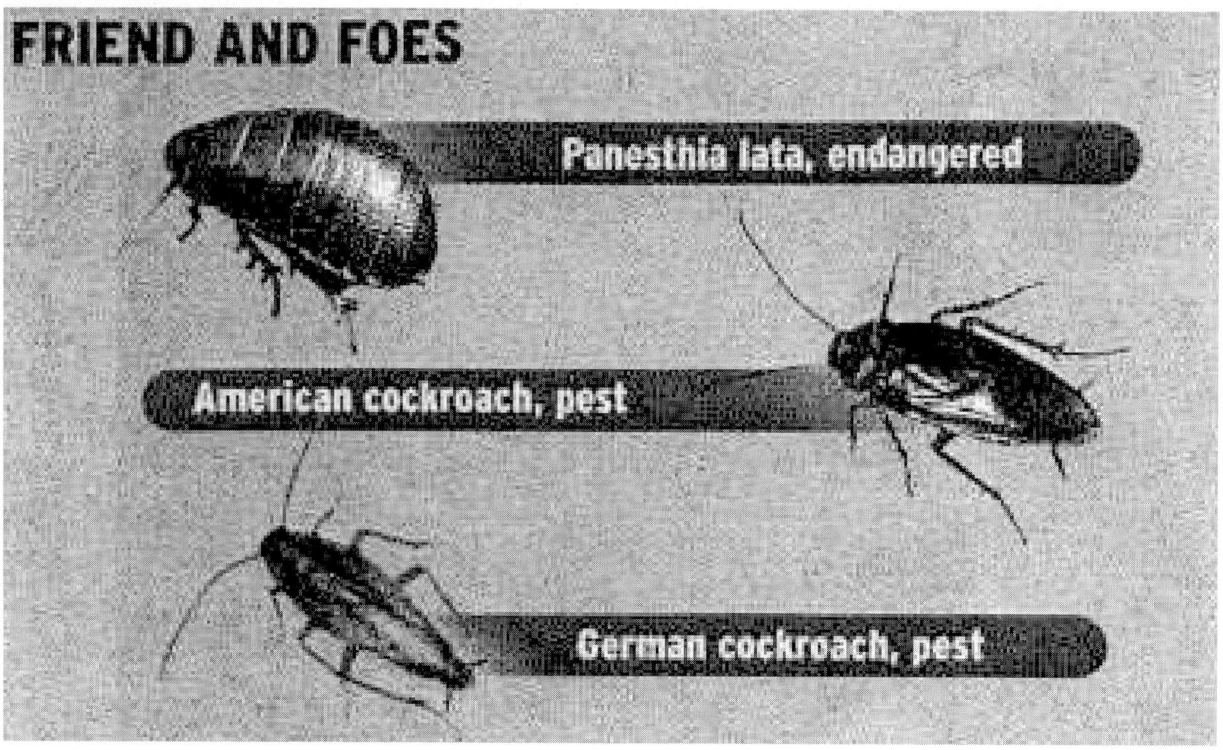

some. "I tend to raise my foot from the floor when I see a cockroach in the kitchen but this is really a nice-looking sort of cockroach."

Lord Howe Island is a rich source of study for scientists. The wood feeding cockroach is one of 23 threatened species including several species of plants, a woodhen, worm and snail.

Mr Murphy can see new avenues for promoting the island's biodiversity make-up: "Lord Howe Island – leading the way in invertebrate conservation."

First documented in 1868, the cockroach was all but wiped out after the black rat was introduced to Lord Howe in 1918. An introduced grass has also contributed to the plummeting numbers.

The cockroach's preferred habitat of decayed logs is also increasingly under threat, although there are plans to regenerate areas of the island in order to boost numbers.

Two colonies of the creature remain on two smaller islands away from Lord Howe but exactly which two is being kept secret for fear of looting. "There could be international smugglers out there who will come and take them," Mr Murphy said.

Figure 11: Extract from the *Sydney Morning Herald* 24 October 2003 (p. 3) (Courtesy, Stephanie Peatling and the *Sydney Morning Herald)*. This article from the popular media demonstrate how the TSC Act can be used to encourage community support for invertebrate conservation issues.

While there are difficulties in applying threatened species legislation to invertebrates, the invertebrate science community has yet to recognise that this legislation can make a valuable contribution as one of a number of useful tools for invertebrate conservation efforts, through promoting community support and involvement, incorporating consideration of invertebrates into land use planning decisions, encouraging scientific research, identifying critical areas for protection, and controlling threatening processes.

At the Royal Zoological Society Forum in October 2003, of which this book is a record, several participants asked whether the effort spent on single species conservation through threatened species legislation had been a failure and should be diverted instead to broader landscape protection measures. It is our belief that success in biodiversity conservation will rely on the continued application of a range of approaches at different scales, including targeted species-specific conservation measures, as well as strategic threat abatement actions and broader landscape protection and management. Concentrating on broad landscape planning initiatives, to the detriment of maintaining active programs to address threatened taxa with special requirements, will not optimise biodiversity conservation. Heywood and Iriondo (2003) made this point strongly, arguing that a purely landscape-level approach to biodiversity conservation was as flawed as a purely single-species approach, while Hnatiuk (2003) warned that relying on the simplified modelling of real world complexity that is an essential part of landscape-level planning would, in conjunction with other shortcomings in environmental management, result in unwanted outcomes including species loss, environmental degradation and declining agricultural production. The issue of broad landscape-level planning as a surrogate for biodiversity conservation is particularly relevant to invertebrate conservation. Protecting a vegetation-based community through broad landscape planning will, by default, probably also conserve some invertebrate species, but not in any strategic, comprehensive or dependable way. The spatial distribution of invertebrate species and assemblages often has little or no relation to the distribution of broad vegetation communities, so that relying on a landscape conservation framework based on vegetation communities as a surrogate for invertebrate conservation is highly unreliable (Hutchings and Ponder 1999). Many invertebrate species have very precise environmental requirements and occur at spatial and temporal scales that cannot be reliably addressed from a broad landscape approach (Yen *et al.* 1990; New 1995). *Paralucia spinifera* is a case in point, with habitat suitability linked to the temporal occurrence of both the species' food plant and ant associate and decided at the spatial scale of sunlit patches in eucalypt woodland (NPWS 2001d). The plight of *Thersites mitchellae* further demonstrates this point. The species survives today in tiny remnants of its previously extensive habitat, the majority of which are simply too small to be reliably identified at the resolution of landscape-level planning (NPWS 2001b).

Yen *et al.* (1990) considered the introduction of threatened species legislation in Victoria (the *Flora and Fauna Guarantee Act* 1988) to be 'an exciting phase of invertebrate conservation' in that state, seeing it as a transition from invertebrates being merely fortuitous passengers in conservation efforts undertaken for other reasons to being the prime object of targeted conservation programs. Threatened species legislation provides opportunities for invertebrate conservation that were previously available only for high profile, charismatic vertebrate species, and as such is of significant value in efforts to redress the continuing imbalance between invertebrates and vertebrates in the conservation priorities of both government and the community. The potential disappearance of a couple of snails and a small butterfly may not seem important to many, but they are a warning that many invertebrate communities are under threat, whose functional loss may indicate the commencement of ecosystem collapse.

We close with a local newspaper quote from a local government Shire Councillor which highlights both how listing under the TSC Act has significantly raised the profile of an invertebrate species (*Thersites mitchellae*) in land use planning decisions on the NSW far north coast, and the low value still placed on this critically endangered species by some sections of the community:

Councillor X "urged her colleagues not to be influenced by the presence of 'one wretched thing that crawls along the ground' during debate about the location of a future district town centre" (*Tweed Sun* Wednesday 16 April 2003 p. 5).

....and, with apologies to Coleridge and Shelley, our own poetic view of the voice in land use planning that listing under the TSC Act has given this hitherto unregarded species:

Ode to the Snail

Long live the snail
May its viability never fail
May its significance never pale
May it make consultants quail
So that all developments entail
Robust ameliorative measures

MJ Murphy

Acknowledgements

Katrina McKay prepared Figure 5. Kerry Cooper drew the cartoon character 'Mitch the rainforest snail' seen in Figure 6 and accompanying the concluding poem. Figure 10 is reproduced with the permission of the *Sunday Telegraph* and Figure 11 is reproduced with the permission of Stephanie Peatling and the *Sydney Morning Herald*. Dan Lunney, Winston Ponder and an anonymous referee provided helpful comments on the draft manuscript. The authors also acknowledge the significant support and involvement of local communities on Lord Howe Island, the NSW far North Coast and NSW Central Tablelands in the conservation and recovery of threatened invertebrate species.

References

Abbott, R. T. 1989. *Compendium of landshells*. American Malacologists Inc., Burlington MA.

Billing, J. 1999. The management of introduced rodents on Lord Howe Island. Unpublished report by Lord Howe Island Board, Lord Howe Island.

Bishop, M. J. 1978. A revision of the genus *Thersites* Pfieffer (Pulmonata: Camaenidae). *Journal of the Malacological Society of Australia* **4(1-2)**: 9-21.

Bishop, M. J. 1981. The biogeography and evolution of Australian land snails. Pp. 925-54 in *Ecological Biogeography of Australia*, edited by A. Keast. Dr W. Junk Publishers, The Hague.

Braby, M. F. 2000. *Butterflies of Australia* (Vol. 1 & 2). CSIRO Publishing, Collingwood, Victoria.

Brazier, J. 1889. Mollusca. In Etheridge, R. The general zoology of Lord Howe Island; containing also an account of the collections made by the Australian Museum collecting party, Aug.-Sept., 1887. *Memoirs of the Australian Museum* **2**: 1-42.

Cassis, G. (ed.). 1999. Workshop: putting invertebrates on the agenda: opportunities for the future. Pp. 437-45 in *The Other 99%. The Conservation and Biodiversity of Invertebrates*, edited by W. Ponder and D. Lunney. Transactions of the Royal Zoological Society of New South Wales, Mosman, NSW.

Cayzer, L. W., Crisp, M. D., and Telford, I. R. H. 1999. *Bursaria* (Pittosporaceae): a Morphometric Analysis and Revision. *Aust. Sys. Bot.* **12**: 117-143.

Clarke, G. M. and Grosse, S. P. 2002. Genetic Analysis of Populations of the Bathurst Copper Butterfly, *Paralucia spinifera*. Unpublished report prepared for the NSW National Parks and Wildlife Service Central Directorate Threatened Species Unit, Hurstville, NSW.

Colgan, D. J. and Ponder, W. F. 2001. Preliminary Assessment of the Genetics of *Placostylus bivaricosus* on Lord Howe Island. Unpublished report prepared for the NSW National Parks and Wildlife Service Northern Directorate Threatened Species Unit, Coffs Harbour, NSW.

Dunn, K. L., Kitching, R. L. and Dexter, E. M. 1994. The Conservation Status of Australian Butterflies. Unpublished report to the Australian National Parks and Wildlife Service, Canberra, ACT.

Eco Logical Australia Pty Ltd. 2003. Habitat Modelling for the Bathurst Copper Butterfly (*Paralucia spinifera*) (Project No. 01-12). Unpublished report prepared for the NSW National Parks and Wildlife Service Central Directorate Threatened Species Unit, Hurstville, NSW.

Greenslade, P. 1999. What entomologists think about listing species for protection: a survey of butterfly specialists in Australia. Pp. 345-49 in *The Other 99%. The Conservation and Biodiversity of Invertebrates*, edited by W. Ponder and D. Lunney. Transactions of the Royal Zoological Society of New South Wales, Mosman, NSW.

Gunning, R. V. 1999. New South Wales *Threatened Species Conservation Act* 1995; the invertebrate experience so far. Pp. 350-53 in *The Other 99%. The Conservation and Biodiversity of Invertebrates*, edited by W. Ponder and D. Lunney. Transactions of the Royal Zoological Society of New South Wales, Mosman, NSW.

Guy, S. 2004. New Caledonia: *bon appetit*. *Qantas: The Australian Way* (Qantas in-flight magazine) **March 2004**: 36-40.

Hedley, C. 1891. The land and freshwater snails of Lord Howe Island. *Records of the Australian Museum* **1**: 134-144.

Hedley, C. 1893. On the relation of the flora and fauna of Australia to those of New Zealand. *Natural Science* **3 (1893)**: 187-191.

Heywood, V. H. and Iriondo, J. M. 2003. Plant conservation: old problems, new perspectives. *Biological Conservation* **113**: 321-335.

Hnatiuk, R. 2003. Classification of Plant Communities. Chapter 14 (pp. 212-229) in *Ecology: An Australian Perspective*, edited by P. Attiwill and B. Wilson. Oxford University Press, South Melbourne, Victoria.

Horwitz, P., Recher, H. and J. Majer. 1999. Putting invertebrates on the agenda: political and bureaucratic challenges. Pp. 398-406 in *The Other 99%. The Conservation and Biodiversity of Invertebrates*, edited by W. Ponder and D. Lunney. Transactions of the Royal Zoological Society of New South Wales, Mosman, NSW.

Hutchings, P. 2004. Invertebrates and threatened species legislation. Pp 88-93 in Threatened Species Legislation: is it just an Act? edited by P. Hutchings, D. Lunney and C. Dickman. Royal Zoological Society of NSW, Mosman, NSW.

Hutchings, P. and Ponder, W. F. (eds) 1999. Workshop: criteria for assessing and conserving threatened invertebrates. Pp. 297-315 in *The Other 99%. The Conservation and Biodiversity of Invertebrates*, edited by W. Ponder and D. Lunney. Transactions of the Royal Zoological Society of New South Wales, Mosman, NSW.

Hutton, I. 1990. *Birds of Lord Howe Island*. Published by the author, Coffs Harbour, NSW.

International Union for the Conservation of Nature Species Survival Commission. 2003. *2003 IUCN Red List of Threatened Species*. IUCN Species Survival Commission, Gland, Switzerland.

Iredale, T. 1944. The land Mollusca of Lord Howe Island. *Australian Zoologist* **10**(3): 299-334.

Kitching, R. L. 1999. Adapting conservation legislation to the idiosyncrasies of the arthropods. Pp. 274-82 in *The Other 99%. The Conservation and Biodiversity of Invertebrates*, edited by W. Ponder and D. Lunney. Transactions of the Royal Zoological Society of New South Wales, Mosman, NSW.

Lunney, D. 1999. Policy lessons from the 1%. Pp. 290-296 in *The Other 99%. The Conservation and Biodiversity of Invertebrates*, edited by W. Ponder and D. Lunney. Transactions of the Royal Zoological Society of New South Wales, Mosman, NSW.

Mawson, P., R. and J. D. Majer. 1999. The Western Australian Threatened Species Scientific Committee: lessons from invertebrates. Pp. 369-73 in *The Other 99%. The Conservation and Biodiversity of Invertebrates*, edited by W. Ponder and D. Lunney. Transactions of the Royal Zoological Society of New South Wales, Mosman, NSW.

Murphy, M. J. 2002a. Mollusc conservation and the New South Wales *Threatened Species Conservation Act* 1995: the recovery program for Mitchell's Rainforest Snail *Thersites mitchellae*. *Australian Zoologist* **32:** 1-11.

Murphy, M. J. 2002b. Observations on the behaviour of the Australian land snail *Hedleyella falconeri* (Gray, 1834) (Pulmonata: Caryodidae) using the spool-and-line tracking technique. *Molluscan Research* **22:** 149-164.

Nally, S. 2003. Community involvement in the conservation of the endangered purple copper butterfly *(Paralucia spinifera* E. D. Edwards & Common, Lepidoptera: Lycaenidae). *Records of the South Australian Museum Monograph Series* **No.7:** 217-224.

New, T. R. 1995. *An Introduction to Invertebrate Conservation Biology*. Oxford University Press, New York, USA. 194 pp.

New, T. R. 1999. Descriptive taxonomy as a facilitating discipline in invertebrate conservation. Pp. 154-158 in *The Other 99%. The Conservation and Biodiversity of Invertebrates*, edited by W. Ponder and D. Lunney. Transactions of the Royal Zoological Society of New South Wales, Mosman, NSW.

NSW National Parks and Wildlife Service. 2001a. *Lord Howe Placostylus* Placostylus bivaricosus *(Gaskoin, 1855) Recovery Plan*. NSW National Parks and Wildlife Service, Hurstville, NSW.

NSW National Parks and Wildlife Service. 2001b. *Mitchell's Rainforest Snail* Thersites mitchellae *(Cox, 1864) Recovery Plan*. NSW National Parks and Wildlife Service, Hurstville, NSW.

NSW National Parks and Wildlife Service. 2001c. *Declaration of Critical Habitat for Mitchell's Rainforest Snail* Thersites mitchellae *in Stotts Island Nature Reserve*. NSW National Parks and Wildlife Service, Hurstville, NSW.

NSW National Parks and Wildlife Service. 2001d. *Bathurst Copper Butterfly* (Paralucia spinifera) *Recovery Plan*. NSW National Parks and Wildlife Service, Hurstville, NSW.

NSW National Parks and Wildlife Service. 2002. Predictive Habitat Modelling of the Lord Howe Placostylus *Placostylus bivaricosus* on Lord Howe Island. Unpublished report prepared by the NSW National Parks and Wildlife Service Northern Directorate Threatened Species Unit, Coffs Harbour, NSW.

NSW National Parks and Wildlife Service. 2003. *Mitch the Rainforest Snail Teacher Resource Book*. NSW National Parks and Wildlife Service, Hurstville, NSW.

NSW Scientific Committee. 1996. Final determination to list *Paralucia spinifera* as an endangered species under Schedule 1, Part 1 of the *Threatened Species Conservation Act* 1995.

NSW Scientific Committee. 1997a. Final determination to list *Placostylus bivaricosus* as an endangered species on Schedule 1, Part 1 of the *Threatened Species Conservation Act* 1995.

NSW Scientific Committee. 1997b. Final determination to list *Thersites mitchellae* as an endangered species on Schedule 1, Part 1 of the *Threatened Species Conservation Act* 1995.

NSW Scientific Committee. 2000. Final determination to list predation from the Ship Rat *Rattus rattus* on Lord Howe Island as a key threatening process on Schedule 3 of the *Threatened Species Conservation Act* 1995.

NSW Scientific Committee. 2001a. Final determination to list clearing of native vegetation as a key threatening process on Schedule 3 of the *Threatened Species Conservation Act* 1995.

NSW Scientific Committee. 2001b. Final determination to list loss and/or degradation of sites used for hill-topping by butterflies as a key threatening process on Schedule 3 of the *Threatened Species Conservation Act* 1995.

NSW Scientific Committee. 2004. Final determination to list the Lord Howe Island wood-feeding cockroach *Panesthia lata* as an endangered species on Schedule 1, Part 1 of the *Threatened Species Conservation Act* 1995.

Parrish, R., Sherley, G. and Aviss, M. 1995. *Giant Land Snail Recovery Plan* Placostylus *spp.,* Paryphanta *sp. Threatened Species Recovery Plan Series No. 13*. Threatened Species Unit, Department of Conservation, New Zealand.

Pickard, J. 1983. Vegetation of Lord Howe Island. *Cunninghamia* **1(2):** 133-266.

Ponder, W. F. 1997. Conservation status, threats and habitat requirements of Australian terrestrial and freshwater Mollusca. *Memoirs of the Museum of Victoria* **56(2):** 421-430.

Ponder, W. F. 2000. Understanding molluscan diversity in our region – past progress and future directions. *Molluscs 2000 Symposium Abstracts*. Pp. 69-71. Malacological Society of Australasia, Sydney, NSW.

Ponder, W. F. and Chapman, R. 1999. Survey of the Land Snail *Placostylus bivaricosus* on Lord Howe Island. Unpublished report prepared for the NSW National Parks and Wildlife Service Northern Directorate Threatened Species Unit, Coffs Harbour, NSW.

Ponder, W. F., Colgan, D. J., Gleeson, D. M. and Sherley, G. 2003. Relationships of *Placostylus* from Lord Howe Island: an investigation using the mitochondrial cytochrome c oxidase 1 gene. *Molluscan Research* **23(2):** 159-178.

Salas, M., Bonnaut, C., Le Bel, S. and Chardonnet, L. 1997. Activity and food intake of captive *Placostylus fibratus* (Gastropoda: Bulimulidae) in New Caledonia. *New Zealand Journal of Zoology* **24:** 257-264.

Sands, D. P. A. 1999. Conservation status of Lepidoptera: assessment, threatening processes and recovery actions. Pp. 382-87 in *The Other 99%. The Conservation and Biodiversity of Invertebrates*, edited by W. Ponder and D. Lunney. Transactions of the Royal Zoological Society of New South Wales, Mosman, NSW.

Sands, D. P. A. and New, T. R. 2002. *The Action Plan for Australian Butterflies*. Environment Australia, Canberra ACT.

Sherley, G. 1994. Translocations of the Mahoenui Giant Weta *Deinacrida* n. sp. and *Placostylus* land snails in New Zealand: what have we learnt? Chapter 11 (pp. 57-64) in *Reintroduction Biology of Australian and New Zealand Fauna*, edited by M. Serena. Surrey Beatty & Sons, Sydney, NSW.

Sherley, G., Stringer, I. A. N., Parrish, G. R. and Flux, I. 1998. Demography of two land snail populations (*Placostylus ambagiosus*, Pulmonata, Bulimulidae) in relation to predator control in the far north of New Zealand. *Biological Conservation* **84:** 83-88.

Smith, B. J. 1992. *Non-marine Mollusca*. Zoological Catalogue of Australia. Volume 8. CSIRO Publishing/ABRS: Melbourne Victoria.

Smithers, C. McAlpine, D., Colman, P. and Gray, M. 1977. *Island Invertebrates*. Pp. 23-26 in *Lord Howe Island*. Published by the Australian Museum, Sydney, NSW.

Stanisic, J. 1994. The distribution and patterns of species diversity of land snails in eastern Australia. *Memoirs of the Queensland Museum* **36(1):** 207-214.

Stanisic, J. 1997. An area of exceptional land snail diversity: the Macleay Valley, north-eastern New South Wales. *Memoirs of the Museum of Victoria* **56(2):** 441-448.

Stanisic, J. 1998. Survey for Land Snail *Thersites mitchellae* in Northern New South Wales. Unpublished report prepared for the NSW National Parks and Wildlife Service Northern Directorate Threatened Species Unit, Coffs Harbour, NSW.

Stanisic, J. 1999. Survey for Land Snail *Thersites mitchellae* in Northern New South Wales II. Investigation of Wilsons Creek locality. Unpublished report prepared for the NSW National Parks and Wildlife Service Northern Directorate Threatened Species Unit, Coffs Harbour, NSW.

Stanisic, J. 2000. Survey for Land Snail *Thersites mitchellae* in Northern New South Wales III. Investigation of Stotts Island locality. Unpublished report prepared for the NSW National Parks and Wildlife Service Northern Directorate Threatened Species Unit, Coffs Harbour, NSW.

Stork, N. E. 1999. Estimating the number of species on Earth. Pp. 1-7 in *The Other 99%. The Conservation and Biodiversity of Invertebrates*, edited by W. Ponder and D. Lunney. Transactions of the Royal Zoological Society of New South Wales, Mosman, NSW.

Sutherland, L. and Ritchie, A. 1977. Defunct volcanoes and extinct horned turtles. Pages 7-12 in *Lord Howe Island*. Published by the Australian Museum, Sydney, NSW.

Van Praagh, B. D. 1997. Conservation of native earthworms and the role of the Giant Gippsland earthworm as a flagship taxon. *Memoirs of the Museum of Victoria* **56**: 597-603.

Yen, A. L., New, T. R., Van Praagh, B. D. and Vaughan, P. J. 1990. Invertebrate conservation: three case studies in south-eastern Australia. Pp. 207-224 in *Management and Conservation of Small Populations*, edited by T. W. Clark and J. H. Seebeck. Chicago Zoological Society, Illinois, USA.

Yen, A. L. and Butcher, R. J. 1997. *An Overview of the Conservation of Non-marine Invertebrates in Australia*. Endangered Species Program, Environment Australia, Canberra, ACT.

The rivers to the sea: experiences with two endangered aquatic fish species

Bill Talbot[1], Sharon Molloy[1,2], Rebecca Chapman[3] and Marcus Riches[4]

[1] NSW DPI - Fisheries Ecosystems Group, Port Stephens Fisheries Centre Private Bag 1, NELSON BAY NSW 2315. Bill.Talbot@fisheries.nsw.gov.au

[2] NSW DPI - Fisheries Ecosystems Group – Habitat Rehabilitation, Port Stephens Fisheries Centre Private Bag 1, NELSON BAY NSW 2315.

[3] NSW DPI - Animal and Plant Biosecurity Branch, PO Box 21, CRONULLA NSW 2230.

[4] NSW DPI - Fisheries Ecosystems Group – Habitat Protection, 5 Regatta Avenue, BALLINA NSW 2478.

ABSTRACT

There are currently 20 species, populations or communities of fish, aquatic invertebrates and marine vegetation listed on the threatened species schedules of the NSW *Fisheries Management Act 1994*. Two of these species are the eastern freshwater cod *Maccullochella ikei,* and the grey nurse shark *Carcharias taurus*. In the 1970s, well over a decade prior to their listing as threatened, there were concerns over the population status of both fish, and various actions were initiated to help preserve and maintain stocks. For example, both species were totally protected from fishing in 1984, captive breeding techniques for eastern cod were investigated and underwater surveys of grey nurse shark populations commenced. Following listing, however, there has been a significant increase in emphasis on actions to help the recovery of both species, protect their habitat and increase knowledge of their biology and distribution. In the case of eastern cod, these actions have included surveys to document the distribution of remnant and stocked populations, tagging studies, genetic and ecological research, habitat restoration works and significant input into water sharing rules for relevant catchments. For grey nurse sharks, critical habitat areas have been declared at key aggregation sites, controls placed on fishing in and around these areas, additional diver surveys completed, tagging and tracking studies commenced to provide detailed information on movements and improved population estimates, and samples collected for genetic research. The changes in management strategies resulting from the listing of these threatened species in NSW may also have flow-on benefits for other aquatic species.

Key words: endangered, fish, grey nurse shark, eastern freshwater cod.

Introduction

In New South Wales, the primary responsibility for the conservation of biodiversity is shared between two agencies: the Department of Environment and Conservation (DEC) (formerly the NSW National Parks and Wildlife Service) through its administration of the *Threatened Species Conservation Act 1995*, and the Department of Primary Industries (formerly NSW Fisheries) through its administration of the *Fisheries Management Act 1994*. DEC is responsible for the conservation of four vertebrate groups (mammals, reptiles, birds and amphibians), and also for terrestrial invertebrates and freshwater plants. The Department of Primary Industries is responsible for the conservation of all 'fish', which includes vertebrate fish and all aquatic invertebrates (both freshwater and marine), and 'marine vegetation' which includes mangroves, seagrasses and macroalgae.

Historically, the main focus of NSW Fisheries had been to regulate fishing and fish farming industries. In the late 1970s through to the early 1990s, amendments to the then *Fisheries and Oyster Farms Act 1935* began to broaden the department's role to include the total protection of certain species of fish from being caught and killed, the declaration of aquatic reserves, controls on dredging and reclamation as well as a range of other measures aimed at the protection of fish habitat. This trend continued with the introduction of the new *Fisheries Management Act* in 1994, which included specific objectives relating to ecologically sustainable development and the conservation of biological diversity.

In 1998, threatened species provisions paralleling those of the *Threatened Species Conservation Act* were introduced into the *Fisheries Management Act*. These provided a comprehensive framework for the protection and recovery of threatened species, populations and communities of fish, aquatic invertebrates and marine vegetation, allowing for the development and implementation of recovery and threat abatement plans and the identification of critical habitat for endangered species. Consequential amendments to the *Environmental Planning and Assessment Act 1979* required specific consideration of the impacts of developments and activities on threatened species, populations or ecological communities during environmental planning and assessment.

Currently 7 endangered species, 2 endangered populations, 2 endangered ecological communities, and 8 vulnerable species are listed under the *Fisheries Management Act*. These range from large, conspicuous 'icon' species such as the eastern

Pp 125 - 134 in *Threatened species legislation: is it just an Act?* 2004, edited by Pat Hutchings, Daniel Lunney and Chris Dickman. Royal Zoological Society of New South Wales, Mosman, NSW, Australia.

freshwater cod *Maccullochella ikei* and grey nurse shark *Carcharias taurus* to smaller, less well known species such as the southern pygmy perch *Nannoperca australis* and Buchanans fairy shrimp *Branchinella buchananensis*. In addition, one species (a seaweed) is listed as 'presumed extinct'.

The threatened species approach to biodiversity conservation has been criticised on a number of counts. For example, it has been criticised on the basis that it requires relatively good species-level knowledge, is resource-intensive and costly, reactive rather than preventative, and tends to be biased towards highly visible, 'appealing' species (e.g. New 1995; Reed and Clunie 1997; Yen and Butcher 1997; Woinarski and Fisher 1999; Ponder *et al.* 2002). This has led some critics to call for a broader approach which operates at the level of communities, ecosystems, or threatening processes (e.g. Franklin 1992; Bowen 1997; Reed and Clunie 1997; Tear *et al.* 1995; Walker 1992). However, as experience with some of the listed fish species in NSW has demonstrated, threatened species listings can provide a powerful tool in leveraging environmental outcomes on complex issues. Outcomes include raising community awareness of aquatic biodiversity and conservation, gaining support for additional research, and delivering specific actions, such as habitat protection, which have significant flow-on benefits for associated species. Within NSW Fisheries, the threatened species legislation has also played a major role in strengthening the department's commitment to biodiversity conservation, with staff from a range of programs (including environmental assessment, habitat protection and rehabilitation, fisheries management, conservation research, and compliance) now playing key roles in threatened species-related work.

This paper summarises the conservation history of two species that have been a major focus of the NSW Fisheries threatened species program to date; the eastern cod and grey nurse shark. Both were listed in 1984 as totally protected under section 19 of the *NSW Fisheries and Oyster Farms Act*, and subsequently as threatened species under the *Fisheries Management Act*. Some background information on each species is given, together with a summary of changes in the management approach as a result of listing, focusing on key issues (e.g. the influence of the eastern cod's requirements in the development of water sharing plans, the declaration of critical habitat for grey nurse sharks and the significant additional research undertaken to address major information gaps). The implications of these actions for broader biodiversity conservation are also discussed.

Case study 1. Eastern cod

Background information

The eastern cod (Figure 1) is a large, predatory, freshwater fish native to only the Clarence and Richmond Rivers in northern New South Wales. It is closely related (and similar in appearance) to the Murray cod *Maccullochella peelii peelii* and the trout cod *Maccullochella macquariensis*, both from the inland Murray-Darling River system, and the Mary River cod *Maccullochella peelii mariensis* from the coastal Mary River system in southern Queensland.

Eastern cod have been recorded up to 41 kg and over 1 metre in length, although such large fish are rare with most individuals normally less than 5 kg and 70 cm in length (Rowland 1996). They are often found in clear, flowing streams with rocky beds and deep holes, generally in areas with plenty of instream cover such as boulders or large woody debris (snags) (Rowland 1993, 1996; Butler 2000; Figure 2).

Eastern cod were once abundant within the Clarence and Richmond Rivers downstream of tablelands waterfalls (Figure 3), and were an important food fish for early settlers in the region. Populations of the species apparently collapsed in the 1920s and 1930s (exacerbated by a series of major fish kills in the late 1930s), and they continued to decline until the 1980s (Rowland 1993, NSW Fisheries 1999). Development in the catchment for such activities as farming, urban development and mining has had significant effects on the Richmond and Clarence river systems, and the decline of eastern cod is probably attributable to a combination of over-fishing and habitat modification (including sedimentation, snag removal, flow modification, water pollution and barriers to migration).

Figure 1. The eastern freshwater cod *Maccullochella ikei.* Photo: NSW Fisheries

Figure 2. Habitat of eastern cod in the Nymboida River, north-eastern NSW. Photo: NSW Fisheries

Eastern cod was first formally described as a separate species in 1985, after a study on Murray cod identified (through electrophoretic and morphometric data and a cross breeding experiment) that cod from north-eastern NSW were a separate species (Rowland 1985).

This work led to the species being protected from harvesting in 1984, under section 19 of the *NSW Fisheries and Oyster Farms Act*. By this time they were already considered to be extinct in the Richmond River system, very rare or absent in the major northern tributaries of the Clarence River catchment, absent from the Orara River, and present only in small numbers in a few other southern tributaries of the Clarence River (principally the Mann and Nymboida Rivers; see Figure 3), where some pristine habitat still remained (Rowland 1993). By the early 1990s, anecdotal reports from anglers suggested

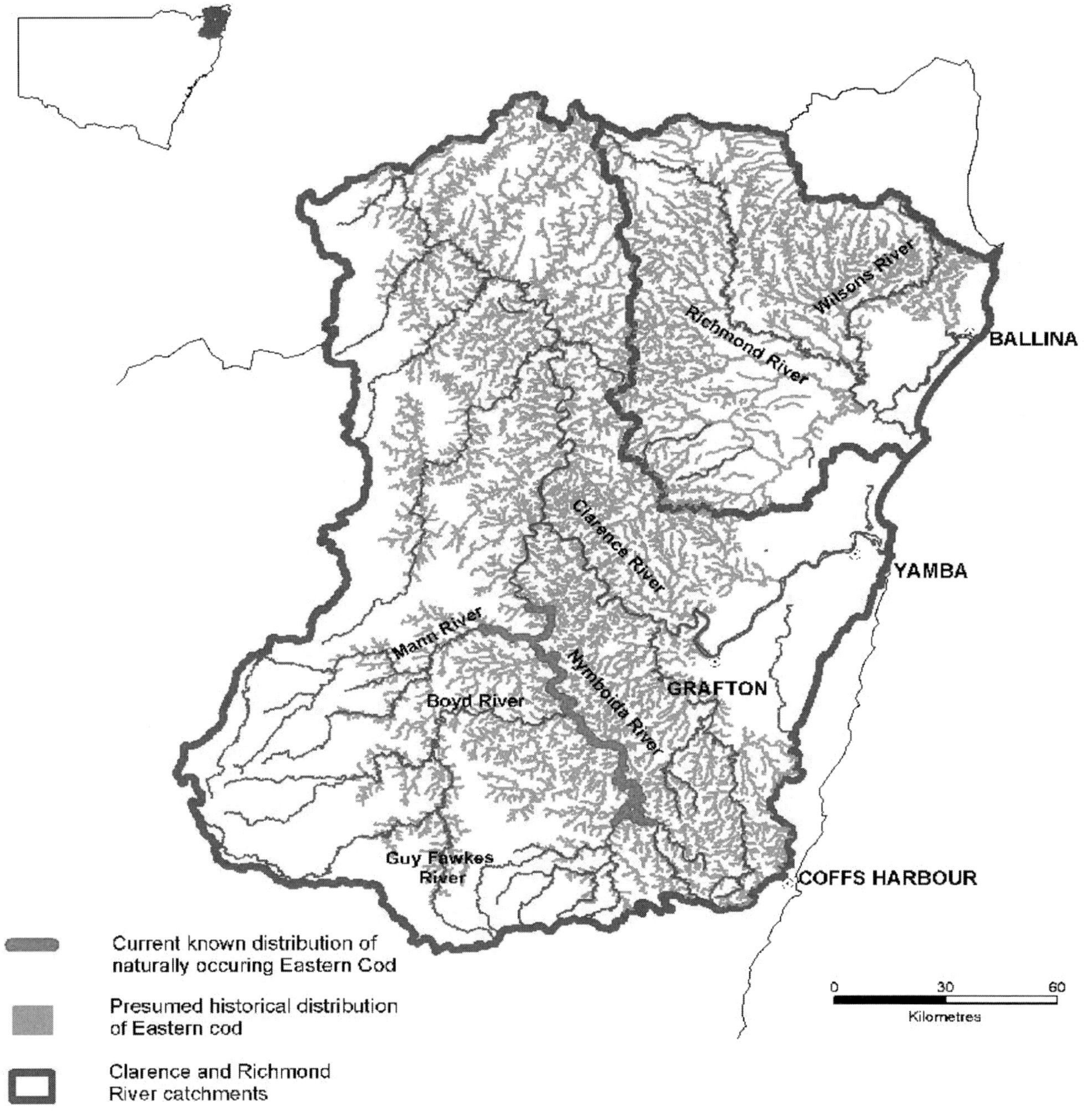

Figure 3. Historical distribution of eastern cod and suspected range of remnant wild population in north-eastern NSW

that their numbers had increased to the point where they were being caught more regularly in a wider area of the Clarence system, but there was little scientific evidence to confirm this. More rigorous surveys suggested they remained highly restricted in distribution.

The eastern cod was listed in 1992 as an endangered species under the Commonwealth *Endangered Species Protection Act 1992*. With the introduction of the threatened species provisions of the *Fisheries Management Act* in 1998, eastern cod was placed directly on Schedule 4 of the Act (endangered species), together with two other freshwater fish species recognised as being of significant conservation concern (trout cod *Maccullochella macquariensis* and Oxleyan pygmy perch *Nannoperca oxleyana*). Eastern cod was also listed as critically endangered on the World Conservation Union (IUCN) 'Red List' of threatened species in 1996 (Baillie and Groombridge 1996).

Management activities prior to listing in NSW

As eastern cod is a large, identifiable fish and a popular angling species, its decline was clearly recognised long before the introduction of the NSW threatened species legislation. Concern over its status led to the initiation of some conservation actions by government agencies and community groups, most notably its total protection from fishing in 1984 and the development of a captive breeding and restocking program shortly afterwards.

Captive breeding techniques for eastern cod were developed by NSW Fisheries between 1988 and 1990 at the Grafton Research Centre, and approximately 30,000 fingerlings were stocked into various locations in the Clarence and Richmond river catchments (including impoundments) at that time. However, no large-scale surveys were carried out to determine the distribution of eastern cod or the survival of the stocked fish.

In 1994 NSW Fisheries issued a permit to a private hatchery at Dorrigo, Booma Fisheries, to undertake further work on the breeding and stocking of eastern cod. This was on the proviso that strict genetic protocols were followed (to prevent loss of fitness through inbreeding depression and maximize available genetic variation in the recovering population) and hatchery-reared fish were used for conservation purposes only. Booma Fisheries stocked 660 fingerlings in 1996 and 30,000 in 1997.

In 1997 a non-profit community group, Project Big Fish, began promoting the conservation of eastern cod with a major focus on raising funds for stocking. NSW Fisheries and Project Big Fish also worked to raise public awareness of the threats to eastern cod through a wide range of activities including signage, media releases and interviews, magazine articles, brochures, newsletters, field days, presentations to schools and community groups and information on the internet.

Some population survey work was also conducted (Faragher *et al.* 1993). However, there was still very little information on many aspects of the species' biology, or even its geographical distribution.

Management activities following listing in NSW

Recovery planning for eastern cod commenced shortly after it was listed as endangered in 1998. In 1999 NSW Fisheries received funding through a threatened species grant under the Natural Heritage Trust to develop the NSW and Commonwealth recovery plan and conduct survey and monitoring work. A recovery planning team was established, and the draft recovery plan was publicly exhibited in November 1999 and finalised in May 2004 (NSW Fisheries).

The draft recovery plan provided for the continuation and expansion of activities that had been initiated prior to listing, such as the captive breeding and restocking program and community education activities. However, it also emphasised the importance of further research to gain a greater understanding of the size, distribution, ecological requirements, historical and existing genetic status of the eastern cod population and the threats to its survival. The plan also recognised the need to initiate management actions to reduce identified threats such as impacts from development and natural resource use on water quality, water flows, and habitat quality (NSW Fisheries 1999).

Research

A threatened species grant obtained under the Natural Heritage Trust enabled NSW Fisheries to commence a series of surveys (still ongoing) at a range of sites in the Clarence and Richmond catchments. The aim of this survey program has been to document the distribution of remnant and stocked populations, estimate their relative abundance, examine population structure, survival and recruitment, assess habitat preferences and develop methods to monitor the status and recovery of eastern cod over the long term (Figure 4). Captured eastern cod

Figure 4. Electrofishing for eastern cod in the Clarence River. Electrofishing works by generating an electric field in the water through a cathode and anode connected to a control box which is mounted on a boat, in a backpack unit, or on the bank. Fish in the vicinity are initially drawn towards the anode, but are then immobilized, allowing them to be collected from the surface. The advantages of electrofishing are that it can be done relatively quickly, used in areas which are difficult to access with other gear types, and the fish recover within a few minutes of collection allowing them to be returned unharmed to the water. Photo: NSW Fisheries

have been fitted with Passive Integrated Transponder (PIT) tags before release (Figure 5), enabling fish recaptured during subsequent surveys or by Project Big Fish volunteers to be 'scanned' to obtain valuable data on movements, growth and abundance. To date 340 cod have been tagged, with preliminary results suggesting strong site associations of fish to specific river reaches or pools and highly variable growth rates (Pollard and Wooden 2002, I. Wooden pers. comm. 2004). Tissue samples (fin clips) have been collected from captured fish to assist in genetic research by Southern Cross University. Some angling has also been conducted, with the assistance of Project Big Fish, to extend the distributional data and genetic sample collection. A PhD student is currently investigating the ecology of eastern cod, and has obtained significant new information on its reproductive biology and behaviour, including video footage of individuals guarding the eggs (G. Butler, Southern Cross University, pers. comm. 2004).

Figure 5. Inserting a Passive Integrated Transponder (PIT) tag into a captured eastern cod. Photo: NSW Fisheries

Additional funding for research and recovery actions has been received from other sources including the Recreational Fishing (Freshwater) Trust Fund and water supply authorities, the latter as a form of environmental offset against potential impacts from high-priority works. In 2003, NSW Fisheries also received funding from the NSW Environmental Trust to undertake habitat restoration works (such as revegetation, improvements to fish passage, and fencing) along a 'demonstration reach' in the Richmond River catchment, and to monitor any changes in eastern cod populations resulting from the habitat restoration work.

Water reforms process

The Clarence and Richmond River systems are largely unregulated, but natural flow patterns have been altered by increasing levels of water extraction for a variety of uses (eg. irrigation, town water supply, stock watering and power generation). The draft recovery plan (NSW Fisheries 1999) recognised that although little information existed on the amount of water extracted or its impact on river health, alteration of natural flows was likely to be a major issue in the management of eastern cod populations. Extraction of water can affect habitat connectivity and quality, both key requirements for the survival of eastern cod. Alteration of flows, particularly during critical lifecycle phases (e.g. pairing and site selection), has been hypothesised to result in reduced fecundity. Furthermore, habitat changes such as reduced water quality, sedimentation of riffle zones and pools and reduced diversity of macrophytes can not only affect eastern cod directly but may also have negative impacts on prey species and general river health and resilience.

In 2000, a new *Water Management Act* was introduced in NSW to provide for the sustainable management of the State's water resources. Local water management committees were established for a number of priority areas throughout the State, with the task of negotiating agreement on water sharing rules as the basis for 10-year water sharing plans.

Protection of the environment is foremost in the objects and principles of the Act, which states that sharing of water from a water source must protect that water source and its dependent ecosystems, and that water sharing plans must establish environmental water rules. However, the precise nature of these rules (e.g. the level at which users must cease to extract water to protect low flows) has been the subject of intense negotiations among the various representatives on the committees, with decisions having major social and economic as well as environmental implications.

A key action in the draft recovery plan for eastern cod was to ensure that the flow requirements of eastern cod were supported during the water reforms process. Given the lack of specific, quantitative information on the biology or habitat needs of eastern cod, NSW Fisheries organised an expert panel of managers, researchers and external experts to define the minimum flow requirements at various times of the year for recruitment, spawning and juvenile dispersal. The panel used expert opinion and available information on the size, timing of reproduction and habitat preferences of eastern cod to recommend:

- a minimum water depth of 30 cm over riffles and a flow velocity not greater than 1.5 m/s (to allow movement of eastern cod between pools, particularly from late June to March during spawning, larval growth and recruitment phases);
- the maintenance of stable water depth of at least 1.5 m in pools (particularly during August-October to protect nest sites and habitat for spawning adults); and
- limiting extraction to mimic medium to high flow events (NSW Fisheries 2000).

Following this process, agreement was reached with the (then) Department of Land and Water Conservation (now Department of Infrastructure, Planning and Natural Resources) to undertake flow modelling to estimate the percentile flows and extraction rules that would correlate with the recommended water depths over riffles and in pools.

NSW Fisheries provided advice to two water management committees (WMCs) dealing with three subcatchments within the range of eastern cod; the Northern Rivers WMC (for Coopers Creek, in the Richmond catchment) and the Upper North Coast WMC (for the Dorrigo Plateau and Orara River, in the Clarence catchment). Although eastern cod are present in all three sub-catchments, the environmental characteristics, current status of eastern cod, and socio-economic issues differ markedly, and this is reflected in the different flow rules that were achieved in each case (Table 1). For Coopers Creek and Dorrigo Plateau, for example, the degraded state of habitat and the existence of significant socio-economic issues meant that full conformity with the eastern cod flow requirements was not achievable.

The Orara River, by contrast, contains some high-quality areas of remaining habitat and had the potential to contribute significantly to the recovery of eastern cod. The WMC was first provided with information on the river system and the linkages between eastern cod and other elements of the ecosystem (e.g. invertebrate food resources), as many members of the committee had little background in, or understanding of, riverine ecology. Subsequently, intensive field work was conducted (in collaboration with staff from other agencies) to validate the earlier flow modelling work and determine the specific river flows needed to meet the requirements of eastern cod. This provided information in a form consistent with flow modelling programs, which was readily understood by the committee. This field-validated information was used in subsequent negotiations when the environmental benefits of improved flows had to be weighed against the potential social and economic impacts.

The eventual outcome was a unanimous decision by the committee to support a set of water sharing rules which included progressive increases in the 'cease to pump' (CTP) threshold over the life of the water sharing plan, the establishment of a 'cod hump' (higher CTP threshold during the critical months when eastern cod migrate) for the first 5 years of the plan, and the establishment of rules for various flow classes. These rules, when implemented, will provide significant improvements in low flow protection, will better mimic natural flow variations, and will benefit not only eastern cod but also the other aquatic species and the broader ecosystem of the Orara.

Monitoring of a range of hydrological and ecological features in the Orara River will be ongoing as part of the NSW water reform process and the Department of Primary Industries eastern cod research program. Performance indicators and associated adaptive management responses will provide a basis for the long-term management of eastern cod populations.

Case study 2. Grey nurse shark

Background information

The grey nurse shark (Figure 6) is a large species of shark native to subtropical and cool temperate waters in the Mediterranean Sea and the Atlantic, Indian and Pacific Oceans. In Australia, grey nurse sharks were historically found around much of the coastline including southern Queensland, New South Wales, Victoria, South Australia and southern Western Australia. The species is now restricted to two separate populations, one off NSW and southern Queensland, and the other off Western Australia (Otway and Parker 2000; Otway *et al.* 2003).

Table 1. Comparison of flow rules contained in water sharing plans for the Orara River, Coopers Creek and Dorrigo Creek.

Water sharing plan	Cease to pump	Do flow rules achieve recommendations of recovery planning?
Orara River	80%ile^ (yr 6-10, #critical month all zones)	Yes (require structural adjustment for change in rules in years 6-10)
Coopers Creek	14-31*-14 ML/d (97-95*-97%ile) (#critical month)	Partial (significant socio-economic impact) (*cod hump – July to September migration period)
Dorrigo Plateau	Zone 1: 75ML/d (98%ile) Zone 2: 23ML/d (99%ile) Zone 3: 17ML/d (97%ile) Zone 4: 13ML/d (98%ile)	No (Significant socio- economic impact, limited habitat availability, possibly outside of natural range).

*Cod hump = jump in Cease To Pump during migration periods to provide for greater depth over riffles (ie. allows for greater fish passage)

^ percentile = % of time that this flow is exceeded

#critical month = the one month in the year when irrigation demand for water is most close to exceeding, or exceeds, the available flows.

Sources: Water Sharing Plan for the Orara River (draft), Water Sharing Plan for the Coopers Creek Water Source 2003 (Gazetted 14 February 2003), Water Sharing Plan for the Dorrigo Plateau Surface Water Source and the Dorrigo Basalt Groundwater Source 2003 (Gazetted 21 February 2003).

Figure 6. The grey nurse shark *Carcharias taurus.* Photo: David Harasti

Grey nurse sharks are regularly seen by divers in a relatively small number of areas along the NSW coast and in southern Queensland, usually in or near deep, sandy-bottomed gutters or caves around rocky reefs or islands fairly close to shore. Despite their fierce appearance, grey nurse sharks are not considered dangerous and generally pose no threat to divers or swimmers unless provoked (Pollard *et al.* 1996).

The life-history and reproductive strategies of the grey nurse shark make it particularly vulnerable to human-induced pressures. Their practice of aggregating at inshore rocky reef sites of the type that are often subject to heavy fishing pressure makes them vulnerable to incidental capture or injury. The grey nurse shark is one of the least fecund of all sharks with females producing an average of one pup per year. This low reproductive rate and their apparently slow growth rate mean that populations, once reduced, will take a long time to recover (Otway *et al.* 2003).

Grey nurse shark populations were reportedly the target of commercial fishing activities in a number of areas in NSW from the mid 1800s up until the Second World War. Between the 1950s and the 1980s grey nurse shark numbers showed a serious decline in NSW waters, largely attributed to the impact of commercial fishing and recreational spear and game fishing, as well as shark control activities such as beach meshing (NSW Fisheries 2002). In October 1997, the grey nurse shark was listed as a threatened species under the *Commonwealth Endangered Species Protection Act 1992.* In NSW it was listed as "vulnerable" under the *Fisheries Management Act* in May 1999 and subsequently upgraded to "endangered" in August 2000. Current threats to grey nurse sharks include death and injury resulting from incidental capture by fishers, as well as illegal fishing activities such as shark finning (i.e. where the fins are cut off and retained for sale and the carcass dumped at sea) (NSW Fisheries 2002).

Management activities prior to listing in NSW

During the 1950s and 1960s grey nurse sharks were falsely viewed as maneaters, and many were deliberately killed during highly publicised operations by spearfishers and scuba divers using powerheads (Ireland 1984). In addition, from 1961 to 1980 over 400 grey nurse sharks were recorded as being caught by game fishing clubs in NSW (Pepperell 1992).

A decline in the proportion of grey nurse sharks caught by gamefishers was clearly evident over the period from 1960 to the 1970s (Environment Australia 1997). Shark meshing captures also showed a significant decline during this period (Figure 7). Gamefishers voluntarily banned the capture of grey nurse sharks in 1979 (Marsh 1995). In the late 1970s and early 1980s a number of underwater film makers, scuba divers and spearfishers began calling for protection of the shark. Following consultation with peak recreational fishing, commercial fishing and scuba diving groups, the species was totally protected from being captured and killed in November 1984 (Pollard *et al.* 1996).

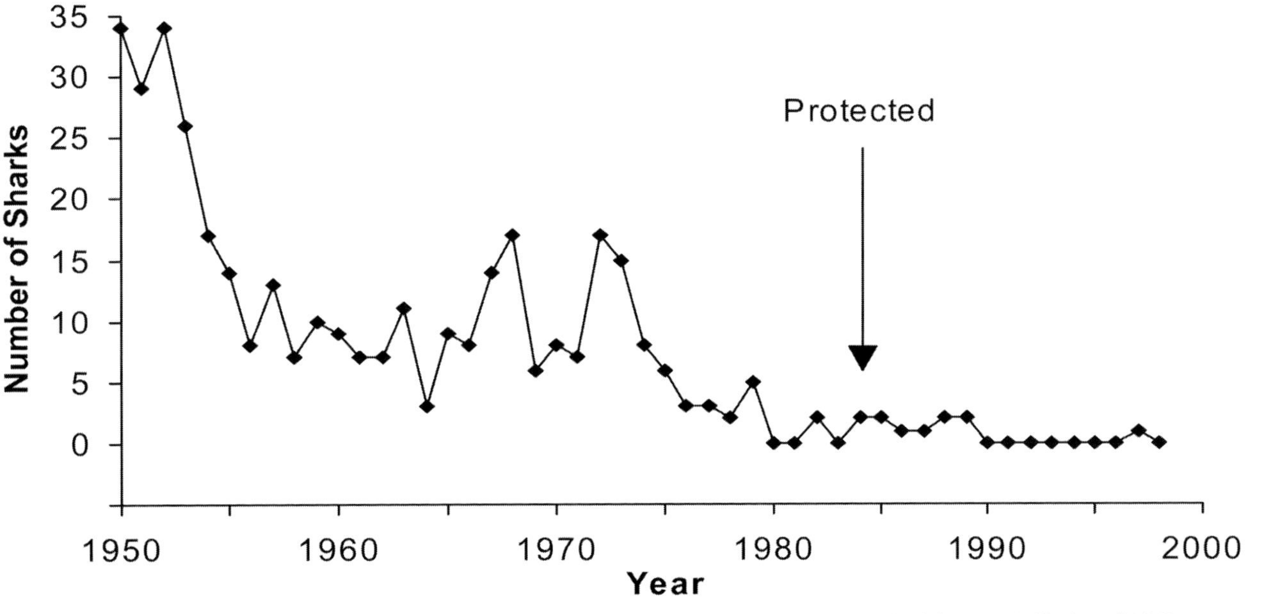

Figure 7. Decline in the numbers of grey nurse sharks caught in shark meshing nets in the Newcastle/Sydney/Wollongong regions from 1950-1998 (from: Otway and Parker 2000)

Research

Prior to the grey nurse shark being listed as threatened in NSW, much of the scientific information about the species' biology (e.g. behaviour, reproduction, age and growth, migration patterns, diet and feeding habits) had been derived from studies in the USA and South Africa.

It was not clear whether this information could be applied in the Australian context. In particular, spatial and temporal patterns in the abundance of grey nurse sharks in south-eastern Australia were not well understood. The data that did exist came primarily from records of captures in the NSW beach meshing program, gamefishers' log books and three diver surveys. These surveys provided some information on local abundance and movement, but they involved studies on a small spatial scale over relatively short periods of time (Otway and Parker 2000).

Although grey nurse sharks were listed as threatened in NSW in 1999, the species had already received national listing two years earlier. Consequently, in 1998 NSW Fisheries was able to begin a large-scale study funded by a threatened species grant under the Commonwealth's Natural Heritage Trust program. This study involved replicate diver surveys of all known grey nurse shark aggregation sites along the entire NSW coast, the mapping of the key sites from south of Sydney to the Brisbane area and an investigation of the potential to declare the important sites as marine protected areas to facilitate the recovery of the species.

Three NSW coastwide surveys were carried out in November/December 1998, March/April 1999 and June/July 1999. The maximum number of grey nurse sharks observed in any one survey was 207. These results indicated that the total number of grey nurse sharks was very low and suggested that the population in NSW had not recovered since the species was first protected in 1984 (Otway and Parker 2000).

Management activities following listing in NSW

Following publication of the report by Otway and Parker (2000) the conservation status of grey nurse sharks was re-assessed and the listing upgraded from vulnerable to endangered in August 2000. NSW Fisheries released a draft grey nurse shark recovery plan for community comment in May 2002 (NSW Fisheries 2002). The draft plan detailed actions required to promote the recovery of the species in NSW waters including a recommendation to declare all known major grey nurse shark aggregation sites as critical habitat areas.

In December 2002, 10 sites were declared as critical habitat and new arrangements were put in place to protect the grey nurse shark. Within critical habitat areas and surrounding 800 metre buffer zones, controls were placed on those forms of fishing thought to pose a significant risk of incidental capture and on diving activities likely to cause negative impacts on the sharks. For example, bans were placed on commercial set-line fishing and recreational demersal (bottom) fishing using bait from anchored boats. Bans were also placed on diving between sunset and sunrise, chasing or harassing the sharks and the use and/or wearing of electronic shark repelling devices.

Within six months of the implementation of the new arrangements, the Minister for Fisheries called for a review of grey nurse shark protection. This move was prompted by a number of factors - divers observed that the number of incidentally hooked grey nurse sharks had remained high; radio tracking studies showed sharks roamed further from aggregation areas at night than had been demonstrated in earlier work, and well outside the critical habitat areas; preliminary research results (detailed below) provided a population estimate with numbers lower than previously thought and demographic modeling indicated that the east coast population could become extinct within a few decades if fishing mortality continued at the estimated current level.

Research

The survey report of Otway and Parker (2000) provided valuable information about grey nurse sharks at key aggregation sites including their distribution, habitat use and abundance relative to historical records. This work did not, however, provide some of the key information needed to determine the true conservation status of the species or to support decision-making on the management of the grey nurse shark population, such as the total population size or key demographic parameters.

The diver surveys developed by Otway and Parker (2000) continued for a further two years. During this time the total number of grey nurse sharks observed did not exceed 292 (Otway *et al.* 2003). In 2002 a tagging study was incorporated into the diver survey program using "roto" tags (similar to cattle ear tags) (Figure 7). This work began to provide information that was highly significant for the management of the recovery program. Standard mark-recapture techniques were used to provide the estimate that there were less than 500 individuals in the total population (Otway and Burke 2004). This work also demonstrated the considerable length of time that particular sharks spent at aggregation sites. In addition, by observing the number of grey nurse sharks that suffered imbedded hooks and hook injuries since

Figure 8. Measuring a grey nurse shark during tagging. Photo: NSW Fisheries

tagging, an estimate could be made of the proportion of the population that experienced incidental capture or interactions with fishing gear (Otway *et al.* 2003). Tissue samples collected during the tagging process also allowed the genetic assessment of grey nurse shark populations (Harcourt *et al.* 2003).

More recently, research using a range of "smart tags" (i.e. acoustic tracking tags, listening stations and "pop up" archival satellite tags) has been initiated to answer questions on the localized movement of sharks around aggregation sites and to determine movement patterns when they are away from these sites. An electronic tracking project found that grey nurse sharks display diurnal movements within a radius of 1200 m from a number of aggregation sites (B. Bruce, CSIRO, pers. comm.). In a project funded by Sea World and NSW Fisheries, "pop up" archival tags (i.e. tags that automatically release from the shark, float to the surface and transmit data via satellite) were fitted to three grey nurse sharks in late 2003. These tags collect data on depth, water temperature and approximate migration path for the tagged sharks. This work will provide new and detailed information on grey nurse shark movements. Preliminary results from this work strongly support current scientific opinion that the sharks are a highly inshore and coastal-dwelling species. The results do not support the hypothesis of some fishers that there is a separate, un-surveyed population in deeper waters and that the sharks occupy critical habitat sites that lie beyond normal scuba diving depths (N. Otway, Department of Primary Industries, pers. comm.).

Discussion

The listing of threatened species under the NSW *Fisheries Management Act 1994* and subsequent recovery planning programs, while primarily aimed at the recovery of the species concerned, can also make a significant contribution to the protection and enhancement of broader biodiversity and ecosystem processes. Work to protect and restore the habitat of eastern cod, for example, particularly through involvement in the water planning process, will have positive benefits for a range of other native aquatic species and their habitats in these catchments.

The areas declared as critical habitat for the grey nurse shark support a rich diversity of marine fauna and flora, including other threatened, rare or protected species of fish. These areas will provide refuge for a range of demersal fish species as they will be protected from the types of fishing activity that would normally pose significant risk of incidental capture. Such fish include the listed threatened species black cod *Epinephelus daemelii*, a number of species protected from fishing such as the Queensland groper *Epinephelus lanceolatus* and estuary cod *Epinephelus coioides*, and species that are currently recognised as overfished such as the two species of wobbegong sharks *Orectolobus maculatus* and *O. ornatus*.

The recovery planning program also creates a greater awareness amongst the community and other stakeholders of the status of threatened species and their habitat requirements, an increased focus on issues such as research, habitat rehabilitation, and compliance and more funding for aquatic research, public education and habitat restoration. This may well lead to more opportunities and support for the conservation of threatened species and the increased protection of aquatic biodiversity. The preparation of statutory plans for the recovery of threatened species also incorporates mandatory monitoring processes to assess recovery actions, which will lead to greater knowledge about these (and other) species and the habitats they use.

Conclusions

Aquatic systems are negatively impacted by a wide variety of human activities. For example, key threats to Australian marine and estuarine fishes are considered to be overfishing, habitat degradation and the introduction (both deliberate and accidental) of aquatic pest species (Pogonoski *et al.* 2002). Cowx and Collares-Pereira (2002) list similar threats in relation to freshwater fish species, as well as the additional factors of river impoundment, water abstraction and water quality deterioration.

These threatening processes can act alone, cumulatively or synergistically, and on-site as well as "downstream". Understanding these threats well enough to effectively develop, communicate, negotiate and deliver appropriate management responses is a difficult task. This task is confounded by a lack of detailed knowledge of the biological requirements of aquatic species and systems, and limited resources to obtain such knowledge.

It is understandable that the single species approach is often seen as an inadequate surrogate for ecosystem-based approaches, which can potentially protect a range of habitats and components of aquatic plant and animal communities without the need for detailed biological investigations of individual species. In an ideal world, decision-making on natural resource management issues would recognise and value the complex functions, requirements and interactions of all species within an ecosystem. In practice, however, many find it difficult to appreciate the complexity of functioning ecosystems or weigh their value against less 'abstract' social and economic considerations.

Our experience, as outlined in this paper, suggests that for some key issues which can threaten biodiversity, such as water extraction and fishing, single species can provide useful and relatively simple reference points and consequently act as powerful drivers of change. This can be seen particularly in those situations (such as community-based natural resource planning) where decisions on very complex issues must be negotiated by groups of stakeholders with vastly different backgrounds, experience, motives and technical understanding. In such cases, broad conservation outcomes, such as improved environmental flows, protection of key habitat areas, and increased community awareness of the issues, may well be achieved through focusing on the relatively simple requirements of a single well-known species.

References

Baillie, J. and Groombridge, B. (eds) 1996. *IUCN Red List of threatened animals*. IUCN, Gland, Switzerland.

Bowen, B. W. 1997. Landscape-level conservation in the marine realm. Pp. 223-225 in *Principles of conservation biology*, (2nd edition), edited by G. K. Meffe and C. R. Carroll. Sinauer, Sunderland, Massachusetts, USA.

Butler, G. L. 2000. Population Structure and Habitat Preference of the Eastern Freshwater Cod, *Maccullochella ikei* (Pisces: Percichthyidae) within the Mann-Nymboida Catchments, NSW. Resource Science and Management, Southern Cross University, Lismore. Integrated project. Unpublished.

Cowx, I. G. and Collares-Pereira, M. J., 2002. Freshwater fish conservation: options for the future. Pp. 443-452 in *Conservation of freshwater Fishes: Options for the Future*, edited by M. J. Collares-Pereira, M. M. Coelho and I. G. Cowx. Fishing News Books, London, UK.

Environment Australia 1997. Report to the Minister – Public nomination to the *Endangered Species Protection Act 1992*: Great white shark and grey nurse shark – recommended by the *Endangered Species Scientific Subcommittee* in the Threatened Species and Communities Section.

Faragher, R. A., Brown, P. and Harris, J. H. 1993. Population surveys of the endangered fish species trout cod (*Maccullochella maquariensis*) and eastern cod (*M. ikei*). *Report for Australian National Parks and Wildlife Service Endangered Species Program.* NSW Fisheries Research Institute, Cronulla. 22 pp.

Franklin, J. F. 1992. Preserving biodiversity: species, ecosystems or landscapes? *Ecological Applications* **3**(2): 202-205.

Harcourt, R. G., Gillings, M., Briscoe, D., Stow, A. 2003. Are east and west coast populations of the critically endangered grey nurse sharks (*Carcharias taurus*) genetically isolated? Final report to Environment Australia 1020pp. Department of Environment and Heritage. Canberra.

Ireland, D. 1984. The grey nurse shark. *Underwater*, **11**: 10-13.

Marsh, N. 1995. A grey future for the grey nurse shark. *Underwater Geographic*, 39: 39-42.

New, T. R. 1995. Focussing on species for invertebrate conservation. *Victorian Naturalist* 112: 29-31.

NSW Fisheries 2004. Eastern cod recovery plan. NSW Fisheries, Port Stephens. 37 pp.

NSW Fisheries 2000. River flow requirements – Eastern Cod. Preliminary draft - 10 October 2000. NSW Fisheries Office of Conservation, Port Stephens. 8 pp.

NSW Fisheries 2002. Draft grey nurse shark recovery plan. NSW Fisheries, Port Stephens. 100 pp.

Otway, N. M. and Burke A. L. 2004. Mark recapture population estimate and movement of Grey Nurse Sharks. *NSW Fisheries Final Report Series*. NSW Fisheries, Sydney, NSW, Australia.

Otway, N. M., Burke, A. L. Morrison, N. S. and Parker, P. C. 2003. Monitoring and identification of NSW Critical Habitat sites for conservation of Grey Nurse Sharks. *NSW Fisheries Final Report Series No. 47*. NSW Fisheries, Sydney, NSW, Australia. 62 pp.

Otway, N. M., and Parker, P. C. 2000. The biology, ecology, distribution and abundance, and identification of marine protected areas for the conservation of threatened Grey Nurse Sharks in south east Australian waters. *NSW Fisheries Final Report Series No. 19*. NSW Fisheries, Sydney, NSW, Australia. 132 pp.

Pepperell, J. G. 1992. Trends in the distribution, species composition and size of sharks caught by gamefish anglers off south-eastern Australia, 1961-90. *Australian Journal of Marine and Freshwater Research* **43**: 213-25.

Pogonoski, J. J., Pollard, D. A. and Paxton, J. R. 2002. *Conservation overview and action plan for Australian threatened and potentially threatened marine and estuarine fishes*. Environment Australia, Canberra, ACT Australia. 375 pp.

Pollard, D. A., Lincoln Smith, M. P., and Smith, A. K. 1996. The biology and conservation status of the grey nurse shark (*Carcharias taurus* Rafinesque 1810) in New South Wales, Australia. *Aquatic Conservation: Marine and Freshwater Ecosystems* **6**: 1-20.

Pollard, D and Wooden, I. 2002. Supplement to Eastern Cod Recovery Plan: Interim Results of Surveys in the Clarence and Richmond River Systems. NSW Fisheries Centre, Cronulla, October 2002.

Ponder, W., Hutchings, P., Chapman, R. 2002. *Overview of the Conservation of Australian Marine Invertebrates*. A Report for Environment Australia. Australian Museum, July 2002 (internet version, http://www.amonline.net.au/invertebrates/marine_overview/index.html).

Reed, J., and Clunie, P. 1997. Conservation of invertebrates in Victoria using the Flora and Fauna Guarantee Act 1988 - achievements and potential for improvement. *Memoirs of the Museum of Victoria* **56**(2): 617-622.

Rowland, S.J. 1985. Aspects of the Biology and Artificial Breeding of the Murray cod, *Maccullochella peelii* and the Eastern Freshwater Cod, *M. ikei* sp. nov. Ph.D. Thesis, Macquarie University, 253 pp.

Rowland, S. J. 1993. *Maccullochella ikei*, an Endangered Species of Freshwater Cod (Pisces: Percichthyidae) from the Clarence River System, NSW and *M. peelii mariensis*, a New Subspecies from the Mary River System, Qld. *Records of the Australian Museum* **45**: 121-145.

Rowland, S.J. 1996. Threatened fishes of the world: *Maccullochella ikei* Rowland, 1985 (Percichthyidae). *Environmental Biology of Fishes* **46**: 350

Tear, T. H., Scott, J. M., Hayward, P. H., and Griffith, B. 1995. Recovery plans and the Endangered Species Act: are criticisms supported by data? *Conservation Biology* **9**(1): 182-195.

Walker, B. H. 1992. Biodiversity and ecological redundancy. *Conservation Biology* **6**(1): 18-23.

Woinarski, J. C. Z., and Fisher, A. 1999. The Australian Endangered Species Protection Act 1992. *Conservation Biology* **13**(5): 959-962.

Yen, A. L., and Butcher, R. J. 1997. *An overview of the conservation of non-marine invertebrates in Australia*. Environment Australia, Canberra.

Threatened species legislation and threatened species recovery: does the former lead to the latter?

James M. Shields

Forests NSW

Locked Bag 23 Pennant Hills NSW 2120

jims@sf.nsw.gov.au

ABSTRACT

Vision statements and key action statements are primary and useful tools for threatened species recovery and threat abatement. In this chapter I provide a view of the necessary vision and key action statements. Listing a species or biological entity through threatened species legislation may or may not lead to its recovery, and this factor should be carefully taken into account in the process of funding natural resource management actions. Recent experience in NSW (1995 -2003) has provided examples of the application of a comprehensive Threatened Species Conservation Act, and its effect on biodiversity management. Some recovery plans have been written, but few actions have been resourced or delivered positive outcomes for the target taxa. Examples where recovery actions have been successful, apparently failed, or are still in question are discussed. Some required elements for successful recovery action and threat abatement are set out (scientific analyses of the environment, monitoring, transparency, community ownership, redundancy, exit strategy). Clearly, it would be better if more actions were resourced and improved conservation outcomes were delivered. A synthesis of current planning processes (2004) required by legislation, which produces fewer plans, processes and meetings, and more measurable conservation outcomes is a simple way forward. This can be done by using a hierarchical process to restore ecosystems, using fundamental and generally available knowledge and techniques.

Key words: threatened species recovery, successful management, vision statement, ecological communities, National Forest Policy, Hastings River Mouse.

Introduction

Many authors, including myself, query the usefulness of current Threatened Species legislation. In particular, and rightly so, the listing process is questioned. However, the legislation in NSW (which is similar to that of Australia, most other Australian States and the U.S.) and its criteria are carefully crafted. They can be used to put in place a process of recovery and successful biodiversity management, despite any flaws in the Act(s)or mistakes in past applications.

In this chapter, I discuss the conservation outcomes from threatened species legislation in NSW, in particular those occurring from 1995 to the present. These outcomes, which have some deficiencies, are discussed in comparison to a proposed process which uses current regulations to produce true recovery for threatened biological entities based on my experience in writing legislation and policy, and through my occupation as a professional wildlife manager and scientist. This experience includes implementing the regulations that result from legislation, the actions which result from policy; and involvement in court when these regulations have been allegedly and actually breached, as well as participating in management actions for the successful recovery of relevant taxa.

Experience has led me to the conclusion that although current processes effectively ensure that regulations are followed, they seldom lead to recovery of the targeted species, population, or ecological community, or the abatement of threats. This however does not need to be the case. The processes required by legislation can and should be converted into synergistic actions for recovery. The successful use or application of legislation requires a clear vision as to what success might comprise, and a basic plan for the key actions necessary to achieve success (Dickman 1996).

The Vision: Clearly Articulated Statements that Define Success

To be successful and useful, environmental legislation (i.e. rules for social behaviour with regard to natural resource management), must have a clear, succinct vision as to what the natural landscape should look like as a result of applying these rules (Caughley 1994). In the case of Threatened Species legislation, there needs to be a clear, common understanding ("vision") on the part of implementers, regulators and the public as to what is an acceptable degree of success. "Key Actions" also need to be identified and prioritised. This step in the recovery process is discussed in the next section, as key actions arise from the vision for success. Vision and action statements are also necessary to garner mainstream community support for environmental legislation, a key element for success.

One way for legislation to be successful and useful is to specify and implement Targets and Standards, which then provides the vision (Beckworth and Moore 2001). This process has its merits and is discussed at length in the

Pp 135 - 144 in *Threatened species legislation: is it just an Act?* 2004, edited by Pat Hutchings, Daniel Lunney and Chris Dickman. Royal Zoological Society of New South Wales, Mosman, NSW, Australia.

documentation of changes to recent legislation in NSW, Australia and elsewhere (O'Connor *et al.* 2003, Song and McGonigle 2001). However, specified Targets and Standards have seldom been implemented successfully, and more rarely do they provide common understanding or vision for success. Elaborating this vision for success is best done through clear declaration of a simple message (a vision statement) to the target audience (in this case, natural resource managers, the corporate world, and citizens of NSW). This is clear from first principles, but there are many examples from business, social work, and land management which show the value of vision statements in successful enterprises (McNamara 2003).

Although simple vision statements can clarify goals for biodiversity management, such statements have not been written for threatened species in NSW or elsewhere in the world (there are no vision statements for threatened species, for instance, listed on the IUCN website; there is a vision statement from the IUCN Species Information Section, run by no doubt by Information Technology professionals, not ecologists). Vision statements occur as part of the planning process in Wildlife Refuges managed by the US Fish and Wildlife Service (*e.g.* USFWS 2000, Kern National Wildlife Complex Planning Update). Threatened species are a part of the planning process for USFWS Wildlife Refuges, and there is some clear vision as to what a successful outcome might be. However, the vision is concerned with land management, not threatened species recovery planning, nor does it contain provisions for implementation.

Some attempts have been made (e.g. the NSW Biodiversity Strategy) to synthesize planning efforts, which could have provided vision statements. The principles in the NSW Biodiversity Strategy have not been adopted or incorporated into threatened species management. The authorship list of this volume, attendance at this conference, and the discussion notes attest to the deficiency of this aspect of the legislation in NSW with regard to vision, action or outcomes.

Government policy statements can specify vision and action. Their publication is not an obligatory part of the legislative process or policy production, although there are examples where these concepts have been addressed by governments. Most notable in Australia is the National Forest Policy (1992) which outlined a vision, actions and outcomes that intended to resolve dispute about reserve size and resource allocation. It specified Comprehensive Regional Assessments (for biodiversity, social values, and economic resources) and subsequent Regional Forest Agreements. The National Forest Policy clearly stated that there would be a "Comprehensive, Adequate and Representative Reserve system, which gave rise to the production of specific criteria for forest ecosystems and their management (*Nationally agreed criteria for the establishment of a Comprehensive, Adequate and Representative Reserve System for forests in Australia.* A report by the joint ANZECC/MCFFA National Forest Policy Statement Implementation Sub-committee (JANIS 1997)). It also specified that there would be a viable and sustainable forest industry in Australia. Although the policy is not necessarily satisfactory to any of the stakeholders, the process was applied, and the tenure and management of forest ecosystems have as a consequence changed radically across Australia.

The National Forest policy illustrates how environmental legislation can work on public land (albeit at significant cost), because very few people are directly adversely affected - problems arise when legislation is applied with the same rigour on private land, with many people and industries affected. Community support consequently recedes, and society is faced with enforcing command-control legislation on an unwilling population.

Vision statements that are produced within the realm of policy, and certainly legislation, must follow a regimented format and be as unequivocal as possible. Therein lies a difficulty: where ecological matters are concerned, good vision statements are inherently equivocal, and there is great value to be taken from the consideration of variability. Of course, a useful vision statement must carefully consider the text of the relevant legislation or policy for which it is trying to provide common understanding. In NSW, the threatened species legislation has two key components: threats (Key Threatening Processes) and recovery (Listing and Recovery Process). Given these two components, vision statements for the threatened species legislation in NSW (and thus Australian and US) acts might be constructed as follows. Two versions of the same vision are suggested for different target audiences.

Vision statements for threats:

Target Audience 1. - *Public-Corporate version:* Key threatening process cease to affect biodiversity across the landscape [of NSW] within 50 years as a result of sound planning, to the maximum extent possible.

Target Audience 2. - *Ecological-Land management version:* Key threatening processes cease to operate within a 50-year time frame, based on Threat Abatement Plans that pragmatically address the needs of biodiversity management, prioritised by the threats, in particular to Endangered Ecological Communities, to the extent that the maximum number of these threatening processes can be removed from the Schedules of the act.

Vision statements for recovery:

Target Audience 1. - *Public-Corporate version:* Endangered species, populations and communities recover or are stabilised across the landscape within the next 50 years as a result of sound planning to the maximum extent possible.

Target Audience 2. - *Ecological-Land management version:* "Threatened species, populations and communities are stabilised and enhanced from their current status within a 50-year time frame by recovery plans for Endangered Ecological Communities and their component species and populations, to the extent that the maximum number of species, populations and communities can be, effectively, removed from the Schedules of the Act.

The two different versions represent an attempt to communicate the same message to two different audiences; one is the general public combined with corporate or business interests, and the other is comprised of regulators, experts (scientists, field staff, technical staff), and land managers. The vision statements put forward may not be comprehensive but they exist, and can form the basis for planning and prioritising actions. The wordy, convoluted version directed towards scientists and public servants includes the key scientific principles necessary to allow objective action based on the scientific method, and takes into account the provisions of the NSW *Threatened Species Conservation Act 1995* (TSC Act).

Vision statements for threatened species recovery and threat abatement have not been written in NSW (or elsewhere). The strategic planning steps that are currently in place (establishment of goals, outcomes, priorities, targets, and standards) therefore have no successful endpoint to which they can aspire or be held accountable - they are structured to most simply meet the requirements of various regulations. A vision and its expression are clearly different from regulations. They are also just a necessary for realistically achieving threatened species recovery and threat abatement.

The next step is consideration of key actions. The vision statements need to consider matters of transparency, pragmatism and ecological principles. By targetting endangered ecological communities (EECs) in reference to recovery, the broadest possible planning approach allows for maximum efficiency and synergism. For threats, maximum efficiency can be gained through a consolidated approach, which prioritises other elements of the biodiversity planning processes. Useful tools have been developed recently to allow the on ground application of consolidated biodiversity planning for EECs (Gibbons and Briggs 2004, Oliver and Parks 2003). Useful tools for addressing threats have been developed for the NSW Fox Threat Abatement Plan (2001) (Meek and Kirwood 2003).

The vision statements all contain the concept of "the maximum extent possible", which is certainly equivocal. This phrase is necessary to provide a common understanding of "success". The words "maximum extent" indicate that all efforts and resources necessary should be devoted to achieving the vision. The word "possible" provides a link to reality, and leads to an objective process of resource allocation. The phrase "maximum extent possible" would not be an acceptable part of a regulation, but is crucial for providing vision.

In the recovery vision statements, I would include the word "effectively" to qualify "removed from the schedules" to cover the likely scenario where an entity remains on the list despite having recovered to the maximum extent possible. Many species and populations will remain at risk, under even the best management regimes (Caughley 1994). A combined vision statement, in its simplest form might be: "To achieve a stable set of regional ecosystems with the minimum number of threatening processes or threatened biological entities". Hopefully, that would serve to focus actions and thought on achieving success - the *removal* of entities and processes from an undesirable status.

Key actions for success

Key actions should be elucidated to cover the following elements of the recovery and threat abatement processes. The first key action is gaining mainstream community support for environmental legislation. Given this generality, I suggest the following order for actions.

1. ***Required Processes:*** Any action list must include the required processes that have been put in place by the legislation. These actions dominate the process at present and are discussed in length at the websites of the NSW Departments of Environment and Conservation and Infrastructure, Planning and Natural Resources, and elsewhere (Australia, USA, UK Acts). The first action required is the development of a plan. That is usually where key actions stop, and where the current vision stops.

2. ***Planning:*** Planning has been developed as a means of meeting legal requirements rather than as a part of achieving success. Conceptually, planning comes first, as was noted in the logo of a now defunct NSW government organisation. Planning is based on the principles of the system within which operations occur and the obstacles which prevent success. Basic input by pragmatic planners to achieve success is required, but often absent, during the planning process.

An "in-joke" amongst natural resource planners in NSW government (2000-03) has been: "Plan First - Think Second". In other words, a good idea is frequently a bad plan.

Planning should not necessarily be undertaken solely by research scientists who are expert in the biological entity or process involved, and certainly not in isolation from other disciplines. A variety of scientific disciplines are relevant to the development of a successful plan. Land management is a science and profession in itself, and senior managers or researchers from this field should be involved. Other professionals required for good planning include administrators, field personnel, geneticists, and public relations experts. Planning provides a structure for the two key actions: assessment, and recovery/abatement actions, and their sub-components. Given this structure, each of the actions in the plan moves directly toward achieving the vision.

3. ***Assessment:*** Assessment has two elements. The status of the biological entity needs to be clearly elucidated, for example population size and threats for a listed species, or the extent of fox predation on native vertebrates. An assessment of the knowledge required to manage the entity is also necessary. If there is sufficient knowledge, a set of recommended actions can structure the recovery or abatement actions. If there is not sufficient knowledge, reasonable steps must be taken to acquire this information (*e.g.* research, technical developments). After such knowledge or technical capabilities have been acquired, the appropriate recovery or abatement actions can them be implemented.

There are some examples where this approach has been taken, notably the Fox Threat Abatement Plan for NSW (NPWS 2001) and the draft Bitou Bush-Bone-seed Threat Abatement Plan (DEC 2004). However, further actions are hampered by the lack of a clear vision as to what they are meant to achieve.

4. Recovery/Abatement Actions: Actions to achieve success should be articulated based on the assessment. At this stage, a projected statement of the scope and scale of actions required needs to be made so resource estimates can then be calculated to achieve success. Again, the Fox TAP and the Bitou Bush TAP are examples where this step has been taken, resources have been allocated, and actions are underway to alleviate the threat.

Recovery and threat abatement actions should always 1) adhere to the best scientific analyses of the environment, 2) provide monitoring information 3) be transparent and accountable, 4) maximise community ownership and 5) include an element of a redundancy, or an exit strategy or both.

5. *Providing Estimates of Possibility:* The vision statement directs action towards achieving the possible. To do this, the inevitable task of costing must be completed. Following completion of Actions 1-3, this can be done objectively. The tasks of resource estimation and allocation are crucial to enable the actions to be carried out.

Resource estimates, a euphemism for costs in social and economic terms, should be presented in the plan, with a range of options and time scales. There are two types of costs. There are the direct costs of recovery actions, and there are the indirect costs of opportunities forgone by recovery actions e.g. prevented actions. The opportunity costs of conserving threatened species, populations, or communities, or abating threats, are often orders of magnitude greater than direct costs (Hawken 1993). Both costs must be calculated and presented to society to allow sound decisions about priorities for natural resource management (Daily 1997).

Resource allocation can then become an iterative process amongst plans. To give an example, if the costs for recovery of populations of the Long-footed Potoroo *Potorous longipes* are correlated with abatement of threat from Fox Predation, the total cost of recovery and abatement for both plans can be judged more realistically.

A more complex (and theoretical) example could be an Endangered Ecological Community centred on woodland bird species listed on the schedules of the Act in NSW. The resources for the recovery plan would cover all those species in the listed community as well as the plants and invertebrates that make up the plant associations that provide resources for the birds. This would probably include several already listed Endangered Ecological Communities described in terms of their vegetation associations.

Calculating the costs of NOT undertaking recovery or abatement (alternative costs) may be relevant, or useful in terms of prioritising implementation (Lowe, pers. comm., McNamara 2003).

6. *Implementation:* Simply put, the actions planned to achieve the vision now be carried out. Review and monitoring are part of implementation. Although this factor appears simplistic, it is now often the case that the only recommended actions carried out are periodic reviews of the plan.

Key actions: research, priorities, and overlaps

In my opinion, research is not always a required activity for achieving success. However, monitoring, which involves scientific measurements of suitable parameters, is required because it is the only objective means to determine success, failure, and future needs. There are cases where the focus on research on threatened entities or process has not provided successful outcomes in terms of recovery or threat abatement in NSW and elsewhere. An example could be the Hastings River Mouse *Pseudomys oralis*, which has been studied in depth in NSW since its rediscovery 30 years ago (Meek 2003). A large amount of information has been compiled, and yet at a recent forum a combined group of experts could not provide guidance to recovery actions in the short, medium or long term, other than more research (Shields and Meek 2004). Nonetheless, carefully designed experiments can provide direct input that increases the chance of successful management if there is a knowledge gap which prevents overcoming obstacles to recovery or threat abatement. Research needs should be carefully articulated in the context of successful management.

Scientists and managers are well aware that ecological questions are boundless, budgets are limited, and success in natural resource management is becoming urgent and important for governments and citizens alike. This is a crucial point. Scientists must be held accountable for prioritising research needs, rather than maintaining everything is important, or replying to management questions of extents with, "How long is a piece of string?"

The legislation provides for exclusion of issues that are covered by other processes, of which there are many, such as the State and National programs to reduce greenhouse gases, or manage salinity and water use. Other government and non-government processes exist which consider weeds, pest animals, animal welfare, and the national research needs for biodiversity. All of these processes should be scrutinised to eliminate redundancy and maximise synergism. For example, Noxious Weeds have a control program at present, and Threatening Abatement Plans centred on weeds should be developed around existing programs.

A hypothetical example

Let us examine a hypothetically successful example for recovery and threat abatement, based on the vision and actions discussed above. A pertinent example is the mallee associations in the western districts of New South Wales. A well written and implemented nomination for an endangered ecological community, centred on real data, could set in train the means for writing an over-arching recovery plan that would consider the needs of a wide range of taxa, threatening processes and economic operations that relate to mallee habitat (Cody and Smallwood 1996, Simberloff 1999). The following scenario might ensue. A landholder has 200 hectares of mallee available. In the past, this could first be cleared at a profit (sale of mallee and the roots) and then agricultural activities could commence to produce revenue (grazing or cropping). In our hypothetical world, the regulation protecting the mallee Endangered Ecological Community is used to give a real value to the loss the landholder incurs from not undertaking these activities, and thus *create a property right for the mallee*. To prevent perverse rewards, incentives are directed toward the landholder to undertake management of the biodiversity entity (the Mallee Endangered Ecological Community) and all of its components, and not towards his neighbour who inherited land that had been flogged for the past century. Recovery plans for threatened species and populations and threat abatement plans could direct funds, activities and profit sharing incentives towards the mallee landholder, or private investment could be sought to manage the identified biodiversity asset and its attendant property rights. For the unfortunate neighbour, suitable incentives for biodiversity improvement could be introduced, through the same recovery/threat abatement process, and other natural resource management instruments (the aforementioned overlaps, e.g. sustainable agriculture, salinity, soil, water, green house gas programs). Direct actions on the Mallee property could include priority for fox abatement funding, subsequent inclusion as a site for re-introduction of Malleefowl *Leipoa ocellata*, funds to establish a commercial stand of mallee for harvest and local use, public relations assistance to capitilise on labelling value, and assistance with property planning. Direct actions on the degraded property would work to restore environmental values (soil, water) at a much more basic level, and there would be little funding for ecological resources.

Two concepts are applicable in our example, based around how funds are allocated. There is funding to compensate for actions foregone because of legislation, where a landholder is paid for opportunity lost due to regulations. Alternatively there is funding where a landholder is paid to manage land for natural values. Compensation comes from government, but funding to manage for natural resources may come from a variety of sources.

The status quo in NSW 2004: using threatened species legislation for evil rather than good?

Examination of the first principles with regard to threatened species, and legislation regarding thereto, is a useful step. Let us examine how threatened species recovery is addressed currently by the relevant legislation in NSW (and many other places around the world).

Unclear visions and unfocused actions?

The vision and actions to halt decline to threatened species, communities, and populations are currently embedded in the legislation, and they are straight forward: The aim is to stop threatening activities through regulation. The actions are that people and corporations who do not comply with the regulations may be prosecuted. These principles are easy to write into legislation, and easy to enforce in a command control system of regulation (Hollings and Meffe 1996, Briggs 2003). For instance, habitat assessment for threatened fauna can be written into the requirements for development applications. If the assessment is not done, development may stop or the developer may be prosecuted. The issues involved with command regulation discussed at length by Farrier and Whelan (2004).

The visions and actions to achieve recovery are not so clear, and are obscured in legislation (Beckwith and Moore 2001). One vision is that populations will increase as threatening processes are abated. Actions are not specified in legislation for recovery beyond the requirement to prepare a Recovery Plan. Actions in recovery plans are seldom articulated beyond the research or information gathering (survey) phase. The exact steps to be taken to achieve recovery or abate threats are either not known or considered to be too expensive for detailed planning to be worthwhile.

Recovery Plans and Threat Abatement Plans are nonetheless developed and duly published. In real terms two conditions must be met for them to be effective: public acceptance and adequate resources (Hollings and Meffe 1996). Currently acceptance by government is provided for through the legislation - any Minister whose agencies, actions or land are affected must approve the plan. Private citizens are compelled to comply by the regulations of the Act. Vegetation management plans for private property need to consider the conditions set out in recovery or abatement plans, but there is no command requirement to do so at present (September 2004). However, acceptance by the public at large can only be achieved by successful cooperation, education, incentives and persistence (McDonald 2003) which is, at present, seldom realised. Adequate resources are the second condition which must be met. All actions must be paid for. At present, government funds the vast majority of recovery actions - in the case of birds, almost 100% of costs have been met by the States or the Australian Government (Smythe *et al.* 2000). In many cases in NSW, direct costs are calculated, but no resources are allocated. Indirect costs are almost never calculated, although Forests NSW has made an attempt to do so in recent years (SFNSW 2002). In short, complying with the requirements of threatened species legislation is regarded as the equivalent of recovery or abatement action. It is not. Compliance and successful recovery or abatement action are not necessarily the same thing.

Let us examine the role of recovery in the context of the actual processes of listing, planning and on-going management.

Threatened species legislation processes

The normal process by which the *Threatened Species Conservation Act (1995)* is enacted are described below. The purpose of this discussion is to accurately describe what happens when a species, population, community or threatening process enters the regulatory process.

Nomination

The first step in the process of threatened biodiversity management and threat abatement in NSW, and in most other legislation, is the nomination of the species, population, community or threatening process to a scientific committee that administers the schedules of the Act. The presumed motivation for this step is the likely threatened or threatening status of the species, population, community or threatening process. However, in real terms, nominations are often instigated by a perceived threatening development or other human activity.

The initial phase can consume a large amount of resources; the process of nomination must demonstrate that the species, population, community or threatening process meets the criteria in the Act. This work includes a literature review, and in some cases, original research. If clear and well-documented nominations are based on threatening processes affecting the species, population, community or threatening process, the process moves ahead quickly. In reality, the nominations are often superficial and require considerable input from the Scientific Committee during the next phase (Determination).

It is interesting to look at the basic motivation of the nominator and the basic thrust of the nomination. The original NSW legislative text notes that spurious nominations are not to be considered. This, however, is the only social criteria for excluding consideration; otherwise, the Act specifies that social and economic factors are not to be taken into account. Many nominations are of the "not in my backyard" nature (sometimes literally), where the nominator notes a decline in the species, population, or community involved, or the appearance of a threat, and prepares a nomination based on personal interest in the land area or processes. A smaller number of nominations come from experts who have access to data which indicates a decline is imminent or has happened, or that a threatening process exists. A large number of nominations are prepared and submitted by NGOs as a part of their core activity. The Department of Environment and Conservation (NSW) prepares some nominations as part of the core activity of threatened species management. The Scientific Committee also has the power and capability to prepare nominations itself, and it does so when considered necessary (for complex or urgent issues). Finally, nowhere in the nomination process is the recovery of the species explicitly considered, beyond the general knowledge that regulations will now apply to this entity. It is not possible or necessary to include recovery considerations at the nomination stage, but it is necessary to be aware that this has not occurred.

Determination

After nomination, the status of the species, population, community or threatening process is determined by the Scientific Committee. That is, a decision is made to accept or reject the nomination, and subsequently include it (or not) on the Schedules of the Act. This can be a time consuming process, where the evidence put together by the nominator is taken into account along with other information that the Scientific Committee accumulates. In some cases, years may go by before a nomination is determined. In making a determination, the Scientific Committee is not required to consider the steps or actions necessary for recovery or abatement. It is not required to take into account economic or social consequences of any determination. Again, it is not necessary at this stage (for successful threat abatement or recovery), but it is necessary to be aware that that it has not occurred.

Planning

The result of listing on the Schedules of the TSC Act are that a plan must be prepared for the recovery of the entity under consideration or the abatement of the listed Key Threatening Process. These plans are the responsibility of the Department of Environment and Conservation, through a recovery team. The actual preparation of the plan is normally done by consultants under contract to DEC, and the resultant document is approved by the recovery team, put on display as a draft for public comment, and signed off by the relevant Minister.

Implementation

These plans become statutory documents, which must be followed by government agencies and the general public. In particular, the provisions of these plans must be followed by Local Government, in the process of land management, business and recreational activities. This involves consideration in the development approval process, but does not include "routine agricultural activities". There are, however, many "loop-holes" in the requirements for compliance with recovery plans, particularly where industrial or suburban development is involved. The combination of loop-holes, lack of vision, and lack of clear action plans for recovery result in little implementation activity. However, most plans are reviewed regularly by the Recovery Team, which is also a requirement of the legislation. Along with the other recent and relevant policy developments described below, the processes described above constitute the recovery and threat abatement actions at present in NSW. [NB. In NSW, legislation is in the process of amendment at the time of writing September 2004].

Other Recent and Relevant Policy Developments

Vision and action plans for recovery and abatement management can come from other government, NGO and private initiatives. Over the past decade, a variety of processes have been underway which affect the recovery of threatened species and the management of natural resources (Allan and Curtis 2003, NSW Department of Urban Affairs and Planning 1999). A major recent change was with regard to forest management. On the east coast of NSW, land tenure, funding for land management, and regulations for threatened species have changed greatly (Lindenmayer and Recher 1998). In particular, large areas of former State Forest are now part of the nature reserves and National Parks managed explicitly for conservation goals by the Department of Environment and Conservation. The process of conversion to nature reserve status and National Park continued in 2004, with large areas under moratorium from logging until they can be converted into nature reserve or National Parks, or the conflict with NGOs dissipates (Nicholson 1999). In areas that remain as State Forest, operations are conducted under the auspices of a Threatened Species Licence, which contains provisions to protect threatened species at the regional, landscape, and local-site level.

In brief, there is no logging of old growth forest, rainforest, riparian areas, rocky outcrops, cliffs, drainage lines, or heath, and all of these areas are protected by buffer zones. Winter flowering Euclapyts, Banksias, Allocasuarinas and Xanthorrhoea are all afforded specific protection. It is noted that the definition of rainforest includes stands with eucalypt emergents, *e.g.* is a broad, rather than narrow, interpretation of rainforest. Hollow trees and recruits *i.e.* trees that may grow to produce hollows are protected within logged stands, except for plantations, and hollow trees are further recruited in the prescriptions for other landscape level resources (riparian areas, rainforest buffers, as listed above). Stags, *i.e.* dead trees, and down wood, *i.e.* fallen trees and branches, have special prescriptions in the license, as do individual threatened species (NSW Department of Urban Affairs and Planning. 1999). The conditions of this licence are monitored vigorously by the Department of Environment and Conservation, NGOs and individual citizens.

Off Crown land, a variety of recent changes to native vegetation regulation have occurred. Generally, it is now illegal to clear native vegetation without a permit, in particular listed ecological communities, for agricultural or other rural land use, based on requirements from Native Vegetation Acts (1997, 2003). Consent from the appropriate government agency (Catchment Management Authorities, 2004 onwards) may be required when a listed entity will suffer significant impact from a development. There are a variety of loop holes in these regulations - for instance, most non-rural development is exempted. Further, no detailed agreements or regulations have been developed as was done for the forest industries.

Recovery and threat abatement actions

Let us examine three sets of recovery efforts. Recovery actions have been limited by funding or lack of knowledge, as noted above. In some cases, however, actions have been taken that significantly reduce threats and provide for recovery of the target taxa. In other cases, actions have achieved little, while a large number are "still working".

Most recovery actions are related to stopping development which would destroy habitat, *e.g.* the species or other entity must be considered in the development application process, and by other government processes affecting land management. Precautionary actions exist for forestry operations, where new prescriptions must be developed for newly listed species, populations or communities. In some cases, habitat restoration work is undertaken (usually weed removal, tree planting or feral animal control). However, the actions and funding are limited, and there is seldom synergism between elements of the recovery program. Three examples follow.

Success: Gould's Petrel *Pterodroma leucoptera leucoptera* - is an example of successful recovery action. The elements of vision and action were addressed in the planning processes, and then the actions, including research, were implemented. In the first instance, the problem was assessed (Priddel *et al.* 1995a) in terms of population status. Aside from some birds on the nearby Boondelbah Island, the species was limited to one breeding site, Cabbage Tree Island, New South Wales. Mortality was investigated (Priddel and Carlile 1995b), key threatening processes were identified - the introduced rabbit, weeds - and subsequently eliminated (Priddel *et al.* 2000). The original population on Cabbage Tree Island showed a positive response to these recovery measures. After appropriate technological developments to secure permanent nest sites (Pridel and Carlile 2001a) and determine the feasibility of translocation (Priddel and Carlile 2001b) a second population was established on a nearby island to provide long term security. An ongoing education and publicity effort vastly aided acceptance (Priddel and Carlile 1999). An ongoing monitoring program is part of the Recovery Plan (NPWS 2000). In effect, all the necessary steps outlined in the vision and action statements proposed at the outset of this paper have been put in place.

Failure?: Long-footed Potoroo *Potorous longipes* - The Long-footed Potoroo was described only recently (Seebeck and Johnston 1981). It is a 2 kg potoroid, confined to northeastern Victoria and possibly southeastern NSW. Since its discovery and description in the early 1980s, recovery efforts, particularly in NSW, have shown few positive results. Originally known from a few (16) hair tube records and one jaw bone in NSW (Dovey 1987), it has not been recorded in recent times (over the past five years). The national recovery plan has been adopted for the species by NSW, but the status of the plan in that state at present is unclear. In NSW, large areas (approximately 30 000 ha) of land have been converted from multiple use forest to nature reserve and National Park with direct reference to this species (based on precautionary modelling procedures in the Eden Regional Forest Agreement), but the effect of this tenure change has not been monitored (DEC website 2004). Further, information available to the public online does not reflect this change in land tenure in NSW, and indicates that the species is primarily threatened by forest management practices across its range (Massicot 2002).

With regard to other objectives of the plan, the captive population in Victoria has ceased to reproduce (National Recovery Team *in litt.* 2004). There is no plan for positive management (translocation, reintroduction) although new, genetically compatible (Luikhart *et al.*1997) and relatively abundant populations have been discovered in Victoria. As noted in the National Recovery Plan, this species is in the peculiar position of having increased its known range since listing (due to increased survey effort). The major recovery action is further reservation from logging (National Recovery Team, *in litt.* 2004).

Research into longevity, reproductive rate, diet, disturbance, and effects of predator management has not proceeded or has not produced useful tools for successful recovery.

In NSW, the species is in much the same state now as it was 15 years ago. That is, its status and threatening processes are unknown. The public and the scientific community have little interest in the issue at present in NSW. Virtually none of the principles or practices suggested in the vision and action sections of this paper have been accomplished. The situation has arisen because there was insufficient or incorrect assessment of the species status, insufficient or incorrect assessment of threats, and inadequate or inappropriate actions have thus been the result. Where positive action has seen positive results (population size doubled after predator control at Bell Bird in Victoria, National Recovery Team *in litt.* 200), there has been little recognition by the public or the scientific community. There has been little persistence and less cooperation in managing this species in NSW, which are key elements of recovery. This has led to limited acceptance of the plan, in financial, scientific or public relations terms.

Still working: Regent Honeyeater *Xanthomyza phrygia* is a large member of the honeyeater family, Meliphagidae, which nomadically follows flowering events in forests and woodlands from South Australia to Queensland (Longmore 1991). As these forest and woodland habitats have diminished, it has declined dramatically over the past two decades. The need for a comprehensive assessment of the species and threats to population survival have been addressed (Menkhorst 1993, 1997). Longevity, foraging requirements, captive husbandry, genetics, re-introduction techniques, and habitat restoration tools have been studied and developed, primarily in Victoria (NRE 1998a).

After a decade of recovery work, source populations are now known, there is a successful captive breeding population in existence, and habitat restoration work has created at least 2 000 ha of new habitat, with at least 1 000 ha in a single block (Menkhorst *et al.* 2000, National Recovery Team *in. litt.* 2003).

However, population abundance in the wild has yet to respond to these efforts in a detectable, positive manner. The wild population has yet to make use of new habitat, possibly because some of these areas are remote from source populations. Genetic work has shown the species is essentially one large gene pool (Schodde *et al.* 1992), and mark-recapture work has shown that all major populations intermingle (Menkhorst *et al.* 2000). With this knowledge, it is clear that principles of conservation biology could be achieved using translocation of captive bred birds. That is, there is not a risk of genetic pollution of local populations resulting from translocations or re-introductions.

Recently, Regent Honeyeater habitat has acquired protected status as Endangered Ecological Communities or their equivalent which have been described and enacted into regulation in NSW, Victoria and the Australian Capital Territory (Muir *et al.* 1995, NSW TSC Act website 2004, Australian EPBC website 2004, Environment ACT website 2004). The Regent Honeyeater has been incorporated as part of the Recovery Planning process for these communities in the ACT. An over-arching and synergistic process is beginning to emerge. However, there is little public acceptance and less funding available for constructive action (Menkhorst *et al.* 2000).

The major work of the recovery team for the Regent Honeyeater remains to be done - more habitat needs to be restored, and possibly aggressive action needs to be taken to establish populations in areas of suitable habitat. There has been adequate assessment of the problem, but action and acceptance are still lacking. In particular, funding needs to be secured, preferably from a long term, vested-interest partner with large economic resources. The apiary industry, for instance, has a long term interest in establishing nectar resources across the range of the Regent Honeyeater. Through serendipity, the colours of the target species and the usual logos of the apiary industry are well coordinated from a marketing point of view (bright yellow and black).

Conclusions

Although threatened species legislation empowers biodiversity managers, this legislation has thus far rarely been used effectively in NSW. Instead, a diverse range of activities has been funded, few of which synergistically move toward the recovery of species, populations or communities, or the abatement of threats. By developing a vision for success, and carrying out logical recovery or abatement actions, the situation could be reversed. An over-arching vision, which provides focus to the actions, is a key element which has not been developed. Articulating clear vision statements will allow government and the private sector to plan and deliver funding for effective recovery and threat abatement actions. Continuing to manage biodiversity by increasing the number of listed entities is not a satisfactory end point.

The real task of threatened species or biodiversity legislation is to remove entities from a threatened status, not simply to list these entities on schedules of Acts. To remove threats, it is necessary to address three equally important topics with vision and action statements:

Acceptance - acquire acceptance for the vision and planned actions.

Assessment - properly assess the status of the entity.

Actions - acquire resources, support and continuity for all necessary recovery actions.

Although the proposed processes may not guarantee success, they will guarantee that legislation and regulations assist in the positive management of biodiversity resources, rather than simply add more entities to "the schedules of the Act".

Chris Dickman asked me on the day of the conference if I would retain any entities on the lists, and my answer is: "Some may need to remain on lists in perpetuity, due to inherent risks compounded by human development. However, all threatened species, population and communities can be managed to achieve success."

Acknowledgement

I would like to thank Daniel Lunney and the two referees who commented so effectively on the draft. Elisabeth Larsen provided detailed comment and conceptual input to the manuscript. I would also like to acknowledge the Scientific Committee for the Threatened Species Conservation Act in NSW. It has been a privilege to work with its members over the past 9 years. Bill Shields, Jake Tanner, and Bethany Ostrey made significant contributions to the delivery and creation of this work. Forests NSW and its staff, particularly Doug Binns, Paul Meek and Justin Williams, have contributed significantly to work on threatened species, communities and populations, and to threat abatement. Paul Mahon and Andrew Leys from DEC NSW have contributed to, but are not responsible for, some of the concepts in this paper.

References

Allan, C. and Curtis, A. 2003. Regional scale adaptive management: Lessons from the North East Salinity Strategy (NESS). *Australasian Journal of Environmental Management* 10, 76-84

Beckwith, J. A. and Moore, S. A. 2001. The influence of recent changes in public sector management on biodiversity conservation. *Pacific Conservation Biology* 7, 45-54

Botkin, B.D. 2001. *No Man's Garden: Thoreau and a New Vision for Civilization and Nature.* Island Press, Washington D.C.

Briggs, S. 2003. Command and control in natural resource management: Revisiting Holling and Meffe. *Ecological Management and Restoration* 4, 161-162

Caughley, G.C. 1994. Directions in conservation biology. *Journal of Animal Ecology* 63: 215-44.

Cody, M. and Smallwood, J. (eds) 1996. Long-Term Studies of Vertebrate Communities. Academic Press, Inc.

Commonwealth of Australia. 1992. *National Forest Policy Statement: a New Focus for Australia's Forests.* Australian Government Printing Service, Canberra.

Daily, Gretchen. 1997. *Natures Services.* Island Press, Washington D.C.

Department of Conservation and Environment. 1991. Regent Honeyeater Management in Victoria. Procedural Document 02-20-0642-1, Department of Conservation and Environment. Heidelberg, Victoria.

Department of Natural Resources and Environment, Victoria; National Parks and Wildlife Service, New South Wales; Parks Victoria, State Forests of New South Wales (2000) Long-footed Potoroo (*Potorous longipes*) Recovery Plan, DEH Website, 2004.

Dickman C. 1996. Incorporating science into recovery planning for threatened species. Pp. 63-73 in *Back from the brink: refining the threatened species recovery process.* Edited by S. Stephens and S. Maxwell. Surrey Beatty and Sons, Chipping Norton, NSW.

Dovey, E. 1987. *Use of databases, indirect survey and predictive techniques in the discovery of the Long-footed potoroo in NSW.* Thirty third Scientific Meeting of the Australian Mammal Society. Australian Mammal Society News, Abstract p.16.

Gibbons, P. and Briggs, S. 2004. Biodiversity Incentives Framework : A framework for allocating incentive payments for biodiversity on private land. Draft. Department of Environment and Conservation, CSIRO Sustainable Ecosystems, Canberra.

Gibbons, P., Briggs, S. and Shields, J. 2001. How many credits is your property worth ? *Australian Landcare, December 2001*, 20-21.

Gibbons, P., Briggs, S.V. and Shields, J.M. 2002. Are economic instruments the saviour for biodiversity on private land? *Pacific Conservation Biology* 7, 223-228.

Hawken, P. 1993. *The Ecology of Commerce.* Harper Collins.

Holling C. S. and Meffe G. K. 1996. Command and control and the pathology of natural resource management. *Conservation Biology* 10, 318-327.

JANIS 1997. Nationally Agreed Criteria for the Establishment of a Comprehensive, Adequate and Representative Reserve System for Forests in Australia. Natural Resource Management Ministerial Council, Canberra.

Lindenmayer, D.B. and Recher, H.F. 1998. Aspects of ecologically sustainable forestry in temperate eucalypt forests – beyond an expanded reserve system. *Pacific Conservation Biology* 4: 4-10.

Longmore, W. 1991. *Honeyeater of Australia and their Allies.* Angus and Robertson, Sydney.

Luikart, G J Painter, R. H Crozier, M Westerman, W. B Sherwin. 1997. Characterization of microsatellite loci in the endangered long-footed potoroo *Potorous longipes Molecular Ecology* 1997 6:5 p. 497

Massicot, Paul. 2002. Long-footed Potoroo. Animal Info Website © 2002

McDonald T. 2003. Persistence and cooperation: Undervalued keys to restoration? *Ecological Management and Restoration* 4, 82.

McNamara, C. 2003. Strategic Planning (in nonprofit or for-profit organizations). In: *Field Guide to Nonprofit Strategic Planning and Facilitation.* 296 pp. Free Management Library, Authenticity Consulting, Minneapolis, Minnesota.

Meek, P. and Kirwood R. 2003. Generating Conservation Kernels to Select Areas to Control Foxes and the Implications for Pest Management Practice in State Forests. *Ecological Management and Restoration* 4: 66-72.

Meek P. and Shields, J. 2004. *The Hastings River Mouse Scientific Forum.* 4-5 March, 2004, Cumberland State Forest, Sydney. Proceedings edited by P. Meek and J. Shields. Forests NSW, Locked Bag 23, Pennant Hills, NSW.

Menkhorst, P., Schedvin, N., and Geering, D. 1999. Regent Honeyeater (*Xanthomyza phrygia*) Recovery Plan 1999-2003. DEH Website 2004.

Muir, A.M., Edwards, S.A. and Dickens, J.M. 1995. *Description and conservation status of the vegetation of the box-ironbark ecosystem in Victoria.* Flora and Fauna Technical Report Number 136, Department of Conservation and Natural Resources, Victoria.

Nicholson, E. 1999. Winds of change for silvicultural practice in NSW native forests. *Australian Forestry* 62: 223-35.

Natural Resources and Environment. 1998a. *Threatened Vertebrate Fauna in Victoria.* Department of Natural Resources and Environment, Melbourne.

Natural Resources and Environment. 1998b. *Victoria's Biodiversity: our living wealth.* Department of Natural Resources and Environment, Melbourne.

NPWS (National Parks and Wildlife Service). 2001. *Threat Abatement Plan for Predation by the Red Fox* (Vulpes vulpes). National Parks and Wildlife Service. Hurstville.

DEC (Department of Environment and Conservation). 2004. Draft *Threat Abatement Plan for Invasion of native plant communities by bitou bush/boneseed.* National Parks and Wildlife Service. Hurstville.

NSW Department of Urban Affairs and Planning. 1999. *Report for the Comprehensive Regional Assessment Project NA57/ ESFM: Application of protective measures/ forest practices into a quantitative database.* Resources Assessment and Conservation Division, NSW Dept of Urban Affairs and Planning, Sydney.

O'Connor C. Marvier M. and Kareiva P. 2003. Biological vs. social, economic and political priority-setting in conservation. *Ecology Letters* **6** No. 8 Pp 706-711(6)

Oliver, I. and Parkes, D. 2003. *A Prototype Toolkit for Scoring Biodiversity Benefits of Land Use Change.* Department of Infrastructure, Planning and Natural Resources, Sydney.

Priddel, D. and N. Carlile. 1999. Reclaiming a Petrel's Paradise. *Nature Australia* **3**: 60-63.

Priddel, D. and N. Carlile. 2001. A trial translocation of Gould's Petrel *(Pterodroma leucoptera leucoptera)*. *EMU*: 2001: **101**: 79-88

Priddel, D. and N. Carlile.1995. An Artificial Nest Box for Burrow-Nesting Seabirds. *EMU* 1995: **95**, 290-294.

Priddel, D. and N. Carlile.1995. Mortality of Gould's Petrel *(Pterodroma leucoptera leucoptera)*.at the Nesting Site on Cabbage Tree Island, New South Wales. *EMU* 1995: **95**, 259-264

Priddel, D., N. Carlile and R. Wheeler. 2000. Eradication of European rabbits (*Oryctolagus cuniculus*) from Cabbage Tree Island, NSW, Australia, to protect the breeding habitat of Gould's Petrel *(Pterodroma leucoptera leucoptera)*. *Biological. Conservation* **94** (2000) 115-125.

Priddel, D., N. Carlile., C. Davey and P. Fullagar. 1995. The Status of Gould's Petrel *(Pterodroma leucoptera leucoptera)* on Cabbage Tree Island, New South Wales. *Wildl. Res.* 1995, **22**, 601-10.

Schodde, R., Mason, I.J. and Christidis, L. 1992. Regional age and sexual differentiation in the Regent Honeyeater *Xanthomyza phrygia*. *Corella* **16**: 23-28.

Seebeck, J.H. and Johnston, P.J. 1980. *Potorous longipes* (Marsupialia : Macropodidae); a new species from eastern Victoria. *Australian Journal of Zoology*, 28, 119-134.

Simberloff, D. 1999. The role of science in the preservation of forest biodiversity. *Forest Ecology and Management* **115**: 101-11.

Smyth, A.K., Lamb, D., Hall, L., McCallum, H., Moloney, D. and Smith, G. 2000. Towards scientifically valid management tools for sustainable forest management: species guilds versus model species. Pp 118-24 in *Management for Sustainable Ecosystems*, edited by P. Hale, A. Petrie, D. Moloney and P. Sattler. Centre for Conservation Biology, the University of Queensland, Brisbane.

Song, S. J. and McGonigle, R. M. 2001. Science, power and system dynamics: the political economy of conservation biology. *Conservation Biology* **15**. 980-989.

State Forests of NSW. 2002. *Seeing: Social, Environmental and Economic Report 2001/02*. State Forests of NSW, Locked Bag 23, Pennant Hills, N.S.W.

List of Bibliography of the Hastings River Mouse:

Blackwell, G., Graham, K., Hochuli, D. and Meek, P. D. 2003. Habitat use by the Hastings River Mouse (*Pseudomys oralis*): are management prescriptions accurate? Poster Paper at the Ecological Society of Australia, 2003 Conference, Armidale NSW.

Fox, B., Read, D., Jefferies, E and Luo, J. 1994. Diet of the Hastings River Mouse (*Pseudomys oralis*). *Wildlife Research* **21**: 491-505.

Jerry, D. R., Dow, T. A., Elphinstone, M. S. and Bavestock, P. R. 1998. Historical and contemporary maternal population structuring in the endangered Hastings River Mouse (*Pseudomys oralis*). *Conservation Biology* **12**, 1017-1022.

Graham, K. 2003. Responses of invertebrate and mammalian fauna to hard and soft edges in Eucalypt forests, Chapter 4. Honours Thesis, University of Sydney.

Keating, J. 2000. Factors affecting the distribution of two rare small mammals at Grady's Creek Border Ranges national Park, North-East NSW. Honours Thesis, Southern Cross University, Lismore.

King, G. C. 1984. Habitat utilised by *Pseudomys oralis*. *Australian Mammalogy* **7**, 139-47.

King, G and Mackowski, C. 1986. Two new localities for *Pseudomys oralis* Thomas (Rodentia:Muridae) in NSW. *Australian Mammalogy* **9**: 63-5.

Meek P. D, 2002a. Radio tracking and Spool-and-line study of the Hastings River Mouse *Pseudomys oralis* (MURIDAE) in Marengo State Forest NSW. State Forests of NSW Unpublished Report.

Meek P. D, 2002b. The nest of the Hastings River mouse *Pseudomys oralis*. *Australian Mammalogy* **24**: 225-227.

Meek P. D., McCray, K. and Cann, B, 2003. New records of Hastings River Mouse *Pseudomys oralis* from State Forest of NSW pre-logging surveys. *Australian Mammalogy* **25**: 101-105.

Meek, P. (unpublished) Home range and habitat use of *Pseudomys oralis* I. radio tracking.

NSW NPWS 2003. Draft Recovery Plan for the Hastings River Mouse (*Pseudomys oralis*). NSW NPWS, Sydney.

Pyke, GH and Read DG, 2002. Hastings River mouse (*Pseudomys oralis*): A biological review. *Australian Mammalogy* **24**: 151-176.

Read, D. G. 1993a. Prescriptions for the identification of habitats of the Hastings River Mouse *Pseudomys oralis*. *Forestry Commission NSW Technical Paper No.* **58**.

Read, D. G. 1993b. Body size in Hastings River Mouse *Pseudomys oralis* from old and new locations. *Australian Zoologist* **29**: 117-23.

Read, D. G. and Tweedie, T. D. 1996. Floristics of habitats of *Pseudomys oralis*. *Wildlife Research* **23**, 485-93.

Smith, A, Ferrier, S, Hines, H. and Quin, D. 1996. Modelling the geographical range of the endangered Hastings River Mouse (*Pseudomys oralis*) (Rodentia:Muridae) in North-East NSW. Unpublished Report to the Hastings River Mouse Recovery Team, AUSTECO, Armidale.

Smith, A. and Quin, D. 1996. Patterns and causes of extinction and decline in Australian Conilurine rodents. *Biological Conservation* **77**, 243-267.

Smith, A, Phillips, C., and Townley, S. 1996. Diet and habitat preferences of the Hastings River Mouse (*Pseudomys oralis*) (Rodentia:Muridae). Unpublished Report to the Hastings River Mouse Recovery Team, AUSTECO, Armidale.

Smith, A. P. and Quin D. G. 1997. Microhabitat requirements of the Hastings River Mouse (*Pseudomys oralis*) (Rodentia: Muridae). Unpublished report to the Hastings River Mouse Recovery Team. AUSTECO Environmental Consultants.

Tasker, E. and Dickman, C. 2004. Small mammal community composition in grazed and frequently-burnt Eucalypt forests of the northern tablelands of NSW. Pp 721-740 in *Conservation of Australia's Forest Fauna* (second edition), edited by D. Lunney D. Royal zoological Society of NSW, Mosman, NSW.

Tasker, E. 2003. Disturbance and the Hastings River Mouse: landscape-scale processes and management of the endangered species. Abstract from the Ecological Society of Australia Conference, Armidale 2003.

Townley, S. 1997. Diet and social organisation of the Hastings River Mouse *Pseudomys oralis* (Rodentia:Muriadae) in north-eastern NSW. Unpublished Draft report to NSW NPWS.

Townley, S. 2000. The ecology of the Hastings River Mouse *Pseudomys oralis* in North eastern NSW and Queensland. PhD Thesis, Southern Cross University, Lismore.

Tweedie, T. and York, A. 1993. Survey Guidelines for the Hastings River Mouse (*Pseudomys oralis*). SFNSW Technical Paper 62. State Forests of NSW, Sydney.

The neglected 74% - the non-threatened vertebrates - and a reflection on the limitations of the process that fashioned the current schedules of threatened species in New South Wales

Daniel Lunney[1], Alison Matthews[1], Hal Cogger[2], Chris Dickman[3]

[1]Department of Environment and Conservation (NSW), PO Box 1967, Hurstville NSW 2220

[2]John Evans Memorial Fellow, Australian Museum, c/- 2A Amethyst Ave, Pearl Beach NSW 2256

[3]Institute of Wildlife Research, School of Biological Sciences, University of Sydney, NSW 2006

ABSTRACT

In this chapter we apply an ecological approach to considering the vertebrate species listed on the current schedules of the NSW *Threatened Species Conservation Act 1995*. We do so in light of our knowledge of the threats that have impacted on the vertebrate fauna of NSW, but mainly as a reflection on our role as the Scientific Committee (1992-95) in formulating the schedules of threatened fauna under the *Endangered Fauna (Interim Protection) Act 1991*. We reflect on the limitations of this process, including constraints on time, a lack of adequate knowledge for most species and differences in expert opinion. The final 1992 list included 236 threatened vertebrate species, or 26% of the fauna of NSW. Threatened species have since been elevated by instruments of government and land-use decisions to such a degree that other species are apparently no longer of interest. Thus 74% of our vertebrate fauna has become neglected. Many species are missing out on research effort, and other conceptual approaches, such as threatening processes, are not given priority. In our opinion, the apparent immutability of the scheduled species is illogical given the uncertainty of the listing process, and we see a pressing need for a dynamic database of species status so that the schedules reflect new knowledge and changes in the status of species as they occur. Threatened species deserve special attention, but their relative importance in the wider conservation agenda needs re-examination if we are to fulfil the broader aims of conserving biodiversity.

Key words: threatened species, Scientific Committee, knowledge, fauna, frogs, legislation.

Introduction

When the NSW *Threatened Species Conservation Act 1995* was promulgated in December 1995, it contained two schedules of threatened vertebrate fauna. Schedule 1 contained taxa considered to be endangered, and Schedule 2 contained those judged to be vulnerable. These schedules were those listed under the *Endangered Fauna (Interim Protection) Act 1991*, the precursor to the *Threatened Species Conservation Act 1995*. The procedures and outcomes of that process have been presented elsewhere (Cogger and Lunney 1992; Lunney *et al.* 1996, 1997, 2000). The aims here are to reflect on the adequacy of that process, the apparently static nature of the schedules, and the consequences for the 74% of the vertebrate fauna not currently included in them. One could argue that the legislative process has fallen short of its own aims of conserving biodiversity by putting so much weight on conserving threatened species, thus relying on threatened species as a surrogate for other species. The process has not readily allowed for rapid changes to be made to any data set so that species can be removed from the schedules and others added on the basis of regular monitoring. The existing process now drives a major industry in environmental assessment, and the legislation has a great social impact because landowners must take account of threatened species in planned developments. An examination of some of the assumptions of that process is a requirement in an adaptive management environment, and the process under examination here is the selection of the species that were included in the NSW schedules of vertebrate fauna.

Saving species from the brink of extinction through legislative clout has provided a powerful instrument in helping to conserve our biological heritage. We remain committed to its retention and refinement. However, as ecologists, we see great merit in other approaches to conserving fauna, such as including all species in any study, survey or research program on the basis that different species yield different information about threatening processes, as well as habitat and area-based approaches (*e.g.* Margules and Pressey 2000; Pressey *et al.* 2002), evaluation of outcomes of threat mitigation, and long-term studies to conserve all our faunal diversity. The increasing priority afforded to threatened species, whether through legal obligation or a public sense of responsibility to conserve threatened species, tends to devalue the crucial role that the other 74% of vertebrates play in the conservation agenda (not to mention the invertebrates, *e.g.* Hutchings 2004). Further, as zoologists who have

Pp 145 - 157 in *Threatened species legislation: is it just an Act?* 2004, edited by Pat Hutchings, Daniel Lunney and Chris Dickman. Royal Zoological Society of New South Wales, Mosman, NSW, Australia.

spent much of our working lives becoming familiar with our fauna, and the people who have studied them, we are acutely aware of the limitations of our knowledge in producing definitive lists of threatened species. In this chapter, we revisit the schedules that we, as the initial Scientific Committee (Cogger, Dickman, Lunney), prepared in 1992, and we recall the uncertainty that accompanied our decision-making process. We support those legislative advances that incorporate populations and ecological communities in the *Threatened Species Conservation Act 1995* and the inclusion of threatened ecosystems in the Commonwealth's *Environment Protection and Biodiversity Conservation Act 1999* but we focus here on the species schedules because of the weight that they are now given.

The *Endangered Fauna (Interim Protection) Act 1991*, and the neglected 74%

The NSW *Endangered Fauna (Interim Protection) Act 1991* provided a rare opportunity to move quickly to place species on the first-ever comprehensive NSW list of threatened vertebrate fauna. A modest list of endangered fauna had been part of the NSW *National Parks and Wildlife Act 1974*, but that schedule (Schedule 12) was drawn up without guidance in the Act as to criteria and did not provide for any regulatory mechanism beyond the protection afforded to all recognised vertebrate fauna (mammals, birds and reptiles), although there was an additional penalty attached to the harm of endangered fauna over protected fauna. One aim of the schedule was to draw attention to certain species and the concept of extinction. This was relevant to the NSW *Environmental Protection and Assessment Act 1979*, which required that endangered species be taken into account in relation to various matters that fell within its ambit. In contrast, the brief for the schedules of threatened fauna prepared under the *Endangered Fauna (Interim Protection) Act 1991* provided specific criteria for inclusion and a timetable for action – one month to produce the list, followed by a period of public review leading to finalisation of the list. The Scientific Committee was independent of parliament and ministers and the decisions of the Committee were final; they were not merely recommendations as had been the case with endangered species listed under schedule 12 of the NSW *National Parks and Wildlife Act 1974*.

The *Endangered Fauna (Interim Protection) Act 1991* passed through the NSW parliament in December 1991. The directions for action came to NSW National Parks and Wildlife Service (now part of the NSW Department of Environment and Conservation) in mid-January 1992 and the Committee of three, as specified under the Act, was promptly formed along with a small working team. A list of species was promulgated in a large daily newspaper (the *Sydney Morning Herald*) on 28 February 1992 as part of a statutory public review process. This publication activated the remainder of the Act, and the process of formally dealing with threatened species under their own legislation commenced for the first time in NSW. The review period was completed by August 1992 and the final decisions appeared in the *Government Gazette* on 18 December 1992. Over the following three years the Committee received a few submissions for additions, but the listing procedure had become lengthy, and the few additions decided upon by the Committee were overtaken by the passing of the *Threatened Species Conservation Act 1995*. These few extra species, such as the Southern Hairy-nosed Wombat *Lasiorhinus latifrons*, were included in the new schedules, and they appeared in the *Threatened Species Conservation Act 1995* which passed through the NSW parliament in December 1995. Since the promulgation of the *Threatened Species Conservation Act 1995*, nine species of vertebrates have been added to the schedules and none removed (as of December 2003).

After the lists of NSW threatened species were promulgated at the beginning of 1992, the Committee members felt that we could be challenged legally in court to explain why some species were left off the schedules and why others were placed on the schedules. Such challenges could arise, for example, if a threatened species was located in the middle of a proposed development and was likely to impede it, or when an activity looked as if it would be widely damaging, even though no threatened species occurred on the site. In short, the demand to go to court could come from either side of the development debate. Further, we were acutely aware that the *Endangered Fauna (Interim Protection) Act 1991* had been passed through the NSW parliament by the Labor opposition and the independents against the wishes of the coalition government. While we could reasonably expect support from the government because parliament had passed the bill, we did not know the extent of government support if contentious cases reached the court. This issue of zoology in court had already been a matter of concern for the Royal Zoological Society of New South Wales (Lunney 1992). We did not relish the thought of court appearances because of the arguments that would inevitably arise about our treatment of uncertainty in assembling the lists and about whether the criteria in the Act could really be applied to a particular species if very little was known about it. However, these problems are largely overcome when rankings are derived from collective expert opinion, which is itself based on objective criteria; anyone wishing to challenge that ranking must be able to demonstrate that the available data negate that opinion. Rather, opponents to the listing of a particular species are forced to argue that the methodological approach (*e.g.* Millsap *et al.* 1990) is fundamentally flawed, an argument that is hard to sustain given the widespread acceptance by governments and biologists of this approach.

The requirement to produce the lists quickly might have resulted in some errors, but it did forestall any dithering. The brief period legislated for the preparation of the NSW lists in early 1992 has escaped many commentators. So, too, has the requirement in the legislation to consider all terrestrial vertebrate fauna (birds, frogs, mammals, reptiles) and provide the reasons for non-inclusion of a species as thoroughly as the reasons for inclusion. This required the Committee to prepare the first-ever list of NSW terrestrial vertebrates (a daunting task in itself) and then to justify the status assigned to every species, not just those likely to be included. What is so astonishing to us now is the apparent immutability, or even sacredness, of the schedules as they are embedded in the workings of so many instruments of government and land-use decisions. At the time, we were not aiming to produce a once-only definitive list, based on universally-agreed criteria, of species about which we had perfect knowledge. Problems arise as the scheduled 26% of threatened species drift away from the other 74% of the fauna. We have to devise ways to conserve species not listed under legislation, but which also are in need of special attention, and to address the threats to them and their habitats.

Threatened species deserve close attention but, in our opinion, they represent only one limited component of an effective biodiversity conservation program. Further, the nature of the legislative process means that it rarely keeps pace with rapid changes in the status of species and ecosystems. Consequently, the concept of threatened species legislation and its application need regular revisiting if we are to become more effective in conserving biodiversity. With 74% of the vertebrate fauna of NSW not listed on the schedules of threatened species in 1992 (Lunney *et al.* 1996, 2000), there is a case for considering the status of these outsiders. One thesis pursued here is that they have become a neglected portion of our vertebrate fauna. This matter is highlighted when you consider the nature of the system involved in placing species on, or removing them from, the schedules. There is a pressing need for a dynamic database of species' status so that the schedules rapidly reflect new knowledge and the consequent changes in a species' status. These matters have been identified as being among the nation's highest faunal research priorities (ANZECC/BDAC 2001). Priority 2 of that report – identifying Australia's species – states that the highest priority research includes the need to 'identify, characterise and record the distribution of native species that are or may be threatened'. Research priority 3 – monitoring changes in Australia's biodiversity – is given the highest priority to 'develop and validate methods for direct measures and indicators to monitor the distribution and abundance of species'. Both research priorities stress the need for biodiversity condition to be part of the state of the environment reporting. We argue that monitoring the lists is an essential activity which needs to be supported by a sustained research program with the capacity to regularly revise the schedules. That they need to be revised is apparent to us as we revisit the procedures and lists we prepared in 1992.

In our view, the current threatened species paradigm for conserving biodiversity subverts the original aim of listing threatened species as symbols of the loss of biodiversity and as case studies of the many processes that lead to extinction. It appears to have become a dominating end in itself in that threatened species provide touchstones in a variety of crucial documents that determine national, state and local priorities for biodiversity conservation. The legal requirement to conserve threatened species is respected in such documents, but the need to see this approach as indicative of the broader picture does not emerge from our reading of them. This paradigm manifests itself through the emphasis on survey, monitoring and research of threatened species in such major programs as the Comprehensive Regional Assessments (CRA) under the National Forest Policy Statement (Commonwealth of Australia 1992), and in regulations and court action, or threats of action, that focus on threatened species in development applications. Threatened species have been elevated to such a degree that other species are virtually overlooked. For example, data audits have in the past resulted in threatened species being the only species entered on the Atlas of NSW Wildlife. Threatened species are also the only species subject to pre-logging surveys in the Integrated Forestry Operation Approvals (IFOA) process, which is part of the Regional Forest Agreement (RFA) process, yet threatened species are only just over a quarter of the vertebrate fauna. We point out, that because threatened species usually occur in low numbers, most are not suitable for long-term monitoring programs. Such programs require species that can be found in sufficient numbers to enable an effective monitoring program to be designed (*e.g.* Lunney and Matthews 2004). Moreover, it is critical that we do not treat threatened species simply as those species most deserving of our conservation effort, but rather as early indicators of failing systems – both ecological and social.

Our contention is that it is ecologically more informative and rigorous to conserve biodiversity as a whole by examining and minimising existing and potential threatening processes, such as land clearing, inappropriate fire regimes, logging, roading, toxic waste, salinity, overgrazing, pest species, and disruption to the natural flows of water. Lunney *et al.* (2003) made this point for bats, where national priorities concentrate almost exclusively on a small number of threatened species (9 of 90 taxa) and omit general bat research. These authors suggested that a better approach would be to focus on the threatening processes that affect all bat species, such as logging, management of old mines and identification and protection of roosts, in order to simultaneously determine strategies for protecting those bats that are threatened, as well as instituting measures that will prevent others from declining significantly. In our view, the effort devoted to the threatened species agenda can work, albeit unintentionally, to the detriment of a more holistic approach in conservation as well as to threatened species themselves. For example, the presence of a threatened species was not intended to block all developments on all sites. If threatened species are used regularly to stop developments, then efforts will be made to limit species being included on the schedules, and the process of identifying and conserving biodiversity will begin to fail. The converse can also apply. For example, some people may fear removing a species from a schedule. It seems that the more the schedules are used as planning tools the further we move from the intent of using them to implement more effective conservation programs.

At issue for the neglected 74% of the vertebrate species not on the schedules is that these species do not attract attention for funding, which is directed almost entirely towards threatened species. How then can a comprehensive review be carried out? Consider the following story. In 1989, a small group of us in the research branch of the NSW National Parks and Wildlife Service applied to the Australian National Parks and Wildlife Service (a forerunner to today's Commonwealth Department of Environment and Heritage) for a grant to study the Spotted-tailed Quoll *Dasyurus maculatus* and examine its status in NSW. The application for funding was automatically rejected because this species was not on the national list of threatened species. It was subsequently listed on the national register and began to receive special attention under the Commonwealth *Environment Protection and Biodiversity Conservation Act 1999*. In 1992, the Spotted-tailed Quoll was listed in NSW under the *Endangered Fauna (Interim Protection) Act 1991*, and holds this status under the replacement Act, the *Threatened Species Conservation Act 1995*. In NSW, it is now a species that gains much attention and is the subject of some careful research (*e.g.* Belcher 2004; Körtner *et al.* 2003). On 14 May 2004 it was listed nationally as endangered, and is now recognised as being just one step from extinction. This is a dramatic change in 15 years. How naïve of us to apply for funding for research on a non-threatened species, even if it was to review its status! The first lesson of this story is the recognition of a barrier between those species on the lists and those not on the lists. The next lesson is that the Spotted-tailed Quoll was undoubtedly a species in trouble in 1989. Concerted action for its conservation had to wait until official listing, yet with its congener, the Eastern Quoll *D. viverrinus*, already extinct on the mainland (it is extant only in Tasmania), there was a strong case for research into its status in 1989. The most powerful lesson is to heed the call for species monitoring at the highest level (that is, in validation of management programs, not just compliance), supported by dynamic databases (in which newly-acquired knowledge about a given species is quickly incorporated into the database and an automatic (quantitative) reassessment of its conservation status generated), dedicated staff and a regard for the schedules of threatened species as part of a commitment to adaptive management. Threatened species legislation has helped focus attention on the plight of *Dasyurus maculatus*, and this is an example of the positive value of such legislation. The converse is that unlisted species are out of sight as far as their conservation is concerned.

We thus urge the adoption of a dynamic system of revising the schedules and a commitment to its rolling application. Revision of the schedules needs to be undertaken at least every five years. The process can look at trends in a community, and monitoring can pick up species not on the list. The lists need to be ranked on objective criteria, the ranking to be dynamic as required by good governance, and have provision for alerting relevant authorities. For example, one of us (Cogger) has proposed what seems to be a simple system to trigger the federal environment minister to take action if a development application affects a species that has been flagged under objective criteria, even if the values are estimates. Criteria with a red flag include, for example, those species in which the total area once occupied by the species is known or is suspected to have declined by more than 75%, with other criteria marked by red or yellow flags. The alert for the minister to consider action in relation to a particular development would be the presence of at least one red flag or two yellow flags. The red, amber and green light approach was a feature of the NSW State of the Environment report 2003, and it has been the subject of comedy in the USA in relation to terrorist alerts. However, the idea is sound. The colour of the flags could change, and we hope that this is the only impediment to a trial of their use. Without a categorical early warning system, it is utterly predictable that tempers will become frayed if a government acts too late, and without having given a developer the information in good time. Such a system, with its unambiguous flags, is easily adaptable to internet use and has the advantage of simplicity. It does, however, require a dynamic database which includes geographic range and habitat information as well as population levels.

The uncertainty in the 1992 lists

This section critically examines how the official lists of threatened vertebrate species in NSW were prepared. We aim to loosen the perceptibly growing rigidity in the application of the schedule, as well as to challenge the growing dominance of threatened species in programs to conserve our fauna.

When the 1992 lists were prepared we were acutely aware of the lack of adequate information on which to base them. In part, we avoided the worst aspects of that problem by consulting experts for each species, or group of species. We asked each expert to tick a score sheet for each species. The criteria were ecological, most were numerical, and the method followed that of Millsap *et al.* (1990), modified for NSW. Cogger *et al.* (1993) had already found that it was a valuable method in the preparation of the national reptile action plan. We then asked each expert to examine the reasons listed in the Act, such as 'whether the habitat of a species has been drastically reduced or modified' (1 of 5 criteria to qualify for endangered), or 'whether the population of a species is still abundant but is under threat from severe adverse factors throughout its range' (1 of the 5 criteria to qualify for vulnerable).

Each expert was to tick 'yes' or 'no' as to whether the species they were commenting on fitted the criteria and should be listed as either endangered or vulnerable, and if there was no tick, then the species would remain protected (under the NSW *National Parks and Wildlife Act 1974* – see Jarman and Brock 2004) but not on the list of threatened species. We selected experts on the basis of our personal knowledge of the individual as well as our reading of the scientific literature and our own knowledge of the vertebrate fauna. We did not confine ourselves to one expert, but selected many experts (89 people for 883 species), and for those species where the decision was likely to be disputed or close to the line, we selected up to 11 experts. The number who responded and provided assessments for each species is given in Lunney *et al.* (2000). We also asked each expert whether they felt that their knowledge base for deciding was A (adequate), B (limited), or C (inadequate). Lunney *et al.* (2000) also recorded when there was a difference of expert opinion about whether a species should be listed or not.

The responses showed there was adequate knowledge for only 6% of species, limited knowledge for 42%, and inadequate knowledge for 52% of species, with noticeable differences in the proportions within each vertebrate class (Figure 1). Despite these knowledge limitations, 26% of species were placed on the list of threatened species. It follows that we were operating under a high level of uncertainty with respect to knowledge in preparing the lists. The second element of uncertainty arises when one considers the responses to the option given to each expert as to whether a species should be listed or not.

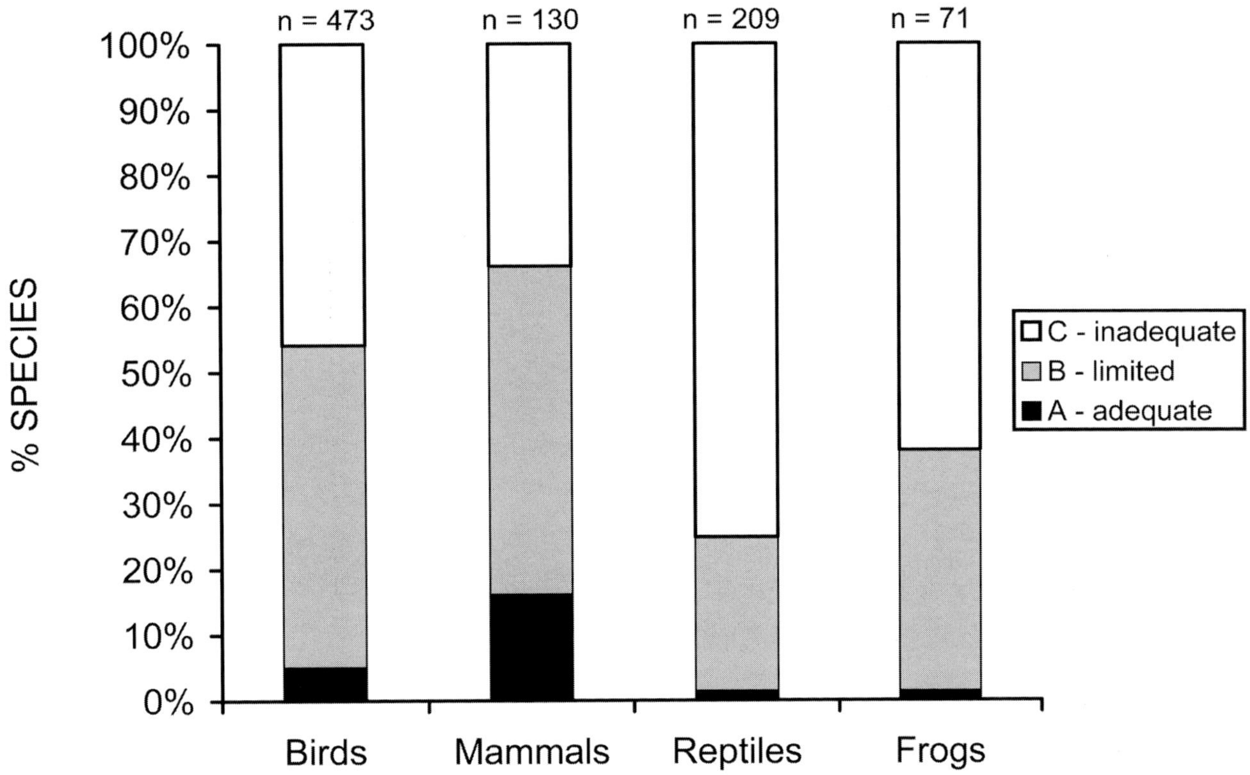

Figure 1. Levels of knowledge in the 1992 evaluation of vertebrate fauna. Percentage of birds, mammals, reptiles and frogs with levels of knowledge from 1992 evaluation. Reptiles and frogs had poor level of knowledge compared with birds and mammals. Overall 'adequate' levels of knowledge on which to assess status of species was low.

Results for each species, and any disagreement (or not) among the experts, were marked in separate columns in the relevant table in Lunney *et al.* (2000). The number of species for which there was disagreement among the experts is shown in Table 1 here. This table was split into the four classes of terrestrial vertebrates. It was also divided into the three categories of knowledge within each vertebrate class. One of the columns shows the percentage of species in each class so that comparisons can be made among the classes. From this table it might be noted that as the level of knowledge declined, the level of disagreement rose (*i.e.* expert opinion became more subjective, being a 'best guess' based not only on available knowledge but also on extrapolation from knowledge of related taxa, anecdotal information, and personal knowledge of habitat preferences). The overall trend is shown in the combined total of all species. The extent of disagreement among the experts was most marked for frogs and reptiles, evident for birds, and least for mammals. For mammals, the percentage disagreement was highest in the limited category. Thus there is a class difference among the vertebrates in relation to knowledge and expert opinion, in large part due to a combination of taxonomic diversity and the size of the research base (number of researchers). Notwithstanding such differences, the standout message for us in this table is that variance in expert opinion is inversely related to the knowledge base. It follows that, since adequate knowledge is the smallest category, and inadequate knowledge the largest, then disagreement is going to be widespread and pervasive.

As zoologists, we can point to the major gaps in knowledge of our fauna, even of the vertebrates, which are numerically insignificant compared with our invertebrate fauna. From a policy perspective (whether legal, bureaucratic or NGO threatened species activist), matters to be challenged include the legislative inflexibility of the schedules, the rigidity with which they have been applied to funding decisions, the setting of priority actions for conservation, and the limited role of experts, especially when forced to cope with incomplete or inadequate knowledge on which to make a categorical decision about a species' status. However, in defence of the expert system, if the species had been allocated randomly from a group of non-experts, the level of disagreement would be 100%. Overall, it was 26%, but with frogs at 57% when the level of knowledge was inadequate, even a non-frog expert could conclude that frogs are a group that should be revisited. We believe that the only way in which these problems can be adequately addressed is through linking dynamic databases on taxa and ecosystems to legislation, either by incorporating such databases into the legislative process (which is probably impractical) or by making them formal instruments to guide regulations and ministerial decision-making.

We also note that if the only species listed were those with a unanimous vote for listing, then the lists would have been halved. Conversely, if we had listed all species that at least one expert thought warranted inclusion, the list would have been half as long again. The final list of 233 species was about half way between the two outer limits (127 and 353 species) of what could have been

Table 1. The number of species for which expert opinion differed as to status in relation to the level of knowledge for the species.

Note: the total numbers of species (883) are those assessed in 1992 and listed in Table 1 of Lunney *et al.* (2000); by November 2000 there were a total of 891 species known from New South Wales (Lunney *et al.* 2000).

Level of knowledge	Number of species	Expert opinion differs	
		n	%
Birds			
A – adequate	24	0	0
B – limited	232	40	17
C – inadequate	217	54	25
Total	473	94	20
Mammals			
A – adequate	21	1	5
B – limited	65	19	29
C – inadequate	44	10	23
Total	130	30	23
Reptiles			
A – adequate	3	0	0
B – limited	49	12	24
C – inadequate	157	59	38
Total	209	71	34
Frogs			
A – adequate	1	0	0
B – limited	26	9	35
C – inadequate	44	25	57
Total	71	34	48
All species			
A – adequate	49	1	2
B – limited	372	80	22
C – inadequate	462	148	32
Total	883	229	26

deemed acceptable, giving a range from 14% to 40% of the vertebrate fauna listed as threatened. *Inter alia*, this reflected the application of the criterion in the Act that, 'the Committee is to have regard to …(e) any other matter which the Committee considers relevant', as well as the substantial use we made of the numerical scoring procedure. However, since we did not use numbers as a cut-off point, then the decision to list or not list reflects our expert judgment in relation to each individual score and the weight of expert opinion. We thus have two layers of expert opinion, with the Committee's expertise being a combination of the numerical and voting scores provided by the experts, guided by our own local knowledge of the fauna, the experts and the relevant scientific literature.

A variety of decisions could have been applied to the data we obtained in constructing the lists, ranging from the conservative to the speculative. For example, we could have required unanimity among the experts to proceed to a listing, used just one vote for listing, or given most weight to the Committee's own expertise. In practice, the Committee adjudicated on the range of opinions among the experts. When we now examine the differences among the vertebrate classes, we note that the reptiles show an 8-fold difference among 80 species that would have been listed with a decision to give the highest rank to listing, compared to 10 by the most cautious decision (Table 2). The Committee settled on 27 species of reptiles. The decision-making for the frogs is also instructive. By using the conservative criteria, i.e. the lowest alternative, only seven species would have been listed, and none would have been called endangered. In the end, four frog species were listed as endangered and the dramatic rise to fame of the Green and Golden Bell Frog *Litoria aurea* was underway (*e.g.* Goldingay 1996; Greer 1996; Pyke and Osborne 1996; Pyke and White 2001; Christy 2003).

Table 2. Comparison of the number of species listed with possible outcomes if alternative criteria were used.

Highest alternative – includes species where at least one expert nominated listing. Lowest alternative – includes species only where all experts agreed on listing. Endangered category here includes extinct species.

Listed category	Listed 1992	Highest alternative	Lowest alternative
Birds			
Protected	363	331	425
Vulnerable	78	86	33
Endangered	32	56	15
Total	473	473	473
Mammals			
Protected	53	38	68
Vulnerable	40	39	33
Endangered	37	53	29
Total	130	130	130
Reptiles			
Protected	182	129	199
Vulnerable	23	69	10
Endangered	4	11	0
Total	209	209	209
Frogs			
Protected	52	32	64
Vulnerable	15	24	7
Endangered	4	15	0
Total	71	71	71
All species			
Protected	650	530	756
Vulnerable	156	218	83
Endangered	77	135	44
Total	883	883	883

The special case of frogs

In 1991, frogs were not considered to be fauna under the NSW *National Parks and Wildlife Act 1974*. Fauna until then comprised just three classes of vertebrates (birds, mammals and reptiles). In fact, reptiles have enjoyed legislative recognition only since 1974 with the passage of the *National Parks and Wildlife Act 1974*. Thus, in 1992 the NPWS did not have a corporate working knowledge of frogs, and it had to rely on the skills and collections at the Australian Museum, as well as local expertise, expressed, for example, in the books of Barker and Grigg 1977 and Barker *et al.* 1995 (which was then in the pipeline). The surprise inclusion of frogs as fauna under the legislation has been commented on previously (*e.g.* Cogger and Lunney 1992; Lunney and Ayers 1993), but what emerged was a new way of looking at the world. However, recent reviews on frogs have changed our perception on which species should be on the lists.

In 2001-2003, under the auspices of WWF (Australia) and Rio Tinto through their Frogs! partnership program, a review was conducted of the frogs of Australia by the program's Scientific Panel, on which one of us (Cogger) sat. A modified version (to reflect the special biological and ecological features of frogs) of the Australian Reptile Action Plan (Cogger *et al.* 1993) was adopted. Subsequently a small group of specialists was convened to complete the final iteration of a Millsap analysis of all Australian frogs in May 2003, producing results and a ranking of threatened frogs that varied in fairly minor respects from those determined by the Scientific Panel alone. The latter results are cited in Table 3. However, one needs to bear in mind that these scores cannot be compared directly with the 1992 NSW scores. As already indicated, the Millsap criteria were revised for the 2003 scores to better match frog biology and ecology. Further, being a national ranking, the criteria used were applied at a different geographic scale to those used in the NSW analysis. Nevertheless, retaining the Millsap scoring system makes it easier to analyse the new results using Millsap's suite of statistical tests, as well as to make comparisons with earlier data sets. The NSW subset of the 2003 WWF review was constructed to make a comparison with the 1992 status listings for frogs in NSW (shown in Table 3).

The frogs are ranked in Table 3 according to their Millsap score in early 2003. Their 1992 NSW status is presented in the next column. Some discrepancies become immediately apparent. The changes in status in NSW since 1992 are presented in the final column, and some movement towards the WWF 2003 position can be seen. What stands out as particularly odd in this table is the 1997 Australian Frog Action Plan (Tyler 1997). One notices that the current rankings are not well aligned with the 1997 Frog Action Plan, which was based, ultimately, on the subjective views of an unknown number of people (after lots of input from others). Further, it was not based on any published suite of objective criteria, but the Frog Action Plan states (p.4) that the IUCN criteria that then applied were adopted, and the definitions of these categories were repeated in the Action Plan. The main issues are that (1) the reasons for the decisions to place a species in a particular category were not transparent (*i.e.* the application of the IUCN criteria to each taxon

Table 3. Threatened frogs in New South Wales. Rank – from a 2003 assessment by the WWF Frogs! Panel using Millsap biological scores. 1992 NPWS Millsap Assessment (Lunney *et al.* 2000) used to derive threatened species listing in NSW under the *Endangered Fauna (Interim Protection) Act* 1991 and then transferred under the *Threatened Species Conservation Act*, 1995. 2003 NSW – identifies changes to the NSW threatened species listing since 1992.

Rank (NSW only)	NSW Frog Species	1992 NPWS Millsap Assessment	1997 Aust Frog Action Plan	2003 WWF Frogs! Panel Millsap Score (biological criteria only)	2003 NSW (changes since 1992)
1	*Pseudophryne corroboree* ^	Vulnerable	Endangered	65	Endangered
2	*Adelotus brevis* *	At least one expert nomination	unlisted	53	Endangered population β
3	*Litoria castanea* ^	Endangered	Endangered	52	
4	*Litoria spenceri* ^	Endangered	Endangered	51	
5	*Litoria piperata* ^	Vulnerable	Vulnerable	50	
6	*Pseudophryne pengilleyi*	not considered	of concern; insufficiently known	49	Vulnerable
7	*Mixophyes iteratus* ^	Vulnerable	Endangered	44.3	Endangered
8	*Mixophyes fleayi* ^	Vulnerable	Endangered	44	Endangered
9	*Crinia tinnula* ^	Vulnerable	of concern; insufficiently known	38	
10	*Litoria aurea* ^	Endangered	Endangered	37	

Rank	NSW Frog Species	1992 NPWS Millsap Assessment	1997 Aust Frog Action Plan	2003 WWF Frogs! Panel Millsap Score	2003 NSW
11	*Assa darlingtoni* ^	Vulnerable	Unlisted	36	
11	*Heleioporus australiacus* ^	Vulnerable	of concern; insufficiently known	36	
11	*Mixophyes balbus* ^	Vulnerable	Vulnerable	36	Endangered
11	*Litoria booroolongensis* *	At least one expert nomination	of concern; insufficiently known	36	Endangered
12	*Pseudophryne australis* ^	Vulnerable	of concern; insufficiently known	35	
13	*Litoria subglandulosa* ^	Vulnerable	of concern; insufficiently known	34	
14	*Philoria loveridgei* ^ φ	Vulnerable	unlisted	33	
14	*Philoria sphagnicolus* ^ φ	Vulnerable	of concern; insufficiently known	33	
15	*Litoria olongburensis* ^	Vulnerable	Vulnerable	32	
16	*Philoria kundagungan* ^	Vulnerable	unlisted	31	
16	*Litoria freycineti* *	At least one expert nomination	of concern; insufficiently known	31	
17	*Litoria revelata* *	At least one expert nomination	unlisted	27	
18	*Litoria raniformis* ^	Endangered	Vulnerable	24	
18	*Litoria pearsoniana* *	At least one expert nomination	of concern; insufficiently known	24	
18	*Notaden bennettii*	No expert nomination	unlisted	24	
19	*Crinia sloanei* *	At least one expert nomination	unlisted	23	
10	*Pseudophryne bibronii*	No expert nomination	of concern; insufficiently known	21	
21	*Limnodynastes interioris* *	At least one expert nomination	unlisted	20	
22	*Cyclorana verrucosa* *	At least one expert nomination	unlisted	19	
23	*Lechriodus fletcheri* *	At least one expert nomination	unlisted	18	
23	*Pseudophryne dendyi* *	At least one expert nomination	unlisted	18	
23	*Litoria citropa*	No expert nomination	unlisted	18	
24	*Cyclorana platycephala*	No expert nomination	unlisted	17	
24	*Geocrinia victoriana*	No expert nomination	unlisted	17	
25	*Limnodynastes terraereginae* *	At least one expert nomination	unlisted	16	
25	*Pseudophryne coriacea*	No expert nomination	unlisted	16	
25	*Uperoleia fusca*	No expert nomination	unlisted	16	
26	*Limnodynastes salmini* *	At least one expert nomination	unlisted	15	
26	*Paracrinia haswelli* *	At least one expert nomination	unlisted	15	
26	*Litoria alboguttata* *	At least one expert nomination	unlisted	15	
26	*Crinia parinsignifera*	No expert nomination	unlisted	15	

Rank	NSW Frog Species	1992 NPWS Millsap Assessment	1997 Aust Frog Action Plan	2003 WWF Frogs! Panel Millsap Score	2003 NSW
26	*Litoria dentata*	No expert nomination	unlisted	15	
26	*Uperoleia tyleri*	No expert nomination	unlisted	15	
27	*Litoria brevipalmata* ^	Vulnerable	of concern; insufficiently known	14	
27	*Limnodynastes fletcheri*	No expert nomination	unlisted	14	
27	*Uperoleia laevigata*	No expert nomination	unlisted	14	
28	*Litoria jervisiensis* *	At least one expert nomination	unlisted	13	
28	*Cyclorana novaehollandiae*	No expert nomination	unlisted	13	
28	*Litoria chloris*	No expert nomination	unlisted	13	
28	*Litoria phyllochroa*	No expert nomination	unlisted	13	
29	*Limnodynastes dumerilii*	No expert nomination	unlisted	12	
29	*Limnodynastes tasmaniensis*	No expert nomination	unlisted	12	
29	*Uperoleia rugosa*	No expert nomination	unlisted	12	
30	*Neobatrachus sudelli*	No expert nomination	unlisted	11	
31	*Mixophyes fasciolatus* *	At least one expert nomination	unlisted	10	
31	*Litoria lesueuri* Ψ	No expert nomination	unlisted	10	
31	*Uperoleia capitulata*	No expert nomination	unlisted	10	
32	*Litoria gracilenta* *	At least one expert nomination	unlisted	9	
33	*Crinia signifera*	No expert nomination	unlisted	8	
33	*Limnodynastes peronii*	No expert nomination	unlisted	8	
33	*Litoria latopalmata*	No expert nomination	unlisted	8	
33	*Litoria tyleri*	No expert nomination	unlisted	8	
34	*Litoria caerulea* *	At least one expert nomination	unlisted	7	
34	*Litoria nasuta* *	At least one expert nomination	unlisted	7	
34	*Litoria peronii*	No expert nomination	unlisted	7	
34	*Litoria verreauxii*	No expert nomination	unlisted	7	Sub-species *L. v. alpina* Endangered
35	*Litoria fallax* *	At least one expert nomination	unlisted	6	
35	*Crinia deserticola*	No expert nomination	unlisted	6	
35	*Limnodynastes ornatus*	No expert nomination	unlisted	6	
36	*Litoria ewingii* *	At least one expert nomination	unlisted	4	
36	*Litoria rubella*	No expert nomination	unlisted	4	
	Neobatrachus centralis	No expert nomination	unlisted		

Rank	NSW Frog Species	1992 NPWS Millsap Assessment	1997 Aust Frog Action Plan	2003 WWF Frogs! Panel Millsap Score	2003 NSW
	Litoria daviesae #				Vulnerable
	Litoria littlejohni #				Vulnerable
	Neobatrachus pictus ϕ				Endangered

^ Assigned threatened status in 1992

* At least one expert nomination in 1992

Previously undescribed species

ϕ Previously unknown in NSW

β Nandewar and New England Tablelands Bioregion endangered population of *Adelotus brevis*

φ Some populations of *Philoria loveridgei* and *P. sphagnicola* are now assigned to *P. kundagungan*, *P. pughi* and *P. richmondensis* (Knowles *et al.* 2004)

ψ 1992 NSW populations of *Litoria lesueuri* have variously been assigned to *L. lesueuri*, *L. wilcoxi* Günther, 1864 and to hybrid populations by Donnellan and Mahony (2004)

was not published), and (2) the IUCN criteria are global in context, and so do not necessarily reflect national priorities. The use of a scoring procedure such as Millsap *et al.* (1990) can help overcome such difficulties as seen in the Frog Action Plan.

The extent to which the current status of each frog taxon differs from earlier assessments can be seen from the extent to which the status assessments between the columns are mixed up, but with the caveat that a difference of only 2 or 3 points in the scores (on which the rankings are based) is not statistically significant. The taxa worth pointing out are those that have serious discrepancies, such as *Litoria brevipalmata, Adelotus brevis, Pseudophryne bibronii* and *Notaden bennetti.* A further major caveat is that the comparison can be based only on rankings (not threatened status) and these will vary with scale, especially where rankings are not based on the total geographic range of a species. Thus a species whose range in NSW is marginal (*i.e.* it represents only a small part of its total range) will be likely to be assigned elevated threatened status (or ranking) by comparison with the ranking of the same species when total geographic ranges are considered. This is not necessarily a criticism of state rankings, but in most cases a reflection of the interaction of two major criteria – the size of a species' geographic range under consideration and the extent to which that range is itself subject to major threatening processes, such as land clearing and overgrazing.

In relation to the inadequacies of the current threatened species legislation (NSW TSC Act and the Commonwealth EPBC Act), and the manner in which the courts apply them, the concerns now for frogs (and indeed many other vertebrates) are that all legislative instruments fail to take sufficient account of (a) incremental loss of habitat through the tyranny of small decisions; (b) scale of development; (c) edge effects and other effects of fragmentation; (d) the need for buffer zones in many cases, especially those affecting wetlands; and (e) the fact that the number of humans interacting with natural systems post-development is a critical factor in assessing impacts and in reducing their effects through good design and the use of technological solutions (such as fences).

Public attention for threatened species

Threatened species legislation in large part pivots on the view (as encapsulated in the IUCN category definitions) that an endangered species is one category away from the extinct species list and a vulnerable species is heading towards the endangered species list. The message that is loud and clear to the general public is that species can decline to the point of being endangered and that if remedial action is not taken, then the decline will continue to extinction. The Numbat *Myrmecobius fasciatus* is an icon species for Western Australia. Yet few know that it was a NSW resident at the time of European settlement. The Numbat is now one of the state's 27 species of mammals that has slid to extinction since 1788 and is one of 24 mammal species that became extinct in western NSW in the first 60 years of European occupation of that area. Some of the other native mammals, such as the Pig-footed Bandicoot *Chaeropus ecaudatus*, have not only disappeared from western NSW, but from Australia. The cause of the extinction of the mammals of western NSW can be found in the overwhelming changes to the land wrought by sheep and compounded by the rabbits and foxes that followed in their wake. The impact was so profound that a Royal Commission on the Western Lands was formed and in 1901 it handed in its report on the causes and consequences of the use, or rather overuse, of the arid and semi-arid lands of NSW. The views found in that report can be interpreted ecologically to help explain one of the greatest mass extinction episodes of the modern era (*e.g.* Griffiths 2001; Lunney 1994, 2001; Noble 2001).

The catchcry that 'extinction is forever' has a powerful appeal, but the billboard invitation by the Australian Museum in 2003, to come and see the Thylacine 'while it is still extinct', carried an even more powerful message. If as yet undiscovered genetic technology can clone a Thylacine *Thylacinus cynocephalus* from tissue of a specimen in the museum, the cost will be immense. The important message is to conserve a species while it is extant and the extension of that message is to prevent its decline. Reversing a serious and rapid decline is likely to be a monumental task. In seizing public attention, the iconic threatened and extinct species, in our opinion, have done a brilliant public relations job for threatened species. Koalas, for example, have been a great icon for conserving forested land that is not in public ownership (*e.g.* Hamilton *et al.* 2000; Lunney *et al.* 2002). This task would have been less effective if the Koala had not been on the threatened species list. Thus there has been a positive value in working with a threatened species, but the Koala did not have this status in NSW when the work started (Lunney *et al.* 1990).

The most broadly threatened mammalian group in NSW is the rodents (Dickman *et al.* 2000). Many species are threatened or extinct, but some are sufficiently widespread to fulfil another primary goal for conservation – that is, to serve as subjects of an evaluation of threats, including potential for recovery, through experimental design and not just anecdotal information. We must be fundamentally committed to a measurable sense of recovery (or impact), for which experimental approaches such as replicated, balanced designs can be employed. Without such a commitment to monitoring and investigation at this high level of endeavour, there will be too much guesswork, or prejudice, about impacts or recovery. The best species to work with are those, such as the Bush Rat *Rattus fuscipes*, that are sufficiently common to provide estimates of numbers. These are not threatened species; they belong in the 74% bracket. What we are arguing is that good conservation practice demands good science, and good science demands good experimental design, which can rarely be provided by threatened species simply because their populations are usually too small to be studied meaningfully. So, the other 74% is the best group to employ in studies aimed at conserving habitats, defining and curbing the threatening processes, and evaluating recovery.

Conclusion

The monograph by Lunney *et al.* (2000) explained how we arrived at the 1992 lists of threatened and non-threatened native vertebrate fauna of NSW. The aim here is to emphasise the level of uncertainty we encountered as we prepared the lists in 1992, and to present our reflections on those levels of uncertainty. We also consider the deleterious ecological consequences of investing most conservation effort in threatened species, rather than taking a holistic approach that deals with threatening processes for all fauna. The threatened species lists have been a boon for conserving biodiversity, but they should not become an impediment to other intelligent and essential approaches to the difficult task of conserving all our native fauna. Nor should the lists, and the procedures that led to their formulation, remain static. Inadequate knowledge undermines the listings, but how do we increase our knowledge? It is not, we would argue, by looking only at the species on the list. We need to take a more ecological approach and reflect on the broader aim of conserving biodiversity, not just of those species currently on the threatened species schedules. The report, *Biodiversity Conservation Research: Australia's Priorities*, provides one intelligent approach as to what to do, and we note that determining lists of threatened species is an outcome of a monitoring process, not simply a self-contained aim (ANZECC/BDAC 2001). We emphasise this so that others will look critically at the schedules, and the methods, and come up with imaginative solutions for new ways to conserve our fauna, and not just work from an historic list that lacks the latest information. We also need to minimise the time lag between the identification of significant declines (or recoveries) and the updating of legislative schedules to reflect those changes.

The primary conclusion from this analysis is that the current NSW schedules of threatened species were produced with a recognition of the limitations of the methods, and the knowledge and the time available. This recognition was reflected in the inclusion of three measures of variability (level of knowledge, differences in expert opinion, and final judgement by the Committee, which included the numerical scores for each species). These measures can be interpreted in a number of ways, and as few as 127 species, or as many as 353 species could have been listed as a result of the exercise. The final 1992 list was 236 species, or 26% of the fauna. Consequently, we need to consider the lists as approximations, as working lists in need of revision, and not immutable, and certainly not as definitive. Frogs highlight one element of the problem. If one chose to tighten or relax the criteria for inclusion of species on the schedules, then one could simply move a marker up or down a set of species fixed in ranked order of degree of threat. But these rankings will vary over time – often over very short periods – as new information about the status of particular species and their distribution and ecosystems comes to hand through research or observations. The draft 2004 WWF-funded revision of Australian frog rankings shows that a sound, threatened species list is much more than the result of moving a marker up and down. A revision of the 1992 NSW lists is in order.

The second conclusion is that the lists have created a group of outcasts, the non-listed species. We call these 'the neglected 74%', yet it is our judgment that not only should they be included in the primary thrust of conserving biodiversity, but they also offer opportunities, such as populations amenable to effective experimental sampling and manipulation, to aid in the process of identifying threats and evaluating solutions that are not possible with threatened species.

Acknowledgements

We are indebted to many colleagues over many years for discussions on this matter and the internal support from the Department of Environment and Conservation (which includes the NSW National Parks and Wildlife Service) to research this matter, to the Council of the Royal Zoological Society of NSW for its support of the forum that created the opportunity to produce this paper, to members of the WWF Frogs! Scientific Panel and to Irina Dunn, Pat Hutchings and two referees for critical comments on the manuscript.

References

ANZECC/BDAC. 2001. *Biodiversity Conservation Research: Australia's Priorities.* Environment Australia, Canberra, ACT.

Barker, J. and Grigg, G. C. 1977. *A Field Guide to Australian Frogs.* Rigby, Adelaide.

Barker, J., Grigg, G. C. and Tyler, M. J. 1995. *A Field Guide to Australian Frogs.* Surrey Beatty & Sons, Chipping Norton, NSW.

Belcher, C. 2004. The largest surviving marsupial carnivore on mainland Australia: the Tiger or Spotted-tailed Quoll *Dasyurus maculatus*, a nationally threatened, forest-dependent species. Pp 612-23 in *Conservation of Australia's Forest Fauna* (second edition), edited by D. Lunney. Royal Zoological Society of NSW, Mosman, NSW.

Christy, M. T. 2003. Making headlines or saving a species? *Australian Zoologist* 324-28, in the proceedings of the *Zoology and the Media* forum, edited by D. Lunney, A. Matthews and P. Wilson. *Australian Zoologist* 32, No 2, 2003.

Cogger, H. G., Cameron, E. E., Sadlier, R. A. and Eggler, P. 1993. *The Action Plan for Australian Reptiles.* ANCA, Canberra.

Cogger, H. G. and Lunney, D. 1992. The *Endangered Fauna (Interim Protection) Act 1991*: criteria for the selection of threatened, vulnerable and rare species of fauna. Pp. 1-9 in *The Endangered Fauna Act in Action*, edited by the Environmental Defender's Office, Sydney.

Commonwealth of Australia. 1992. *National Forest Policy Statement. A new focus for Australia's forests.* (second edition). Australian Government Publishing Service, Canberra.

Dickman, C. R., Lunney, D. and Matthews, A. 2000. Ecological attributes and conservation of native rodents in New South Wales. *Wildlife Research* 27: 347-55.

Donnellan, S. C. and Mahony, M. J. 2004. Allozyme, chromosomal and morphological variability in the *Litoria lesueuri* species group (Anura: Hylidae), including a description of a new species. *Australian Journal of Zoology* 52: 1-28.

Goldingay, R. L. 1996. The green and golden bell frog *Litoria aurea* – from riches to ruins: conservation of a formerly common species. *Australian Zoologist* 30: 248-56.

Greer, A. E. 1996. Why green and golden bell frogs *Litoria aurea* should not be translocated – an unpublished opinion. *Australian Zoologist* 30: 257-8.

Griffiths, T. 2001. One hundred years of environmental crisis. *The Rangeland Journal* 23: 5-14.

Hamilton, C., Lunney, D. and Matthews, A. 2000. An economic evaluation of local government approaches to koala conservation. *Australian Journal of Environmental Management* 7: 158-69.

Jarman, P. J. and Brock, M. A. 2004. The evolving intent and coverage of legislation to protect biodiversity in New South Wales. Pp 1-19 in Threatened Species Legislation: is it just an act? edited by P. Hutchings, D. Lunney and C. Dickman. Royal Zoological Society of NSW, Mosman, NSW.

Knowles, R., Mahony, M., Armstrong, J. and Donnellan, S. 2004. Systematics of sphagnum frogs of the genus *Philoria* (Anura: Myobatrachidae) in eastern Australia, with the description of two new species. *Records of the Australian Museum* 56: 57-74.

Körtner, G., Gresser, S. and Harden, B. 2003. Does fox baiting threaten the spotted-tailed quoll, *Dasyurus maculatus? Wildlife Research* 30: 111-8.

Lunney, D. (ed.). 1992. *Zoology in Court.* Royal Zoological Society of New South Wales, Mosman, NSW.

Lunney, D. 1994. Review of official attitudes to western New South Wales 1901-93 with particular reference to the fauna. Pp. 1-26 in *The Future of the Fauna of Western New South Wales*, edited by D. Lunney, S. Hand, P. Reed and D. Butcher. Royal Zoological Society of NSW, Mosman, NSW.

Lunney, D. 2001. Causes of the extinction of native mammals of the Western Division of New South Wales: an ecological interpretation of the nineteenth century historical record. *The Rangeland Journal* 23: 44-70.

Lunney, D. and Ayers, D. 1993. The official status of frogs and reptiles in New South Wales. Pp. 404-8 in *Herpetology in Australia: a Diverse Discipline*, edited by D. Lunney and D. Ayers. Royal Zoological Society of NSW, Mosman, NSW.

Lunney, D., Curtin, A., Ayers, D., Cogger, H. G. and Dickman, C. R. 1996. An ecological approach to identifying the endangered fauna of New South Wales. *Pacific Conservation Biology* 2: 212-31.

Lunney, D., Curtin, A. L., Ayers, D., Cogger, H. G., Dickman, C. R., Maitz, W., Law, B. and Fisher, D. 2000. The threatened and non-threatened native vertebrate fauna of New South Wales: status and ecological attributes. *Environmental and Heritage Monograph Series No. 4.* Pp. 1-132. (NSW) National Parks and Wildlife Service, Hurstville.

Lunney, D., Curtin, A., Fisher, D., Ayers, D. and Dickman, C. R. 1997. The ecological attributes of the threatened fauna of New South Wales. *Pacific Conservation Biology* 3: 13-26.

Lunney, D. and Matthews, A. 2004. Conserving the forest mammals of NSW. Pp 988-1021 in *Conservation of Australia's Forest Fauna* (second edition), edited by D. Lunney. Royal Zoological Society of NSW, Mosman, NSW.

Lunney, D., Matthews, A., Stein, J. A. and Lunney, H. W. M. 2003. Australian bat research: the limitations of *The Action Plan for Australian Bats* in determining the direction of research. *Pacific Conservation Biology* 8: 255-60.

Lunney, D., Matthews, A., Moon, C. and Turbill, J. 2002. Achieving fauna conservation on private land: reflections on a ten year project. *Ecological Management and Restoration* 3: 90-6.

Lunney, D., Urquhart, C. A. and Reed, P. (eds). 1990. *Koala Summit. Managing Koalas in NSW.* Proceedings of a conference held at the University of Sydney 7-8 November 1988. Pp. 1-268. (NSW) National Parks and Wildlife Service, Hurstville.

Margules, C. R. and Pressey, R. L. 2000. Systematic conservation planning. *Nature* 405: 243-53.

Millsap, B. A., Gore, J. A., Runde, D. E. and Cerulean, S. I. 1990. Setting the priorities for the conservation of fish and wildlife species in Florida. *Wildlife Monographs Supplement Journal of Wildlife Management* 54: 5-57.

Noble, J. 2001. Guest editorial. *The Rangeland Journal* 23: 3-4.

Pressey, R. L., Whish, G. L., Barrett, T. W. and Watts, M. E. 2002. Effectiveness of protected areas in north-eastern New South Wales: recent trends in six measures. *Biological Conservation* 106: 57-69.

Pyke, G. H. and Osborne, W. S. (eds). 1996. The green and golden bell frog *Litoria aurea*: biology and conservation. Special Edition. *Australian Zoologist* 30.

Pyke, G. H. and White, A. W. 2001. A review of the biology of the green and golden bell frog *Litoria aurea*. *Australian Zoologist* 31: 563-98.

Tyler, M. J. 1997. *The Action Plan for Australian Frogs.* Wildlife Australia, Canberra. 77 pages.

APPENDIX I

The Koala *Phascolarctos cinereus*, shown here wearing a radiotransmitter, is a great icon for the cause of forest dependent fauna, but if threatened species recovery was the only criterion for conserving biodiversity, we would have a warped sense of priorities because we would be neglecting non-threatened species, which constitute most of the fauna.

Photo: D. Lunney.

The Bush Rat *Rattus fuscipes*, shown here inside an "Elliott" live trap, is one of the neglected 74% of vertebrate fauna but it is a valuable species for monitoring impacts because it occurs in sufficient numbers to show changes through time.

Photo: D. Lunney.

The Tusked Frog *Adelotus brevis* was not listed as threatened in 1992 but was ranked second in NSW in a 2003 assessment by the WWF Frogs! Panel using Millsap biological scores. This frog provides one example of how inadequate knowledge has undermined the listing process and why a revision of the 1992 NSW lists is in order.

Photo: D. Lunney.

Threatened species legislation: Does it work for local communities or Local Government?

Judy Lambert

North Head Sanctuary Foundation, PO Box 896 Balgowlah NSW 2093

Email: twswombat@optusnet.com.au

ABSTRACT

The paper examines the Commonwealth and New South Wales threatened species legislation from the perspective of local communities and a local government seeking to protect threatened species. Both the Commonwealth and State legislation are identified potentially to be useful tools, providing ample opportunities for involvement in protecting threatened species and ecological communities and their habitats. However, groups and individuals in the community who have an interest in conserving these key elements of unique natural environments are disengaging. Ministerial discretion is undermining commitment to the Commonwealth's *Environment Protection and Biodiversity Conservation Act 1999* and the NSW *Threatened Species Conservation Act 1995*.

Local Governments, faced with a growing suite of State and Commonwealth demands, have scarcely begun to engage with Commonwealth threatened species legislation, even though Matters of National Environmental Significance addressed by the Commonwealth legislation may be locally relevant.

The protection of endangered species, ecological communities and populations on North Head at Manly is used to highlight some of these concerns as they relate to both Commonwealth and State threatened species legislation. The paper concludes with a call for stronger partnerships between scientists, local government, local communities and the environment movement, to bring governments to greater accountability in using the existing legislation to protect Australia's unique biodiversity, some of which has an iconic status locally.

Key words: Threatened species, community action, Local Government role, legislative opportunities, Ministerial discretion

Introduction

This paper draws on the author's experience as a Ministerial Consultant during the development and introduction of the former Commonwealth *Endangered Species Protection Act 1992*, and with the Environmental Defenders' Office staff during the passage of the Commonwealth *Environment Protection and Biodiversity Conservation Act 1999* (the EPBC Act). The author also draws on experience as one of the leaders of a multi-stakeholder project directed to improved management of the endangered Grassy White Box Woodlands on the western slopes of New South Wales, as an elected local representative (i.e. a councillor) on Manly Council, and as a foundation member of the North Head Sanctuary Foundation.

The North Head Sanctuary Foundation is a community-based not-for-profit organisation whose members include scientists and other professionals working side-by-side with community members with expertise in research, education, and other aspects of natural, Aboriginal, built and cultural heritage. The Foundation was formed in 2002 with the purpose of securing a publicly-owned sanctuary of national significance across the whole of North Head, managed predominantly as a protected environment to conserve its natural attributes, while at the same time celebrating the Aboriginal, immigration and military heritage of this spectacular place, and providing opportunities for learning, research and contemplation.

North Head, at the gateway to Sydney Harbour, is a place of outstanding natural and cultural heritage. It is a tied island largely surrounded by spectacular seacliffs up to 60 metres high and by flooded river valleys. Past uses as a Defence training area and as Australia's oldest and longest serving Quarantine Station have ensured that some of the most extensive and intact tracts of heath and scrubland around Sydney Harbour are found there. The North Head area provides refuge for artificially disjunct species, remnant populations and ecological communities displaced by urbanisation and now recognised as endangered. These natural values, together with its built and cultural heritage associated with its immigration, quarantine and military history see the whole of North Head listed on the Register of the National Estate.

Experiences with North Head and elsewhere indicate that both Commonwealth and State threatened species legislation deserve the label 'a curate's egg' i.e. good in parts. Whether viewed from the perspective of a community advocate for threatened species, a former government policy adviser, or an elected Local Government representative, the Commonwealth and New South Wales Acts are each in part good in protecting threatened species, but they are also deficient in several ways.

Pp 158 - 163 in *Threatened species legislation: is it just an Act?* 2004, edited by Pat Hutchings, Daniel Lunney and Chris Dickman. Royal Zoological Society of New South Wales, Mosman, NSW, Australia.

What Commonwealth legislation has to offer

It is useful, when considering community and Local Government perspectives, to begin with the Commonwealth legislation directed to threatened species conservation.

Through a process of referral, assessment and approval, the *EPBC Act 1999* gives the Commonwealth powers to act on any of seven Matters of National Environment Significance, namely to protect World Heritage properties, wetlands of international significance, species and ecological communities listed as threatened at a national level, listed migratory species, the environment in Commonwealth marine areas, the environment in relation to nuclear actions (including uranium mining), and more recently items of national heritage significance.

A recent review of the Commonwealth-funded Environment Protection and Biodiversity Conservation (EPBC) Unit, done for a community sector client by the author's consultancy business Community Solutions (Lambert *et al.* 2003), clearly highlighted the extent to which the merits of legislation and community willingness to engage with that legislation depend hugely on the will of governments to implement it.

The threatened species aspects of the *Environment Protection and Biodiversity Conservation (EPBC) Act 1999* are potentially quite powerful. The opportunities for community action are numerous, and can be summarised as follows:

- bring actions to the attention of the Minister for Environment and Heritage if they are considered likely to have a significant impact on aspects of the environment, including nationally listed threatened species. Although not formally able to 'refer' matters to the Minister in this process, anyone can ask the Minister to have a project proponent refer an activity for consideration, and community members can also play an important role in scrutinising and commenting on the information provided by a project proponent when they are referring an action to the Minister.
- lobby for a higher level of assessment of any action determined to be a 'controlled action', and make input to Public Environment Reports, Environmental Impact Statements and Public Inquiries triggered by the Minister in relation to 'controlled actions'
- nominate a species or ecological community for listing under the Act
- nominate Key Threatening Processes (e.g. Predation by Feral Cats, Clearing of Native Vegetation), which must be addressed by a Threat Abatement Plan made under the Act
- make representations in relation to the identification of Critical Habitat for a species listed under the Act
- make representations to the Compliance and Enforcement Unit in relation to any perceived breaches of the Act
- in particular circumstances (with 'standing'), take legal action to have the Act enforced

However Ministerial discretion within the Act, and the extent to which the Federal Minister for the Environment has chosen not to use the Act to its full potential or to respond to community representations, are doing much to discredit this legislation. Community sector responses provided during a recent review of services provided by the community-based EPBC Unit (Lambert, Elix and Pearce 2003) reveal a perception of limited usefulness in engaging with the Act. Lack of adequate knowledge and understanding of the Act, lack of resources, cynicism about the effectiveness of the Act in protecting the environment, failure to perceive opportunities to use the Act to protect the environment, and lack of confidence in engaging with the legislative and policy process were all significant factors cited as reasons for a relative lack of community engagement. Environmental non-government organisations and local communities are beginning to disengage – they simply are not willing to invest the time, effort and resources needed to get nominations accepted, seek redress for breaches of the Act, or make referrals under the provisions of the Act if there is no demonstrated political will to act on those representations.

Despite numerous referrals, the Commonwealth only initiated its first enforcement action under the Act three years after the legislation came into force. The *Minister v. Greentree* case heard in the Federal Court of Australia (2003) addressed an issue of impacts on wetlands of international significance.

The Australian National Audit Office in its Audit Report in April 2003 (ANAO 2003), in its performance audit of Environment Australia's[1] administration of the *EPBC Act 1999*, highlighted the areas which are causing community concern.

Of the 603 actions referred to the Minister as 'likely to have a significant impact on the environment', only 29% (175 actions) became 'controlled actions' requiring Ministerial approval, and at the time of the Audit Report 34 of these had been approved, and none refused.

Monitoring and enforcement of compliance with the Act were identified by the Audit Report as being "crucial to the effective operation of the Act", but were described as "still being at an early stage" and "not ideal". Environment Australia's[1] responses to potential breaches of the Act were described as "less than robust".

While the examples of lack of commitment are numerous, perhaps one of the most notorious examples of this lack of robustness is the approval recently given for the Portman iron ore mine expansion at Koolyanobbing in WA. There the Federal Minister approved a development which his own media release (Kemp 2003) at the time of the approval, acknowledged will result in an initial destruction of 30% of individual plants of the endangered species *Tetratheca paynterae*, with a further 20% likely to be affected later.

Precedents like this cause community members working for the conservation of North Head and its endangered species and ecological communities to question the

[1] Environment Australia is the former title for the Commonwealth Department of Environment & Heritage.

robustness of a Ministerial determination that the proposed Mawland Hotel development at the North Head Quarantine Station is "highly unlikely" to have a "significant impact" on nationally listed flora in the area. The determination is apparently based on a commitment by the development proponent to avoid all threatened species and the ecological community within the lease area (Wilmot 2002). This is of particular concern when the proponent of the Mawland hotel development is seeking the right to remove a section of endangered Eastern Suburbs Banksia Scrub through an offsetting process.

Similarly, although the endangered status of the East Coast population of Grey Nurse Shark *Carcharias taurus* was upgraded from 'vulnerable' to 'critically endangered' during 2001 because of a "very severe reduction in population size due to impacts such as fishing", a 20 February 2002 referral to the Federal Minister seeking consideration of the likely impacts of the Mawland Hotel development and associated works on the adjoining Aquatic Reserve which is a potential nursery area for this species was, by 27 February 2002, determined not to be a 'controlled action' requiring approval under the *EPBC Act* (Mercer 2002).

However, all is not negative at the Commonwealth level. In the *Minister v. Greentree* case the Federal Minister for the Environment has successfully taken action in the Federal Court to restrain land clearing in RAMSAR listed wetlands (Federal Court of Australia 2003).

Third party standing provisions, and the removal of the requirement to give an undertaking as to damages under the *EPBC Act*, do enable community groups to take action. The potential for outcomes is demonstrated by one notable case - the *Booth v. Bosworth* Spectacled Flying Fox *Pteropus conspicillatus* case in the Federal Court (Federal Court of Australia 2000). While this case focused on World Heritage (rather than Threatened Species) as a Matter of National Environmental Significance, the result was the granting of a permanent injunction protecting the Spectacled Flying Foxes within the Wet Tropics World Heritage Area against culling by electric grid. In handing down his judgment Justice Branson said "*In weighing the factors which support an exercise of the Court's discretion in favour of granting an injunction… it would be a rare case in which a Court could be satisfied that the financial interests of private individuals, or even the interests of a local community, should prevail over interests recognised by the international community and the Parliament of Australia as being of international importance.*" (para.115)

That the Courts offer a greater opportunity for community action to protect threatened species than does direct referral to the Federal Minister is reinforced by a second case brought by the Humane Society International (Australia). In this case the Grey-headed Flying Fox *Pteropus poliocephalus*, which is listed nationally as vulnerable to extinction and which the Federal Minister thus has responsibilities to protect, was protected from culling through the Court (Federal Court of Australia 2003).

The Commonwealth, in the early stages of the legislation, demonstrated its willingness to list threatened ecological communities. Grassy White Box Woodlands received Commonwealth listing before it was listed in NSW. The former Federal Minister for the Environment, Robert Hill, also showed courage in the face of hostile landholder reaction, when he listed Queensland's Brigalow Woodlands and the Bluegrass Grasslands in April 2001. However, that courage, like other aspects of the application of the *EPBC Act*, has since evaporated. More recently Minister Kemp has listed only two new ecological communities in more than two years as Environment Minister. This despite the fact that there are over 500 threatened ecological communities listed in the States and Territories and 2891 "threatened assemblages" identified in the Terrestrial Biodiversity Assessment completed nationally in 2002 (National Land & Water Resources Audit 2002). Even the public nominations submitted by community groups are processed slowly, and the Minister has fallen behind in meeting statutory deadlines (10 days to Scientific Committee, Scientific Committee advice to Minister within 12 months, Ministerial decision within 90 days after advice).

Local Government & the EPBC Act 1999

If community-based environmentalists are disengaging with Commonwealth threatened species legislation, then Local Government has never engaged. That the *EPBC Act* has scarcely reached the consciousness of many working in Local Governments is reflected in the comment from one committed Local Government Sustainability Officer who said recently at interview (Lambert, Elix and Pearce, 2003) "*I hardly even know about the EPBC Act, let alone using it.*"

Threatened species protection responsibilities arising out of Commonwealth legislation that seems remote from their day-to-day operations are not a high priority for Local Governments faced with ever-growing demands resulting from Commonwealth and State policy and legislation. This situation is exacerbated by a lack of commensurate increase in resourcing from the higher levels of government.

Not so long ago the focus of Local Government responsibilities could be summed up in the 'roads, rates & rubbish' catch-cry. Nowadays Local Government plays an increased role in issues as diverse as environmental management, protection of the built and natural environment, town planning and development controls, housing and welfare, waste and recycling, public health and safety, and provision of recreational, educational and cultural services, among others. Each of these functions brings with it substantial cost, but while in 1975 State Government funding for Local Government in NSW averaged 14.8% of the total expenditure by Local Governments, by 1998 that had fallen to 7.1% (Hale 2003). When this is matched with substantially increased charges to Local Government for Bushfire control, waste disposal and other activities, and in NSW rate-capping which seriously constrains Local Government's revenue-generating capacity, it is not unexpected that Local Government has not generally engaged with a piece of legislation which has not yet demonstrated its relevance to them.

Many in the community expect the Commonwealth legislation to serve as the flagship or role model nationally, so it is not surprising that in the recent review of the EPBC Unit and its work (Lambert *et al.* 2003), respondents from State agencies in several States and from the community were cynical about the future, or focused on State, rather than Commonwealth legislation to protect threatened species and ecological communities.

Protecting species, populations and ecological communities in New South Wales

In moving on to look at the *NSW Threatened Species Conservation Act 1995* (the *TSC Act*), as amended in 2003, this paper uses as a case study, community and Local Government experiences. In particular the work of a local community and its Local Government to conserve the flora and fauna of North Head, Sydney and its surrounding seas are examined.

The northern headland at the entrance to Sydney Harbour, North Head, east and south of the historic stone wall which defines its landward boundary, is an area of some 300 ha, all of which is public land under the care and control of State and/or Commonwealth authorities.

Although there is as yet no published comprehensive flora and fauna study of North Head, the area is known to be home to species, populations and ecological communities listed as vulnerable or endangered under the provisions of the *TSC Act 1995*. Listed are

- Endangered Populations of the Long-nosed Bandicoot *Perameles nasula*, and the Little Penguin *Eudyptula minor*
- the Endangered Eastern Suburbs Banksia Scrub Ecological Community
- the Endangered sub-species of Sunshine Wattle *Acacia terminalis ssp terminalis*
- the Vulnerable species – Camfields Stringybark *Eucalyptus camfieldii* and the Powerful Owl *Ninox strenua*

Several rare plant species are also recorded from North Head.

How well has the *Threatened Species Conservation Act* served these population, ecological communities, and species?

Little Penguins

The Little Penguin *Eudyptula minor* population in the northern parts of Sydney Harbour is the only known breeding colony on the mainland in NSW. After considerable community action, it was listed as an Endangered Population in January 1997. A Recovery Plan was put in place in October 2000, and after further investigation, public consultation, and numerous representations by Manly Council and community members, Critical Habitat for the penguin population was declared in January 2003 (NSW NPWS Jan 2003).

Although Little Penguins are reported to have been present in their hundreds in the North Harbour area in the past, recent NPWS information (NSW NPWS 2003), based on a comprehensive monitoring program, indicates that the population has been "dramatically decreased to around 60 breeding pairs… Numbers are now so low that the population is in danger of becoming extinct".

Yet still, there are activities being approved which further place at risk the future of both this population, and of the endangered Long-nosed Bandicoot population. These activities receive increased public attention since the Little Penguin and the Long-nosed Bandicoot have achieved something of an iconic status for the natural environment of North Head.

A Commission of Inquiry into the proposed Mawland Hotel development at the North Head Quarantine Station (Office of the Commissioners of Inquiry 2002) recommended that proposed ferry movements to and from the Quarantine Station wharf (in the heart of the Little Penguin nesting and rafting area, the Wharf lying within the declared Critical Habitat area) should be the subject of a blanket dusk to dawn curfew, so as to avoid impacts on the penguins as they are coming ashore to feed their young. However, in what appears to be a trade-off between protection of endangered populations of Little Penguins and Long-nosed Bandicoots, and with strong focus on the commercial operations of the proposed Mawland development, approval has been given for the ferries to operate to and from the Quarantine Wharf until 11pm each night. As Little Penguins generally raft together and come assure within approximately one hour of sunset, this ferry operation will cease far too late to offer any meaningful protection.

More recently, as part of the declaration of Critical Habitat, marker buoys have been installed, defining the 50 metre offshore line beyond which anchoring is not permitted during the Little Penguin breeding season. Two buoys placed in key locations have since disappeared and when local community representatives sought to have them replaced, NSW Fisheries Director, Steve Dunn was reported as saying that the buoys had been *"mistakenly placed in an area that was a traditional and legitimate commercial fishing area and needed to be moved"* (*Manly Daily* 9 October 2003).

Community dialogue with various agencies saw replacement of the marker buoys defining the boat exclusion zone, but they have been relocated so as not to interfere with commercial fish net hauling in the area – scarcely a satisfactory resolution of this issue, especially given that NSW Fisheries were consulted as part of the Critical Habitat declaration process.

Long-nosed Bandicoots

Listed as Endangered in 1997, the North Head population of Long-nosed Bandicoots *Perameles nasula,* is a discrete population isolated from other populations once widespread in the Sydney region. The North Head population numbers no more than 100 animals (Banks and Powell 2002). Within this perilously small population, adults dominate slightly, and males dominate at a ratio of almost 4:1 among juveniles (Banks and Hayward 2002), causing the authors to express concern that the population is not growing and would be highly vulnerable to a single catastrophic event.

From these same population studies, Banks has concluded that: *"the Q[uarantine] S[tation] is a core source area for the whole bandicoot population on North Head. Bandicoot numbers are consistently higher there than elsewhere on the headland, and the QS is recognised by most people who have worked in the area as the best area for bandicoots, despite its artificial nature."* (Banks 17 Nov 2001).

Just as with the Little Penguins, the North Head population of Long-nosed Bandicoots is an iconic population on the brink of extinction, for which the *Threatened Species Conservation Act* provides little protection. Despite several years of negotiation, there is still no Recovery Plan in place for this population and Critical Habitat has not been identified.

The Mawland Hotel development proposed for the Quarantine Station will significantly increase traffic on the site, given the proponents plan to have approximately half of their clients arriving by car and the remainder arriving by ferry. The Joint Determining Authorities (Joint Determining Authorities 20 Nov 2002), assessing the development proposal, acknowledge in their opening paragraphs that: *"the determining authorities have concluded as part of this joint determination report that the activity has the potential to significantly affect the endangered Long-nosed Bandicoot and Little Penguin populations"*.

Road kill has been identified as one of the major causes of mortality in the North Head population of Long-nosed Bandicoots (Scott 1995). The Joint Determining Authorities (Joint Determining Authorities 20 Nov 2002, p.71) in their assessment of environmental issues and impacts state: *"In the case of the Long-nosed Bandicoot population, for example, the population viability analysis (Banks 2000) clearly demonstrates the need to apply extreme caution given that any additional road deaths above current background levels will increase the risk of extinction"*. However Environment Minister Debus (Joint Determining Authorities 20 Nov 2002) has given his concurrence to conditions which permit "*2 recorded adult road mortalities above background in any 6-month period*" before additional mitigating measures are required.

Concerns about the increased threat to bandicoots posed by additional vehicle movements if the ferries were required to stop at dusk were used by the Minister as a justification for not pursuing the ferry curfew recommended by the Commission of Inquiry into the Conservation and Adaptive Re-use of the North Head Quarantine Station. In the Joint Determining Authorities report (20 Nov 2002, p.71) it is clearly stated that: "*...the determining authorities do not support the recommendation of the COI to impose a blanket dusk-to-dawn curfew on the operations of the ferry service, given the subsequent impacts this could have on vehicle traffic to the site*".

With such trade-offs between endangered populations occurring it is scarcely surprising that members of the community who have worked long and hard for the conservation of North Head and its natural and cultural values are cynical about the effectiveness of the *Threatened Species Conservation Act 1995* when it is pitted against major tourism development on public land.

Eastern Suburbs Banksia Scrub

One other example deserves mention in highlighting why inadequacies in implementation of the *Threatened Species Conservation Act 1995* are adding to community cynicism.

Eastern Suburbs Banksia Scrub is an ecological community listed as Endangered at both Commonwealth and State levels. Occurring on nutrient poor, ancient Aeolian dune sands, it is estimated that only 138 hectares of this once much more abundant sclerophyll heath/scrub remain, all of it in small (0.06-8.5ha) fragmented remnants (NSW NPWS March 2003). Despite its name, some of the most significant patches of Eastern Suburbs Banksia Scrub are scattered across North Head. Included among these is a small patch of 0.3 ha that has been approved for removal to enable car park construction for the Mawland Hotel development. The Aeolian dunes on North Head are limited in extent, and debate over the extent to which restoration ecology can accomplish replacement of complex ecological communities remains unresolved. Yet removal of a small area of Eastern Suburbs Banksia Scrub has been approved (Heritage Council Approvals Committee, 5 Feb 2003) to enable construction of an on-site car park, the only condition of approval being that "habitat regeneration works on an area elsewhere on North Head up to 20 times the size of the area to be impacted" be carried out.

The land on North Head is all public land, managed by public authorities in the public interest. Given the ongoing and acknowledged threats to endangered species and populations provided in the examples given here, members of the North Head Sanctuary Foundation, like other community members striving for conservation of threatened species, are justifiably disenchanted with the effectiveness of the relevant legislation, whether Commonwealth or State. The legislation does provide a benchmark against which government action can be measured, but the extent to which objectives of conserving threatened species and ensuring their recovery can be circumvented is substantial. That disenchantment is shared by many in Local Government, which frequently looks to State and Federal Governments to lead by example.

So, do local communities & Local Government see threatened species legislation as simply an Act?

Yes and No. Both the Commonwealth and State legislation are seen as part good, part bad.

The community in general would be worse off, and so would the populations, species and ecological communities needing protection, if there were no such legislation in place. The legislation at least provides opportunity for community input.

But, without demonstrated political will to use the legislative provisions to protect threatened flora and fauna, those in the community who are concerned are increasingly likely to disengage with processes they see as delivering feeble outcomes. Local Government, with a plethora of other responsibilities handed down without supporting resources, may ignore the issues for as long as possible.

The answers are not simple. However scientists, local communities and Local Government must each contribute the expertise they can bring and together seize opportunities to highlight the impacts that political reticence has on our native plants, animals and the smaller species that support them. If political representatives are to ensure that threatened species, populations and ecological communities are protected, it is important that those best able to bring local and professional expertise to the debate act as if the threatened species, ecological communities and populations they care about are their own.

At a time when governments are calling for 'partnerships' to protect many aspects of our environment, a coming together of concerned scientists, national, state and local environmental organisations, and Local Governments can and should present challenges to both State and Commonwealth Governments to make greater use of the nomination, assessment, monitoring and compliance aspects of the available legislation.

References

Australian National Audit Office 2003. Referrals, assessments and approvals under the *Environment Protection and Biodiversity Conservation Act 1999*. Auditor-General's Audit Report No.38 2002-2003, Commonwealth of Australia, Canberra. www.anoa.gov.au

Banks, P. 2000. Population viability analysis for the Long-nosed Bandicoot population at North Head, NSW: modelling the effects of increased traffic flow on adult mortality. Unpublished report prepared for the NSW National Parks and Wildlife Service.

Banks, P. 17 Nov 2001. Submission on SIS for Mawlands Development of Quarantine Station, North Head. Submission to Commission of Inquiry into Conservation and Adaptive Re-use of North Head Quarantine Station at Manly. School of Biological Sciences, University of NSW.

Banks, P.B. and Powell, F. Jan 2002. An analysis of the NSW NPWS North Head Long-nosed Bandicoot monitoring program: Unpublished report for the NPWS Central Directorate Threatened Species Unit. School of Biological Sciences, University of NSW.

Banks, P.B. and Hayward, M. July 2002. An analysis of the May 2002 Long-nosed Bandicoot Population Census: Unpublished report for the NPWS Central Directorate Threatened Species Unit. School of Biological, Earth and Environmental Sciences, University of NSW.

Federal Court of Australia 2000. *Booth v. Boswell*. http://www.fedcourt.gov.au/gfx/j011453.pdf

Federal Court of Australia 12 February 2003. *Humane Society International Inc v. Minister for the Environment & Heritage*. http://www.austlii.edu.au/cgi-bin/disp.pl/cases/cth/federal_ct/2003/64.html

Federal Court of Australia 8 Aug 2003. *Minister for the Environment & Heritage v. Greentree*. http://bar.austlii.edu.au/au/cases/cth/federal_ct/2003/857.html

Hale, S. 26 June 2003. Local Government Amendment (Elections Bill). NSW Legislative Council Hansard Article No.35.

Joint Determining Authorities 20 Nov 2002. North Head Quarantine Station Conservation and Adaptive Re-use Proposal. Report prepared by NSW Minister for the Environment, NSW Heritage Office & NSW Waterways. Sydney.

Kemp, The Hon. Dr David 5 Sept 2003. Koolyanobbing Expansion Approved with Conditions. Media Release, Minister for the Environment and Heritage, Canberra. www.ea.gov.au/minister/env/2003/mr05Sep2003.html

Lambert, J., Elix, J. and Pearce, H Sept 2003. Evaluation of EPBC client servicing. Unpublished report prepared for the World Wide Fund for Nature (Australia), Sydney.

Mercer, S. 1 March 2002. Letter to J Lambert, Friends of Quarantine Station, re. Mawland Hotel Management Pty Ltd.

National Land and Water Resources Audit 2002. Australian Terrestrial Biodiversity Assessment. Department of Environment & Heritage, Canberra. http://audit.deh.gov.au/ANRA/vegetation/docs/biodiversity...

NSW Heritage Council 5 Feb 2003. Conditions of Planning Approval of Draft Joint Determining Authorities prepared by Joint Determining Authorities dated November 2002, as amended by the Heritage Council meeting of 5 February 2003 pursuant to section 63 of the Heritage Act 1977. NSW Heritage Council, Sydney.

NSW National Parks and Wildlife Service 2003. Critical Habitat identification report for the Endangered Population of Little Penguins at Manly. Director General's Report, NSW NPWS, Hurstville.

NSW National Parks and Wildlife Service 2003. Critical habitat: Little Penguin population. www.nationalparks.nsw.go v.au

NSW National Parks and Wildlife Service March 2003. Eastern Suburbs Banksia Scrub Endangered Ecological Community Recovery Plan. Draft for Public Comment. Director General's Report, NSW NPWS, Hurstville. www.nationalparks .nsw.gov.au

Office of the Commissioners of Inquiry for Environment and Planning July 2002. Proposed Conservation and Adaptive Re-use of the North Head Quarantine Station. Report to NSW Minister for Planning, The Hon. Dr Andrew Refshauge. http://www.coi.nsw.gov.au

Register of the National Estate Database. North Head, Manly NSW. Australian Heritage Commission, Canberra. www.ahc.gov.au/cgi-bin/register/site.pl?101621

Scott, L.K. 1995. Nutritional ecology and population biology of the Long-nosed Bandicoot, *Permales nasuta*: Implications for conservation. M.Sc. Thesis. University of Sydney, Sydney.

Wilmot, S. Jan 2002. Environment Protection and Biodiversity Conservation Act 1999 Referral. Unpublished submission to the Federal Minister for the Environment on behalf of Mawland Hotel Management Pty Ltd. Gunninah Environmental Consultants, Crows Nest.

Struggling with the Act – some observations by ecological consultants

Martin Denny

President, Ecological Consultants Association of NSW Inc.

88 Showground Road, Castle Hill, NSW 2154

ABSTRACT

This paper explores the influence of the New South Wales *Threatened Species Conservation Act 1995* (TSC Act) and its predecessors upon the assessment of impacts on native fauna. There has been a move away from an overall assessment of impacts upon total biodiversity towards an approach focused on threatened species and communities. For ecological consultants dealing with the TSC Act, interpreting its many requirements can be difficult and can often lead to differences of opinion between the proponents of a development and the Determining Authority. Some of the positive and negative aspects of the TSC Act are described, and it is concluded that there is a large gap between satisfying the requirements of the Act and providing information about the effects of various activities upon the fauna and their habitats.

Key words: *Threatened Species Conservation Act*, Assessment of Significance, species impact statements, ecological consultant

Introduction

This paper presents the combined thoughts and concerns of members from the Ecological Consultants Association of NSW Inc (ECA). The ECA represents professional ecological consultants in NSW and has the following objectives:

- To liaise with regulatory authorities and other bodies on matters relevant to the practice of ecological consultants
- To prepare and/or support standards relevant to the practice of ecological consultants, including a code of ethics, standards for the practice methodology, peer review of work
- To provide or co-ordinate training programs for the betterment of the practice of ecological consultancy
- To undertake any activity that will promote the interest of ecological consultants in NSW[1].

History of the act and its impact upon fauna

Prior to its discovery and settlement by Europeans, Australia could be considered to be a land that was occupied by plants and animals unencumbered by demands from management plans, captive breeding, taxonomic testing and legislative controls.

Until the early 20th century, plants and animals of NSW were described by their scientific names and by their physical, behavioural and ecological characteristics. Animals could roam freely within their favoured biotope and could easily be recognised by their evolved defining characteristics. However, the imposition of legislative controls on fauna with the *NSW Birds and Animals Protection Act* in 1918 changed how fauna were defined in NSW. I will not go into the history of legislation associated with the conservation of fauna in NSW, as this is dealt with by Jarman and Brock (2004) in another paper, but I would like to view the development of the current legislation through the eyes of the fauna affected and to discuss its impact upon our current approaches to impact assessment. This paper only addresses fauna, and leaves any discussion of the effects from the *Threatened Species Conservation Act* on flora to someone more qualified.

Although the *Birds and Animals Protection Act 1918* recognised the need to protect some species of birds and mammals in NSW, the *NSW Fauna Protection Act 1948* divided these species into two groups, Protected and Scheduled. Protected animals became the property of the Crown and could not be killed, except during declared 'open seasons'. Scheduled animals were unprotected and could be hunted and killed at any time (although 'closed seasons' were possible). The list of Scheduled animals includes species now considered threatened (e.g. Grey Falcon, *Falco hypoleucos*, and 'flying foxes'), whilst another 23 species considered unprotected under the Fauna Protection Act are now classed as protected. These species included cormorants, Wedge-tailed Eagle *Aquila audax*, falcons, parrots, Dusky Moorhen *Gallinula tenebrosa* and the Common Wombat *Vombatus ursinus*.

Thus some fauna in NSW were no longer just defined in terms of their biological characteristics. Birds and mammals were now placed into two categories and some species could now be considered as being more important (protected) than others. This approach was repeated in the *NSW National Parks and Wildlife Act 1967*.

[1] Further information can be obtained from www.ecansw.org.au

Pp 164 - 171 in *Threatened species legislation: is it just an Act?* 2004, edited by Pat Hutchings, Daniel Lunney and Chris Dickman. Royal Zoological Society of New South Wales, Mosman, NSW, Australia.

The NSW *National Parks and Wildlife Act 1974* (NPW Act) replaced the 1967 Act and consolidated the above collection of Acts that had protected the flora and fauna (although only some birds and mammals) of this State. From 1974 onwards, native animals were labelled with the terms 'protected' or 'endangered' and their lives were never the same. Animals could also be classed as 'locally unprotected' or 'prescribed'. There were schedules of endangered fauna as well as protected amphibians. The 'taking' – in those days this meant 'killing' - of native animals could be undertaken under licence.

Thus, one no longer viewed a native animal as an integral part of the environment, interacting with the biotic and abiotic aspects of their surroundings. Each species had become labelled as either a protected or an endangered animal, and those species listed in Schedules 12 and 12A of the NPW Act were now considered to be more important than those that were only protected (amphibians left off Schedule 12A were still not considered 'protected'). So began the change of impact assessment from the general to the specific.

The importance of impact assessment was set in place with the introduction of the *NSW Environment Planning and Assessment Act 1979* (EP&A Act). This Act introduced legislative requirements for development applications and environmental impact statements that included an assessment of the harm to the environment (this included most native fauna). Because there was little emphasis on endangered species (this had to wait until the introduction of the NSW *Endangered Fauna (Interim Protection) Act 1991 No.* 66) assessments of impacts upon native fauna were taken in general terms. This is illustrated by the following extracts from Environmental Impact Statements produced in the early 1980s:

1. North Hermitage Colliery Pit Top Proposal at Fernbrook Environmental Impact Statement prepared by Nexus Environmental Studies for Coalex Pty Ltd September 1980. Environmental Report "Impact of the Proposed mine Top on Wildlife in Marrangaroo Valley" by M.J.S.Denny May, 1980:

"The impact of coal mining at the North Hermitage site will be significant for the terrestrial fauna living at the site. Most animals will initially move away from the disturbance and some will find it difficult to recolonize new areas. However, the relatively small area covered by the mine site (about 3.7 ha) will mean that overall numbers of animals in the Lithgow region will not be affected." (p.8)

Part of the assessment included observations of wombat activity along Marrangaroo Creek, as well as descriptions of aquatic macro-invertebrate fauna within the creek. The report concluded that Marrangaroo Creek should be preserved in its present condition 'for the sake of the total ecology of the area'.

2. Ulan Coal Mines Stage 2 Colliery Development and Expansion Environmental Impact Statement Prepared by Kinhill Stearns Engineers for Ulan Coal Mines Ltd February 1983:

"The clearing of vegetation from parts of the three sites surveyed will cause a loss of fauna from the same areas. This will occur through the land clearing process and its associated noise and disturbances, as well as through the consequential changes to habitat. However, none of the species located at the three sites are unique to the site, area or region. Previous studies in the Upper Hunter region show that the species found at Ulan are distributed over a wide region in a range of habitats similar to those identified by the current survey." (p. 103)

The Statement also provided figures showing the distribution of wildlife habitat as well as a general appraisal of the biodiversity of the area studied together with comparisons of survey results from nearby areas of similar habitats (e.g. Goulburn River, Newnes Plateau).

These two extracts show the approach to impact assessment prior to the emphasis on endangered species. Although endangered species (as listed in Schedule 12 of the NPW Act at the time) were found in the North Hermitage survey (Gang-gang Cockatoo *Callocephalon fimbriatum*, White's Thrush i.e. Bassian Thrush *Zoothera lunulata*) there was no requirement to target these animals or their habitats. Rather, the overall species assemblages were assessed, in terms of their contribution to the region and their value to the local area. Also, wildlife habitat was assessed in terms of its value to the region's biodiversity and not to that required by an individual (endangered) species.

The introduction of the EP&A Act provided a basis for future detailed assessments of impacts upon fauna. But more importantly, its introduction became a milestone in the development of an impact assessment industry. From 1979, it became necessary for a developer of a project of sufficient status (i.e. a designated development) to produce an impact assessment statement and possibly a fauna impact statement from 1992 onwards. This required persons of sufficient expertise to undertake investigations to determine the extent of impact upon various parts of the environment, including flora and fauna. Thus, an industry, based upon consultants assessing impacts upon the environment, was born and has blossomed over the last 24 years. Part of that industry includes ecological consultants. In 1979, there were one or perhaps two ecological consultants in NSW; today there are more than a hundred.

The introduction of the now repealed NSW *Endangered Fauna (Interim Protection) Act* in 1991 introduced the concept of a Fauna Impact Statement and set out seven factors required when deciding whether there is likely to be a significant effect on endangered species (the seven factors are given in Appendix 1). Unfortunately, or perhaps unconsciously recognising the importance of all native vertebrate species, when it passed through parliament in December 1991, the *Endangered Fauna (Interim Protection) Act* put 'protected' instead of 'endangered' fauna in Section 4A, i.e. the test using the seven factors had to assess the 'significant effect on the environment of protected fauna', not just endangered fauna.

An example of the effect from this wording can be seen in the Environmental Impact Statement for the North West Transport Links East (prepared by Manidis Roberts Consultants and Snowy Mountains Engineering Corporation Ltd for the NSW Roads and Traffic Authority, 1991). The Flora and Fauna Evaluation prepared by Mount King Ecological Surveys produced a fauna Impact Statement for 131 protected vertebrate fauna species, ranging from the now endangered Eastern Bentwing Bat *Miniopterus schreibersii oceanensis* to the more common Australian Magpie *Gymnorhina tibicen*. Only two species assessed were listed in Schedule 12 of the NPW Act (Green and Golden Bell Frog *Litoria aurea,* and Diamond Python *Morelia spilota*), but it was possible to discuss conservation status within the Statement so importance locally could be addressed. Thus species such as the Brown Pigeon *Macropygia amboinensis*, Gang-gang Cockatoo and Powerful Owl *Ninox strenua* could be highlighted as having declining population status in the Sydney region, and emphasis placed on restoration of their habitat. Although this approach to impact assessment was laborious, it did provide a broad platform to look at and assess the total biodiversity of an area where impacts may occur. The drafting error in the Act was rectified in the *Endangered Fauna (Interim Protection) Amendment Act 1992 No. 97*, and only the impact upon endangered species (as listed in Schedule 12 of the NPW Act) needed to be assessed by the seven-factor test.

Now, there was increased emphasis on endangered species, and less on the general fauna populations i.e. biodiversity of species and populations. Australian fauna had now become a class-ridden society, i.e. endangered or non-endangered, with the invertebrates, being completely ignored.

Finally, in 1995, the NSW *Threatened Species Conservation Act 1995* (TSC Act) was proclaimed. This Act, together with the recent *NSW Threatened Species Conservation Amendment Act 2002*, is one of the major subjects of this forum.

The TSC Act has set the focused approach to impact assessment in concrete, with definitions of threatened species (called and defined as 'endangered species' in the *Endangered Fauna (Interim Protection) Act*), endangered populations and communities, critical habitat and key threatening processes.

The process of assessing whether an activity could be considered a significant impact upon a threatened species etc is clearly spelt out as the 8-part test (an amended seven-factor test). The 8-part test is given in Appendix 1. Recovery plans and threat abatement plans are described and their use defined. The Fauna Impact Statement was expanded to include all threatened species of flora and fauna (including invertebrates and non-vascular plants), endangered populations and endangered ecological communities and was renamed a Species Impact Statement.

In all, the TSC Act plus amendments provides all the requirements for consultants to undertake impact assessment of native flora and fauna. Of all native flora and fauna? Well, perhaps not all – just those species, populations and ecological communities listed at the back of the Act, i.e. the Schedules.

There are benefits to species that are listed in the Schedules of the TSC Act. Such benefits include the possibility of increased research funding to threatened species such as the Powerful Owl and Barking Owl *Ninox connivens*, and the power to stop or modify developments (e.g. the humble Cumberland Land Snail *Meridolum corneovirens*, has wrought havoc to residential development in western Sydney).

The nature of the Act

Whatever one's philosophical approach to impact assessment, the *Threatened Species Conservation Act* has become the ecological consultants bible. Each Section and sentence must be studied, interpreted and acted upon. Each word and definition needs to be puzzled over, and any official interpretation and guidance by NPWS and its replacement, the Department of Environment and Conservation (DEC), in the form of Information Circulars and workshops, is eagerly consumed.

Because the text of the Act, like the bible, can be interpreted in many ways by people of different persuasions and agendas, its application can sometimes be considered a bit like a lottery or the roll of a dice. Whoever determines the assessment produced by an ecological consultant holds the power to accept or reject such an assessment. Often such a person or agency can take a broad approach or can look at the assessment from a strictly narrow fundamentalist interpretation of each word within the Act. This is called the 'black letter' approach to law, and is practised by many within the environmental and legal community.

Thus the Determining Authority, be it the DEC or another government authority, or local government staff, are able to interpret the Act in a manner that may be different from that taken by the proponent. Consequently, the passage of a development through its assessment phase may not depend on what you know, or whom you know, but on the attitude taken by unknown representatives of the Determining Authority. This attitude is acknowledged by the DEC in their guidelines for persons undertaking surveys and assessments of threatened species (NPWS and SMEC Australia 2003):

"Ultimately, it is left to the discretion of the consent, determining or licensing authority to form a view as to whether a proposed development, activity or action is likely to have a significant effect on threatened species, populations or ecological communities, or their habitats" (p. 3-11).

This problem extends to the NSW Land and Environment Court, where the black letter approach dominates in an adversarial environment. Decisions from the Court are more than likely to be based upon precedence, i.e. previous interpretations of the law, rather than on the merits of the case.

Unfortunately, by applying a black letter approach, the Act has become a checklist of do's and don'ts. It is now possible for an applicant without knowledge of the interaction between species and their environment, or of broader biodiversity principles, to pass through the assessment process by ensuring that the detailed requirements of the Act are met. Such an approach ignores the spirit of the TSC Act, particularly Object (a) i.e. "to conserve biological diversity and promote ecologically sustainable development". At present, one could argue that the Act does not meet this objective, by narrowly focusing on threatened species and ignoring the remainder of the biological diversity. There is no need to change the objects of the Act, rather change the Act to suit the objective.

The checklist approach makes life easier for those who have the task of determining whether the assessment process has been correctly undertaken. They can also go through a checklist of items to ensure that all have been covered. Whether they are covered adequately or whether their needs to be a broader assessment covering the full range of plants and animals and their interaction is not brought into the determination.

The checklist approach reaches out to include the contents of Species Impact Statements, the added requirements set by the DEC Director-General for a SIS, the techniques and methodology used during surveys and the qualifications of those involved in the assessment.

Using the Act

For all of the general complaints about the evolution of legislative controls on conservation of the natural environment, the *Threatened Species Conservation Act* and Amendments creates the environment in which ecological consultants must work. So it is important to see what benefits and liabilities stem from the implementation of this Act. First, let us look at some of the benefits.

Primarily, the Act does provide protection to a set of plants and animals within NSW (those found on the Schedules). Together with the NPW Act, all native vertebrate and some invertebrate species in NSW are protected and cannot be harmed without a valid development consent or activity approval and/or a licence from the DEC. The inclusion of a test for significant impact within the EP&A Act and the TSC Act has had a strong effect on how activities are developed. The original intention of the seven-factor test in 1992 and the 8-part test in 1995 (called Assessments of Significance, NPWS and SMEC 2003) was to provide a quick assessment of the significance of impact from a development. If such an assessment led to a Species Impact Statement then detailed surveys and assessments could then be undertaken. However, over the years, results from Assessments of Significance have become critical. Many proponents were quick to realise that the results from such an assessment were not always available for public comment or submissions (although an 8-part test accompanying a Development Application is available to the public), whereas a Species Impact Statement becomes public property. Thus the production of an SIS could be expensive and its assessment drawn-out, i.e. period for exhibition, time to respond to submissions and time for a decision. It is far better to put the resources into the 8-part test and attempt to show that no significant impacts are expected.

This has led to a reversal in the sequence of events, i.e. detailed surveys now come before an 8-part test. The 8-part test is now produced as an extensive assessment of impacts. Initially, an 8-part test would have been only one to two pages long, now it can be at least 50 to 60 pages.

The benefits from this approach are many. For the ecological consultant, it may mean that you have two bites at the cherry, i.e. a survey prior to the 8-part test and then another survey to cover those requirements needed for a SIS. But far more important is the attitude taken by some proponents to ensure that a SIS is not required. This will often mean that advice on mitigation of impacts given by an ecological consultant will be taken. Thus, a proponent may see the advantage of setting aside areas of land proposed for development as a trade-off for ensuring a successful passage of the modified development. This is illustrated in the paper by Whelan *et al.* (2004) when describing the conservation of an area containing *Darwinia biflora*.

Another benefit stemming from the TSC Act is the emphasis on providing an area of compensatory habitat as part of the development. Similar actions are part of other policies within NSW. The provision of compensatory wetlands is part of the NSW Wetlands Management Policy when social and economic imperatives give no other alternative (Department of Land and Water Conservation 2002). Land managers in south-western NSW can set aside areas of land as compensation for the clearing of other parts of their property (Land Use Agreements under the Southern Mallee Project). Land Use Agreements involve landholders developing existing grazing country for dryland cultivation. Any development which takes place on a property must be offset with an equal or larger Private Reserve (where stock and watering points are removed) within the property. Up to September 2003, 22 properties have been granted consents to develop 37,550 ha, with 100,000 ha being placed in Private Reserves for perpetuity.

Such arrangements are now being incorporated into large developments and are defined and described in Environmental Impact Statements and Species Impact Statements, i.e. there is compensation for the loss of habitat during development by providing and protecting another area of similar habitat. Such an approach has been undertaken for the Western Sydney Orbital Tollway (Sinclair Knight Merz and PPK Environment & Infrastructure Pty Ltd 2000).

Finally, the use of the TSC Act and the ensuing publicity associated with controversial developments has raised the public profile of threatened species and communities. I am sure that no one knew about the Cumberland Land Snail, Bathurst (=Purple) Copperwing Butterfly *Paralucia spinifera* or of the Cumberland Plain Woodland before these were the subject to assessment under the TSC Act. Such publicity has also made the public aware that there are constraints on developments that may affect biodiversity, and that the concept of biodiversity is sufficiently important to result in legislation to enforce its protection.

Let us now look at the other side of the coin. The focused view of biodiversity assessment via the TSC Act has, in the view of many consultants, eroded the broader view of flora and fauna impact assessment, and led to trivialising ecological assessment. Species richness, habitat values and conditions are no longer part of the assessment process unless they focus upon threatened species and communities. The Western Australian Environmental Protection Authority has recognised the importance of placing species into an ecosystem context in its Position Statement No.3. "The EPA expects proponents to ensure that terrestrial biological surveys provide sufficient information to address both biodiversity conservation and ecological function values within the context of the type of proposal being considered and the relevant EPA objectives for protection of the environment" (Environmental Protection Authority 2002).

The EPA of WA emphasises that, "It is important to clarify that the intrinsic value of a species should not be judged only by its rarity or how threatened the taxon may be. Even though a species may occur in large numbers, it may be a major component (or 'keystone species') for the sustainability of the system." The position statement also points out that it is important to consider the concept of species as it relates to diversity (e.g. species richness) or endemism, particularly regards areas of high biodiversity ('hotspots'). It is felt that NSW DEC should adopt a similar approach.

The increased emphasis on the application of Assessments of Significance is leading to a confused and vague interpretation of the factors within the test. A submission in 2003 to the DEC by the Ecological Consultants Association of NSW (ECA) raised many questions concerning the interpretation of the yet to be proclaimed 7-part test (the 7-part test is outlined in Appendix 1). Definitions of terms used in the seven factors are inconclusive, particularly as a quantitative term of reference. Terms such as 'local viable population', 'local occurrence', 'substantially', 'extent', 'importance' and 'locality' need to be defined precisely before accurate assessments can be undertaken. Taking one example of the concerns of members of the ECA, Factor (d), Sub-factor (iii) of the proposed 7-part test (see Appendix 1) states that "in relation to the habitat of a threatened species, population or ecological community: the **importance** of the habitat to be removed, modified, fragmented or isolated to the long-term survival of the **species, population** or **ecological community** in the **locality**" must be taken into consideration when assessing impacts.

The four underlined terms in this sub-factor require a clearer definition to avoid varying interpretation. The use of the collective terms "species, population or ecological community" can cause confusion. For example, does "population" refer to an endangered population? Or does population refer to a *population* of the threatened *species* being assessed by the 7-part test. If so, is it a local population? And is it defining the local population as those within the "locality" – another undefined term. Does ecological community mean endangered ecological community, or the ecological community of which a threatened species is part? There needs to be clarification about what type of criteria needs to be considered in assessing 'importance' of habitat to be removed or modified to the species, population or endangered ecological community. There may need to be separate criteria for each category. Also, it may be necessary to define habitat in terms of the particular species under consideration. For example, if the species to be assessed were the Cumberland Land Snail, there would be a significant impact if the herb and shrub layers were removed or modified, but the trees retained.

The term "locality" has been inconsistently defined in the past. The draft guidelines for persons undertaking surveys and assessments of threatened species (NPWS and SMEC Australia 2003) state: "The size of the locality will vary depending upon the location of the study area and the amount of information available. For example, in Sydney the locality may be defined as a 5km radius of the subject site compared with far western NSW where the locality may be a 50km radius of the subject site due to a limited number of surveys conducted" (p. 4-36). This places some expectation upon the consultant to define locality that will be consistent with that used by the DEC. The size of a locality is usually defined in the DEC Director-General's requirements for a Species Impact Statement. Examples from such requirements show that the radius for a locality in western NSW is usually 20 km, and 10 km for the North Coast of NSW. Hence, are we to consider the local population as those individuals likely to occur within a 5, 10, 20 or 50 km radius (or any other nominal radius)? Consequently, the definition of this term will vary in its significance to the ecology of the species, and can be open to exploitation. For example, for a Common Planigale *Planigale maculata* with a very small home range of about 1 ha, the "importance" of 10 ha of Common Planigale habitat to be modified in a locality of about 160 km^2 (10 km radius circle) could be argued not to be significant, as numerous sub-populations may occur within such a wide area which may contain a lot of potential habitat (which for a species found in a wide range of habitats, can be substantial). Thus, the habitat of a small population on the 10 ha of habitat proposed for development could be argued to not be significant in a "local" sense, and could be eliminated. As the title of the paper suggests, many consultants are 'struggling with the Act'. They are struggling because some terms and concepts used in the Act are ambiguous and difficult to interpret. Much of the energy used in attempting to provide a clear assessment of impacts can go in trying to determine what is required under the Act. Clearer guidelines to the Act and better communication between determining authorities and consultants would make this task easier.

The inclusion of key threatening processes, recovery plans and threat abatement plans within the proposed 7-part test present problems in terms of adequacy of available information. Factor (f) of the 7-part test asks "whether the action proposed is consistent with the objectives or actions of a recovery plan or threat abatement plan" (the 7-part test is outlined in Appendix 1). At present, there are few recovery plans available for threatened species or communities, and according to the Amendment Act, the Service (now DEC) has three to five years to provide such plans. However, there is an obligation for a proponent to ensure that any action is consistent with the objectives of these plans. Also, there are many more key threatening processes listed (22) than there are related threat abatement plans available (2). Thus, it will be highly likely that an action will constitute a key threatening process, e.g. clearing of vegetation, but it is not possible to apply any of the objectives of the relevant threat abatement plan, as none are available on vegetation clearing.

Another problem is the increasing complexity of the Act, particularly when combined with the other legislative requirements associated with impact assessment e.g. State Environmental Planning Policies, vegetation clearing requirements (e.g. Regional Vegetation Management Plans), NPWS Translocation Policy, *NSW Fisheries Management Act 1994*, *NSW Rural Fires Act 1997*, Animal Welfare concerns and the Commonwealth *Environment Protection and Biodiversity Conservation Act 1999*.

Each new amendment to the Act adds increasing specialised work for the consultant ecologist. In addition, the Schedules are being continually upgraded with additional species, communities, populations and key threatening processes. Between 1995 and 2002, over 300 listings involving flora and fauna have been added. There were 241 fauna species listed as threatened in the TSC Act in 1995. In 2004, there are 289 fauna species listed as threatened. In March 2004, there were 14 Preliminary Determinations listed that covered nine threatened fauna species, a Key Threatening Process and four endangered ecological communities. A Preliminary Determination needs to take into account any submissions from the public before being listed as a Final Determination. DEC recommends that a proponent take into account those endangered species, populations and ecological communities that are listed as Preliminary Determinations when undertaking Assessments of Significance (NPWS and SMEC Australia 2003).

The increasing specialisation of the assessment process, because of the listing of species that are only known to a small number of taxonomists, has placed a greater strain on the role of ecological consultants. Consultants have had to become familiar with a range of obscure invertebrate species, as well as vertebrate species that have undergone taxonomic changes since being listed (e.g. Gunther's Skink, *Cyclodomorphus branchialis*, is now called the Mallee Slender Blue-tongued Lizard, *C. melanops elongatus*).

As we hear so often from the political arena, The Devil is in the Detail! Nothing could be truer for ecological consultants. The aims, intentions and requirements of the TSC Act are clearly spelt out. However, the difficulty is being able to implement the aims and satisfy the requirements. Vague terms, unrealistic demands, ambiguous expectations and inconsistent determinations all lead to difficulties when undertaking the assessment process.

There is a great need to refer to such groups as the Ecological Consultants Association of NSW for their practical knowledge to ensure that the Act becomes more workable. It is praiseworthy that NPWS, now DEC, has invited the ECA to contribute to the development of the guidelines for the proposed 7-part test, and in assisting in the certification process for persons undertaking Species Impact Statements (this is a requirement under the TSC Act and Amendments).

In praise of the Act

On reflection, particularly after presenting such a negative view of the Act, I came upon a revelation. In the case of the TSC Act, the government and bureaucracy were stepping outside their normal conservative role and have provided a *Threatened Species Conservation Act* that is too far ahead of its time!

All that the Act states and requires is correct and is important for the conservation of flora and fauna. The trouble is that the scientific, consulting and planning communities cannot provide the answers needed to satisfy the requirements of the Act. These communities are struggling to answer such questions as:

- the extent of a local population;
- the specific feeding, shelter and breeding requirements of each threatened species;
- the degree of loss of habitat that may affect a species;
- the impact from the creation of a road;
- the effect from constant or erratic noise upon a breeding population;
- the home ranges of threatened species

and so on. Yet, these are questions that must be answered before a rigorous Assessment of Significance, i.e. using the 8-part or the proposed 7-part test, can be achieved.

There is an opportunity for a conduit of information between consultants and members of the scientific community through conferences etc. Conferences and workshops targeting problems relating to assessment of impacts upon threatened species need to be developed. With the proposed certification of consultants undertaking Species Impact Statements information relevant to impact assessment could be channelled through the DEC. This already occurs through Threatened Species Management Information Circulars and Species Profiles, and could be upgraded by the use of the internet (e.g. regular updates via the email).

So, to be honest, we should not be looking down our noses at the *Threatened Species Conservation Act* and its administration by the DEC. Rather we should be turning to the scientific community and asking why they haven't provided the information needed to help support this innovative piece of legislation. This piece of legislation was welcomed by all interested in flora and fauna conservation, but has fallen into disrepute because of lack of support by a community that fought for its introduction.

References

Department of Land and Water Conservation 2002 *Compensatory Wetlands A discussion paper under the NSW Wetlands Management Policy* Department of Land and Water Conservation and NSW State Wetland Advisory Committee.

Environmental Protection Authority 2002 *Terrestrial Biological Surveys as an Element of Biodiversity Protection Position Statement No.3 March 2002*, Environmental Protection Authority, Perth.

Jarman, P.J. and Brock, M.A. 2004 The evolving intent, coverage and use of threatened species legislation in Australia. Pp 1-19 in Threatened Species legislation: is it just an Act? Edited by P. Hutchings, D. Lunney and C. Dickman. Royal Zoological Society of NSW, Mosman, NSW, Australia.

Lunney, D., Curtin, A.L., Ayers, D., Cogger, H.G., Dickman, C.R., Maitz, W., Law, B. and Fisher, D. 2000 The threatened and non-threatened native vertebrate fauna of New South Wales: status and ecological attributes. *Environmental and Heritage Monograph Series No.4*. Pp 1-132. NSW National Parks and Wildlife Service, Hurstville.

NPWS and SMEC Australia 2003 *Draft Threatened Species Survey & Assessment Guidelines* NSW National Parks and Wildlife Service, Hurstville.

Sinclair Knight Merz and PPK Environment & Infrastructure Pty Ltd 2000 *Proposed Western Sydney Orbital Environmental Impact Statement, Species Impact Statement and Working Papers.*

Whelan, R., Brown, C.L. and Farrier, D. 2004 How and when might the precautionary principle be applied in threatened species conservation? Pp 49-58 in Threatened Species legislation: is it just an Act? Edited by P. Hutchings, D. Lunney and C. Dickman. Royal Zoological Society of NSW, Mosman, NSW, Australia.

APPENDIX I

Factors to be taken into consideration when undertaking Assessments of Significance

1. Seven-factor test (Schedule 2 *Endangered Fauna (Interim Protection) Amendment Act 1992 No. 97*):

In deciding whether there is likely to be a significant effect on the environment of endangered fauna the following factors must be taken into account:

(a) the extent of modification or removal of habitat, in relation to the same habitat type in the locality;

(b) the sensitivity of the species of fauna to removal or modification of its habitat;

(c) the time required to regenerate critical habitat, namely, the whole or any part of the habitat which is essential for the survival of that species of fauna;

(d) the effect on the ability of the fauna population to recover, including interactions between the subject land and adjacent habitat that may influence the population beyond the area proposed for development or activities;

(e) any proposal to ameliorate the impact;

(f) whether the land is currently being assessed for wilderness by the Director of National Parks and Wildlife under the Wilderness Act 1987;

(g) any adverse effect on the survival of that species of endangered fauna or of populations of that fauna.

2. 8-part test (Schedule 5 *Threatened Species Conservation Act 1995 No. 101*):

The following factors must be taken into account in deciding whether there is likely to be a significant effect on threatened species, populations or ecological communities, or their habitats:

(a) in the case of a threatened species, whether the life cycle of the species is likely to be disrupted such that a viable local population of the species is likely to be placed at risk of extinction,

(b) in the case of an endangered population, whether the life cycle of the species that constitutes the endangered population is likely to be disrupted such that the viability of the population is likely to be significantly compromised,

(c) in relation to the regional distribution of the habitat of a threatened species, population or ecological community, whether a significant area of known habitat is to be modified or removed,

(d) whether an area of known habitat is likely to become isolated from currently interconnecting or proximate areas of habitat for a threatened species, population or ecological community,

(e) whether critical habitat will be affected,

(f) whether a threatened species, population or ecological community, or their habitats, are adequately represented in conservation reserves (or other similar protected areas) in the region,

(g) whether the development or activity proposed is of a class of development or activity that is recognised as a threatening process,

(h) whether any threatened species, population or ecological community is at the limit of its known distribution.

3. 7-part test (yet to be proclaimed Schedule to *Threatened Species Conservation Amendment Act 2002*):

The factors to be taken into account when making a determination as to whether an action, development or activity is likely to significantly affect threatened species, populations or ecological communities or their habitats are:

(a) in the case of a threatened species, whether the action proposed is likely have an adverse effect on the life cycle of the species such that a viable local population of the species is likely to be placed at risk of extinction,

(b) in the case of an endangered population, whether the action proposed is likely to have an adverse effect on the life cycle of the species that constitutes the endangered population such that a viable local population of the species is likely to be placed at risk of extinction,

(c) in the case of an endangered ecological community, whether the action proposed:

(i) is likely to have an adverse effect on the extent of the ecological community such that its local occurrence is likely to be placed at risk of extinction, or

(ii) is likely to substantially and adversely modify the composition of the ecological community such that its local occurrence is likely to be placed at risk of extinction,

APPENDIX I

(d) in relation to the habitat of a threatened species, population or ecological community:

(i) the extent to which habitat is likely to be removed or modified as a result of the action proposed, and

(ii) whether an area of habitat is likely to become fragmented or isolated from areas of habitat as a result of the proposed action, and

(iii) the importance of the habitat to be removed, modified, fragmented or isolated to the long-term survival of the species, population or ecological community in the locality,

(e) whether the action proposed is likely to have an adverse effect on critical habitat (either directly or indirectly),

(f) whether the action proposed is consistent with the objectives or actions of a recovery plan or threat abatement plan,

(g) whether the action proposed constitutes or is part of a key threatening process or is likely to result in the operation of, or increase the impact of, a key threatening process.

Plenary

PLENARY

Chris Dickman (University of Sydney): We have planned a two-part plenary. In the first part, we invite you to ask today's speakers questions on any of the topics that they covered. In the second part Rob Whelan, David Farrier and Martin Denny have kindly prepared some summary points and will lead discussion on the questions of uncertainty and also on how we can ensure that we achieve the objects of the threatened species legislation. So, are there any questions?

Mark Burgman (University of Melbourne): Chris, this is a question for you. You made two assertions in your talk that went largely unchallenged. I'd like to put them to you now. One was that your group, the Scientific Committee, was performing well. I'd like to know what was your measure of performance. You could be doing it all wrong and I would have no way of knowing. I didn't see any substantiation.

Secondly, you claimed that all but two of the assessments that you completed were unanimous. I'm deeply suspicious of any committee process that is routinely unanimous, let alone almost exclusively unanimous. I think there must be something wrong or you're submerging dissent under threat or something. I'd like to know what threat mechanism you're using, because I'd like to use it!

Thirdly, did you notice and do you regret having no Indigenous representative on the Scientific Committee?

Chris Dickman: With respect to your first question, the Committee has several functions so there are a number of ways of measuring success. One is that the community gets to know about the legislation and begins to take it on board. If you look at the numbers of nominations that were received by the Committee from 1996 until 2002, there was a steep increase. This indicates increasing public awareness of the legislation and the ability for anybody to contribute information about threatened species, populations or communities.

Another measure of success is that we were attracting increasing numbers of certain kinds of nominations, particularly for endangered ecological communities and key threatening processes, increasing our ability to conserve broadly. If we're serious about conserving taxa, communities and ecological processes across the board, that increase in awareness and increase in submissions has to be good.

From the Committee's perspective, I think we did well in dealing with the deluge of information that came in. We got it right in the sense that there has been only one formal court challenge to the Committee's operations, and that's been dismissed. There are probably other aspects, but I'll come to the second question.

Your second question was about the unanimity of the Committee's decisions. Yes, well, it was just harsh and vicious chairmanship! I think the Committee was able to do a lot of background research, and cover a lot of discussion in meetings. These were not ordinary meetings. They were once a month, sometimes more frequent, and lasted all day, from a 9 o'clock pre-meeting to often 6.00 or 7 o'clock at night. So we really had time to cover any issues.

Also, all Committee members had an opportunity to read the information to be discussed at the meeting several days in advance, thanks to Sue Chate, our Executive Officer, who marshalled all the information before sending it on. So everybody was up to speed with what was going to be covered in the meetings. I think, given the opportunity to consider and discuss the issues, it was actually not that difficult to come to a unanimous conclusion in virtually all cases. So unanimity was achieved through the opportunity to discuss broadly and by the readiness and the preparedness of the Committee members.

People had the opportunity to raise their hands and dissent; there was no need in fact for unanimity to be accomplished. Perhaps towards the end of the day everybody was getting tired!

Your third question was about Indigenous representation: I'm sure it would have been of great benefit to the Committee to have had an Indigenous representative. Unfortunately, that is not part of the TSC Act. The part of the Act that specifies the backgrounds and expertise of the Committee members doesn't provide for Indigenous representation. It doesn't disallow it, but it doesn't say there should be an Indigenous representative either. I think that perspective certainly would have been valuable.

Dan Lunney (Department of Environment and Conservation, NSW): That's left you a bit less than satisfied, hasn't it, Mark?

Mark Burgman: Look, no, there are probably lots of other questions, but my experience of expert judgments in committees is that you can achieve consensus by negotiation or you can have some sort of decision-making process like voting. You can also amalgamate opinion numerically, quantitatively, and arrive at a process of consensus in which you re-examine the data and then re-examine the degree of disagreement.

Pp 172 - 179 in *Threatened species legislation: is it just an Act?* 2004, edited by Pat Hutchings, Daniel Lunney and Chris Dickman. Royal Zoological Society of New South Wales, Mosman, NSW, Australia.

You could then produce a judgment, as courts do that have dissenting opinions, or you could present a summary of the information that says that this is the point at which this person changed their mind. It would make the evaluative process much more transparent. At present you simply provide a point estimate, a yes or no, and we don't know why, but you all 10 or 11 agree.

As a committee, you are then difficult to cross-examine. You are a lot of fairly important people to disagree with, so you become very difficult to challenge. You may start to assume a mantle of authority that is unassailable, even unwittingly.

Chris Dickman: I guess it would have been possible, potentially anyway, to record a lot of the cut and thrust of our discussions. It would have been easy for many of the nominations that were open-and-shut cases; everybody had a predetermined decision on which way they'd go on these, and they all happened to be the same. But there was certainly a lot of debate and cut and thrust about many of the nominations.

I think the difficulty would have been in terms of recording it all. The minutes for each meeting, just the record of decisions and other things that were discussed and decided by the Committee, usually ran to between 12 and 15 pages. So, although it would perhaps have been desirable, full recording was probably unworkable in practice and could not have captured all of the points that were raised.

If we felt there would be some benefit in capturing the diversity of opinions that contributed to a decision on a particular determination, we tried to make sure that the points were covered and that referencing was extensive. In some cases, such as the determination on land clearing, we produced an extensive series of appendices that provided additional information, indicating the contributions of various Committee members. Generally though, it would have been very difficult to capture the entire range of debate and opinions about everything that was considered.

David Farrier (University of Wollongong): Chris, I'd like to follow up Mark Burgman's question. You dealt with one side of consensus. I'm interested in the decisions that you made not to list particular species. Did you always have consensus at that level or was there lots of disagreement at that point?

Chris Dickman: The species that were not listed, the decisions not to proceed, were again in almost all cases unanimous. But like the species or populations or communities that ended up being listed, these cases were often subject to a lot of debate. I think some of the nominations for koala populations, for example, tied the Committee up in knots in trying to determine whether the nominations could be sustained. Did they fit the criteria in the Act?

Some of the difficulties that we had with such nominations for endangered populations were difficulties in determining where a population started and stopped. The question of boundaries is a really difficult one, particularly for mobile animals. There are issues not only of mobility, but also of lack of survey, perhaps lack of access into areas where the species might occur, making it very difficult to determine whether a population could be defined. Even though we didn't proceed with quite a large number of nominations, we considered them in detail, and the decisions not to proceed were virtually unanimous.

Martin Denny (Ecological Consultants Association): Chris, one of the questions I'm asked constantly by frustrated land developers is: "Why can't you delist a species?" Have there been any examples of delisting? That is, where species have been taken completely off the threatened list - not moving through the levels, but completely removed? And how do you go about it?

Chris Dickman: I think the first example of delisting was of a plant species, *Pultenaea campbellii*. The procedure for taking a species off the lists altogether is exactly the same as for putting one on. A nomination can be made by any person. The Committee reviews the evidence, then decides whether to proceed to a preliminary determination or not. In the case of the *Pultenaea*, the Scientific Committee decided to seek additional information on how extensively the species was distributed, whether any new populations had been discovered, and the extent of any threats. We discovered that there had in fact been new surveys that revealed that the *Pultenaea* was more widespead than the Committee had thought when the species was first listed. As a consequence, the determinations, preliminary and then final, were made to delist the species altogether. So, yes, delistings do occur, but admittedly there have not been a large number of cases.

Paul Adam (University of NSW): On the issue of rejection of a nomination and how unanimous the agreement was, one of the important things is that most of the rejections are, in a sense, automatically triggered. The Scientific Committee or individual members of the Committee may decide at meetings that they can't make a decision unless certain information or data are provided, and then a request for this additional information goes back to the nominator.

Most of the rejections under Section 21 occur because the nominator is unable to provide the information that would enable a decision to be reached. Normally this is because the information is deficient; or it doesn't exist. There have been a small number of cases with publicly advertised

PLENARY

rejections, and in these cases all the information that would be appropriate to making a decision is available. In those cases the Committee has come to the conclusion that there is a very strong case not to make a listing. There have been a small number of those cases, but the majority of rejections are because some or all members of the Committee have felt they can't make a decision unless certain information is available, and the nominator hasn't been able to provide it.

Paul Burcher: My question is for Dan Lunney. Given your concern for the 74 per cent of species not listed, what's the status and future of the New South Wales Biodiversity Strategy?

Dan Lunney: The NSW Strategy for Conserving Biodiversity [produced by the NSW National Parks and Wildlife Service in 1999] is a different way of approaching the issue of biodiversity conservation. The topic of threatened species is just one of the strands of thinking. The Strategy is currently under revision, and there is also the national strategy [1996], and local councils also have strategies as do different local community and catchment groups. All these matter, but it is the clout that threatened species legislation delivers that makes it so important across Australia. The selling of the importance of threatened species makes for a very powerful image.

From one way of looking at it, threatened species legislation and the actions that followed in the last decade or so are a response to the massive conservation problems identified through the 1970s and 1980s. In one sense we have legislation that deals with yesterday's problems, but we have yet to catch up with the rate at which we are understanding current problems and to legislate, or put in place other policies to force a broader approach on conservation issues.

Not that these other issues are neglected, but the question you have to ask is, is the current emphasis on threatened species the right emphasis to conserve biodiversity? The NSW Biodiversity Strategy is great from the point of view that it covers so many approaches, it does identify a wide range of options. It identifies actions by priority, but a lot of those priorities slide away when it comes to funding because they do not have legislation to force matters, such as going to court for lack of compliance.

No name recorded: I had a question in regards to private land, when a recovery plan or a listed threatened species is involved. To what extent are the writers of the recovery plan or the Scientific Committee responsible for a socio-economic study? Also, given that it is largely command legislation, but a lot of people believe that incentives are the best way to go, how much is that going to come through in actually writing the plans, and who assesses whether that socio-economic study is adequate or not?

Rob Humphreys (Department of Environment and Conservation, NSW): I'm one of the managers of the Threatened Species Unit in the Department of Environment and Conservation. The responsibility for socio-economic assessments falls on the Department. We recognise that we are not experts on social and economic matters and, depending on the impact of a preliminary assessment, we may or may not go to external expert consultants to do that work for us.

As an example, in the Cumberland Plain Woodland recovery plan, we've spent in the order of $150,000 so far on independent external economic assessments. Those assessments will go out on exhibition with the plan. That will provide an opportunity for people who are affected to provide a comment on whether they agree or disagree with those economic impacts.

A lot of the recovery plans we write, such as for a species which is known only from a very few sites, or is wholly within the National Park estate, might have very limited social or economic impacts, so the amount of social and economic assessment that goes into the plan will vary, depending on the species and its locations.

There were amendments to the TSC Act which came into force at the beginning of this year that will set up a social and economic advisory committee in parallel to the Scientific Committee. This will have various terms of reference and all recovery plans will need to be sent to that committee for their advice.

In terms of impacts on private landowners, we can't write recovery plans or dictate actions which bind private landowners. We can only enter into voluntary-type actions on private land. As far as statutory authorities or government agencies are concerned, we can't put actions into recovery plans unless the ministers responsible for those agencies agree to those actions being included. So, to a degree, we're pretty hamstrung as to what sort of actions we can put into recovery plans.

Gillian Courtice (Curlew Biological Services): I live on the mid north coast. Up there, a lot of the application of threatened species legislation is with private landholders who are slowly developing the coastline with a sort of urban creep that may start off with 25-acre blocks, which go down to five acres and then one and so on.

Since the introduction of the latest Rural Fires Act, we find that the Act is really overstepping or overwhelming threatened species legislation in the applications to Council. I guess the socio-economic aspect of fire protection is overcoming the threatened species legislation to the point where all subdivisions, even if they are large rural holdings, have to be totally underscrubbed and therefore have all habitat removed, whether threatened species are present or not.

This will happen unless you can make a really strong case that the last animal on earth is living on that subdivided block. It seems that urban creep and habitat impact have accelerated enormously since the asset protection zones of the Rural Fires Act have been enforced. I just wondered if any of the speakers had a comment as to how that could be addressed.

Michael Murphy (Department of Environment and Conservation, NSW): On the north coast where I work, I think the new fire legislation is a good thing. What we were seeing in the past was that developments would get the okay, with only the immediate impact of the development being considered. Then they'd come along and say, "Well, now we have to put all these fire breaks in", even though they weren't considered as part of the original development. The way we look at it is that the fire buffers that would be needed if a development goes ahead are part of the footprint of the development. If it comes to DEC for concurrence, it is then considered as part of the overall impact of the development.

Stuart Little (Department of Infrastructure Planning and Natural Resources): I worked on some of that bushfire policy. The approach that we adopted was through the planning and development control provisions and the new planning for bushfire protection guideline. What we tried to achieve was that, first of all, the bushfire protection measures get kept on the side of the development and are therefore not palmed off onto the biodiversity side, or the bushland side. Secondly, that the set-backs be retained on the side of the development wherever possible, for existing developments that have already gone in and for existing properties where the bush is already close to that particular asset.

What gets triggered then is the bushfire hazard reduction assessment, which comes through the new bushfire code. What we are trying to provide, first of all, is a planning process through the bushfire risk management plan that identifies where your asset protection zones are. Then there's a streamlined process through the bushfire assessment code that actually makes explicit provisions for consideration of threatened species, and other biodiversity matters, which should get taken into account when developing the type of bushfire protection measures that are warranted and how far they should go into the bush. There are set distance thresholds provided in that code document.

Judy Lambert (North Head Sanctuary Foundation): There have been a few comments from various speakers, myself included, about what I'll call offsetting, for want of a better term. I'm wondering if anybody would like to comment on this. I believe it's a fairly widely held community view; certainly in my interactions it is. If an ecological community is already sufficiently at risk to be listed as endangered, then surely the remaining remnants that are being used for the so-called offset should already be getting management attention to enhance their degraded conservation value; should they really be allowed to be the subject of an offset trade-off?

Rob Humphreys: I'll volunteer to answer that question. Judy is right. There are a lot of landowners who have threatened species or threatened species' habitat on their land, but the legislation does not require the landowner to carry out any active management of their land. So if a landowner has a threatened species, they can just sit back with their arms crossed and watch it slowly degrade over time, and a lot of threatened species sites on both public and private land are decreasing in value through lack of active management. In some situations - and whether offset or compensatory habitat or ameliorative measures are the appropriate terms, I'm not so sure - if funds can be found to actively manage the remnant sites, then it enhances and increases the value of those remnants.

Tony Auld (Department of Environment and Conservation, NSW): Can I just add to what Rob is saying. People use the word "offset" for several different things, and it leads to confusion that is part of the problem. It is the case that, where you lose a remnant of bushland, the way you compensate is to actually create new bushland in an area that is severely degraded, with no current biodiversity values. That's a different thing to what I think Rob was talking about. If you have a political decision to develop, you try and use whatever resources you can get from losing a particular patch of remnant to protect another patch, or patches, of remnants.

So they are two slightly different things. In the case of the former, the problem is the timelag effect and our ability to reconstruct communities. We should not be calling these offsets - the new community we are trying to create does not have anything to do with the existing community, because it really is not that community and it may not be for centuries.

David Keith (Department of Environment and Conservation, NSW): I'm interested in the issue of uncertainty and building - or the dynamics of knowledge, I suppose, and how it reflects upon the whole process of regulating and conserving threatened species in particular, and biodiversity in general.

We've been doing things like impact assessment now for 25 years formally in New South Wales. I get the impression that we have not learned an awful lot in terms of how we go about mitigating impacts, or making decisions about whether impacts are acceptable or not in terms of community values. For example, when we place conditions of consent on a particular development, my question is: "How well have we done in the last 25 years in assessing the success of those conditions to mitigate the impact that they were proposed to mitigate?"

PLENARY

I guess it's not a new issue. Peter Fairweather commented on this subject, very effectively I thought, maybe 15 years ago, when there seemed to be a fairly woeful standard of environmental assessment with investigations that were purported to be scientific but had no replication and no controls and very little commitment in terms of resources to follow up observations in any sort of sensible scientific design. This is something that I think we have largely neglected today, but I think it is an important part of the whole threatened species issue, and it certainly needs as much scrutinising as the success of the various other stages in the process.

Anne Conway (Countrywide Ecological Service): I had a long history in the bureaucracy, dealing with EISs on both sides of the fence. I think that is a really key question and I think the fundamental failure of the environmental assessment system in New South Wales is that there is no monitoring on several levels. The first being whether in fact the conditions of consent are actually met.

There are many developments, that I am sure people in this room would know, where the approval was given, with conditions of consent, and you know damn well they have not even complied with them, let alone going back 5 or 10 years later and looking at what the environmental impact statement said, and whether the assertions that were made in the environmental impact statements were in fact correct and that there was no impact, or only little impact..

I've had a long interest in some of the really big infrastructure developments. It would be very interesting to go back and look to the early days and look at what was said in the EIS, what the conditions of consent were, what actually happened on the ground, what the EIS said, and what in fact happened. Were there other impacts that were quite significant that were in fact picked up in the EIS and what sort of things that should have been picked up. We could learn from that.

It could be fairly non-threatening to the government if you did it for the early ones. I think it's the sort of thing that the scientific community should be pushing, to actually spend some money and have done, because we will never learn if we don't go back and look at what happened. We will continue to blunder on into this century, making exactly the same mistakes as we made way back when.

Dan Lunney: Thanks, Anne. We shall now shift to the second part of the plenary, with three new convenors to bring a fresh dimension to this matter. They have assembled, as a stimulus for thought and further discussion, some comments and identified some themes that have become obvious during the sessions today. These are presented below, in the comments by Rob Whelan, David Farrier, and Martin Denny, and are followed by responses from forum participants.

Rob Whelan (University of Wollongong): One important point of relevance to the forum is that the first couple of objects within the NSW *Threatened Species Conservation Act 1995* actually don't say very much about threatened or endangered species. They talk about conservation of biodiversity and ecologically sustainable development. Given that these issues are contained in the objects of the Act, then we need to think more broadly about how to achieve them.

David Farrier (University of Wollongong): It seems to me that there are several questions related to listing. The fundamental issue is whether we should list anything at all (i.e. avoid using listing as a conservation process) or whether there are other processes which we can use, broad-brush processes such as land use planning. This would avoid what is often a tortuous process of working out whether or not a species should be listed.

Assuming that we decide that it is useful to list something, there is the issue of what we should list. There have been various points raised in relation to whether we should focus on listing many species or just a few. "Flagship species" have been mentioned. We have not had much discussion of the idea that nominating one species can save a lot of others.

There has been discussion of listing on a broader scale than species. Habitats were mentioned, although it is not clear exactly what 'habitat' would mean in this context. In Queensland, ecosystems are being listed; threatened ecosystems are identified, not under the threatened species legislation but under vegetation clearance legislation.

The next question is, if you do decide to list, whether the actions this triggers can ever be implemented vigorously enough to achieve adequate protection for the listed entity. Often the result of listing is some sort of approval process. It has been argued that we go through the motions, and end up with some sort of amelioration that appears to be sufficient to allow an approval to be granted. It has been suggested today that a lot of the amelioration does not really achieve its objective. I have always been concerned about amelioration: whether it is, in fact, just something that we put up with because we really can't bite the bullet of saying, "This development shouldn't take place". If it shouldn't take place, then perhaps we need to look for ways of softening the blow to the land user by producing some sort of conservation agreement, which will involve flow of funds.

Then the last question: What are the alternatives and do we trust them? I think we really do need to look beyond threatened species legislation. It does not work in a vacuum. There is lots of activity going on at the moment with vegetation clearance legislation in both New South Wales and in Queensland. In Queensland, they are listing threatened ecosystems and also pursuing a parallel process of listing species. Basically you have two regulatory systems – two command regulatory systems – operating alongside each other.

Then there is the question of whether or not we should really start looking at the land use planning system. People have mentioned the NSW *Environmental Planning and Assessment Act.* Our focus is very much on the concurrence process at the moment, but maybe we should step back a little and look at the planning process, the LEP making process, the REP-making process.

Martin Denny: One thing that I did not mention earlier, and it comes in here very importantly, is that in Western Australia (through the EPA) there are statements that the species should be assessed within the ecosystems. They talk about keystone species, and also 'hotspots' – areas with high species richness – neither of which brings in the idea of listing of threatened species.

The other thing I'd like to bring up is why do we have a seven- or eight-part test? We should not have it at all. I'd like to put that up as a suggestion: if a development is to go ahead, it should go straight to a species impact statement. All of the significance of impact should be assessed at that level. We don't need an eight-part test or seven-part test at all. It's only duplicating what's going to come next and it's a waste of time and a waste of money.

Rob Whelan: We have drawn out a number of themes from several of the presentations today that dealt with uncertainty of various kinds. We have summarised some of them above. Several people have referred to the uncertainty of the science. One component of uncertainty is not knowing how an organism is going to respond to a particular development or a particular type of action, nor being able to predict how it will respond to a proposed mitigation action.

We have talked about ignorance of various sorts; not that invertebrates are ignorant (despite their simple nervous system), but that we're ignorant *about* them. In many cases, we don't even know if a particular listed organism actually exists in an area. If we know nothing about an organism, if we don't even have that first hint that there might be an impact, it is unlikely that we could apply the precautionary principle.

A third type of uncertainty has to do with definitions. Paul Adam especially emphasised the importance of definitions, and several people returned to that theme during the day.

There is also the issue of change over time. Even if we agree on definitions, and if our science becomes more complete, the landscape out there is not stationary. It is not static. We also heard about how communities change over time. How can we deal with this sort of uncertainty or variability?

How should we be responding to these components of uncertainty? I suggest that there are two levels of response: internal and external. How we respond internally is how, as a community of scientists, managers and policy-makers, we could help guide the development of legislation so that it can deal with the uncertainties in predicting distribution and abundances of organisms and their ecological responses. These predictions, with their associated uncertainties, are needed in imposing development controls, amelioration actions, recovery planning, or threat abatement planning.

Really interesting questions were stimulated by Mark Burgman's optimism about presenting our uncertainty to the political world and to the community, and educating them about so-called ' facts' and their uncertainty. This issue has been debated at the tea breaks and lunch. People sound very nervous about this subject because of the way in which politicians and special-interest groups will twist that admission of uncertainty. So we might do it with the best intentions, but how is it going to be used externally?

Stuart Little: The first key point looked at was biodiversity and ESD. What I think is becoming clear is that there is indeed a need for a threatened species framework, but that it is driving everything and it is not sitting properly within the context of a wider biodiversity framework. When Dan Lunney spoke about 'the other 74 per cent', it became clear that there is a need for a mechanism that considers ecological processes, and things in the landscape, such as wildlife corridors.

There is also the need to deal better with the source of the problem - the actual threat. The Threat abatement planning process does not appear to be well resourced and there are not many threat abatement plans. The other thing is the development control and planning processes. Again, responses seem reactive to particular development applications.

There has not been a lot of discussion about integrating biodiversity considerations into wider land use decisions, as David Farrier was suggesting, such as regional vegetation management plans, catchment blueprints, and local environmental plans. How do we actually encapsulate biodiversity outcomes at the early stages of planning processes before the area is zoned and individual developments come through?

The final thing is that the relationship between threatened species and conservation in the wider natural environment is driven by regulation. Potential solutions to such things as salinity and land degradation are driven by incentives. There seems to be an absence of regulation on that side and an absence of incentives on the threatened species side. I think we need to balance that up a bit.

Paul Adam: I was going to raise some of the same issues. I would love to think that the solution was going to be in planning. However, I look back at history and conclude that I was over-optimistic. The important question is, given that we have a planning system, how do we redesign it so that it will actually meet these ends?

The other issue I'd throw into the debate is this: As the number of listings increases, we have got to seriously revisit the whole question about what we want recovery planning to do, whether it really is appropriate, and whether the same timetable should be set for everything. We need to make much more strategic use of resources so we are not producing plans and losing sight of those things we urgently need to have a plan for, rather than having a mechanistic planning process.

Jack Baker (Department of Environment & Conservation, NSW): I don't think listing is a way to achieve conservation of biodiversity, but if you don't list things, we won't know what biodiversity we are trying to conserve. It comes first.

Judy Lambert (North Head Sanctuary Foundation): I agree that listing per se probably does not achieve the conservation of biodiversity, but listing, as a recognition of threat of extinction, is a far more saleable tool to both the community and decision-makers than is the conservation of biodiversity. Unless we include it as an essential step in the process, we are probably going to go backwards. One of the approaches may be to focus more on habitats and communities and less on individual species, except where they are really on the brink of falling off the planet.

Rob Whelan: Several speakers mentioned other things that happen as a result of listing. Listing triggers a range of activities, but also the listing of species raises community awareness, such as in Simon Nally's example. It also leads to an increase in knowledge. We know a lot more about many populations because of the excitement and interest that followed the listing. As mentioned in the fish example, alteration to management practices occurred as a result of information gained after listing, but not as an official part of the process. There are benefits of listing that we have yet to quantify.

Helen Mills (Department of the Environment and Heritage): Listing of threatened species and ecological communities is very important for all those reasons of recognition, but we need to stop trying to fund programs solely on the basis of a species being threatened, we need to prioritise. An awful lot of the nominations that come in are triggered by the fact that recovery plans are not being funded. They want to get it listed, so they can get it funded. Many taxa already have recovery plans, but the recovery *actions* are not being funded. Listing may not necessarily help them unless we stop prioritising endangered species for funding above other species and ecological communities.

Alison Colyer (Threatened Species Network): The main thing is to try to engage the community in threatened species recovery work through awareness-raising, through on-ground projects, and so on. We tend to find we use the flagship approach, and that's largely because it is reactive to what the communities that come to us want to do, and partly because it is a great way to engage the community and get them interested.

We tend to use species because this helps people understand broader biodiversity issues and the broader landscape management issues. That is where listed species are vital. Biodiversity is a concept that the general community generally cannot get their heads around, especially ecological communities. It is difficult for the wider human community to understand the complex interactions involved. Listing is therefore vital as a contact point, because we know that the vast majority of threatened species management has to be done on private land with community members.

Martin Denny: The seven-part test (of likelihood of significant impact) gives a chance for negotiation before the development goes ahead. One of the problems with the seven-part test and associated negotiation is that none of the things that the proponents say are locked in concrete. They can say anything they want to and they can go to a seven-part test and it could be assessed in a certain way. If a proposal goes straight to a Species Impact Statement, proposed actions have to be signed off by the proponent.

Rob Whelan: It seems to me that there is a fundamental problem here. The example that I used was of *Darwinia biflora*. If the determining authority, the Council in that case, had been negotiating with the developer about amelioration to remove the impact, the question is: "Have the putative ameliorative actions really removed the impact (even if they are put in place as planned)?" There is no external scrutiny of the degree to which they have achieved protection. We are trusting that the process of assessing likely effectiveness of proposed amelioration actually worked at the Council level. If the proposal does go to the next level – SIS – there is at least an opportunity for people to see whether something is flawed in the argument.

Lyn Dawson (RZS member): We've been talking about scientific certainty. We've been talking about the difficulties of understanding taxonomy of the life cycles of organisms. There's a lot of money going into funding threatened species conservation or threatened species protection. I'm just wondering, "Is there any money going to go into the scientific basis for this?" If so, where is the money going to come from? It is very tedious science, it is slow science, it is boring science, it is not fashionable science. Would anybody like to comment on that?

Liz Denny (RZS member): Sitting here today I've realised that we have done it back to front. In fact, everything should be listed, and the game should be to get it off the list. Yes, it's a wonderful idea! If you listed absolutely everything, and then you say: "Okay, you want to do something? Show the Scientific Committee that they can take this off in this region or this locality." Then everyone would get money for research because developers would want as much research as possible so that they can say, "There's bloody millions of them here." What a good idea!

Kate McIlroy: I want to go back to the discussion that, instead of listing species, we should be listing ecological communities. I can't believe I'm actually going to say this, because I'm not particularly convinced that listing ecological communities is working effectively! Essentially, a listed ecological community always just ends up being a comprehensive species list anyway. You select an area for protection or otherwise based partly on whether or not it's got the percentage of the species in the list.

Generally the species list contains mostly botanical species anyway. Not solely. I can be looking at an area in western Sydney and there might be four macropods, none of which is listed, but they may be regionally significant and an integral part of the community - but this is completely irrelevant to the protection of the Cumberland Plain Woodland Ecological Community in the area.

Claire De Lacey: There seems to be some misconception that the basis of an eight part test lies in science. It doesn't. The eight part test consists of many value-based judgments and that's all it comes down to. Very little science is portrayed in an eight-part test. That's why it makes it so important to assess these things at a planning stage, because otherwise it eventually is simply at the whim of the government of the day.

If they're having a bad hair day or a good hair day, then it shines through in terms of how they manage threatened species. Environmental management is something that we should be looking at in perpetuity and not something that is here today and gone tomorrow. That seems to be what's happening at the moment, with so many value-laden judgments in environmental assessment.

Danny Hirschfeld: I work in Sydney's eastern suburbs. We have about 300 ha of remnant vegetation left. Most people probably would not realise that it is that much. But a lot of it is endangered, most of it Eastern Suburbs Banksia Scrub, an endangered ecological community under both state and commonwealth legislation. Those pieces of legislation have worked, to a large extent. Development applications being assessed with regard to Banksia Scrub, and other endangered items, have resulted in 90 per cent of those areas being conserved. The listing process has been great, also for an endangered wattle. It has contributed to biodiversity conservation in the eastern suburbs.

Rob Whelan: Isn't that a wonderful note on which to end this session? This example is one that has achieved the conservation of biodiversity. Thanks very much on behalf of the three of us for having the opportunity to do this. I think we hand over to Pat Hutchings for the next bit.

Pat Hutchings (Australian Museum): I should like to thank all the speakers. We're going to go straight into the annual general meeting. It will only take 20 minutes. Then we can all go and have a drink… I promise!

Threatened species legislation: just one act in the play

Chris Dickman[1], Pat Hutchings[2] and Daniel Lunney[3]

[1] Institute of Wildlife Research, School of Biological Sciences, University of Sydney, NSW 2006. cdickman@bio.usyd.edu.au

[2] Australian Museum, 6 College Street, Sydney, NSW 2010. path@austmus.gov.au

[3] Department of Environment and Conservation (NSW), PO Box 1967, Hurstville NSW 2220. dan.lunney@environment.nsw.gov.au

ABSTRACT

Governments in Australia and elsewhere in the world use legislation to conserve species and other elements of biodiversity that are threatened by human activities. This forum examined to what extent relying on the legislative process is justified and, in particular, considered the effectiveness of legislation in achieving broad conservation goals. A synthesis of the papers presented at the one day forum run by the Royal Zoological Society of NSW is presented in this final paper. These papers display considerable insight into how threatened species legislation works, from the decision to list a species, to the consequences of listing for government departments, local councils, community groups and consultants. While the papers concentrated on the NSW Acts, they were compared with Commonwealth and other States' legislation as well as overseas examples. For many contributors to this forum, threatened species legislation is deficient in that it focuses on conspicuous, iconic and well-known taxa, and does little to conserve small, cryptic organisms, common but declining taxa, or ecological processes. In addition, threatened entities must often be identified using imprecise criteria that can be interpreted differently, and uncertain or incomplete information, while recovery planning for ever-expanding lists of threatened taxa is difficult and expensive. Scarce resources could be made to go further by allocating them to priority taxa, but criteria for ranking priorities are themselves highly contentious. Given these difficulties, an efficient way forward would be to identify common threats that afflict suites of populations, species and ecological communities and to target these for management. This is possible, but has been given insufficient priority under some existing legislation such as the NSW *Threatened Species Conservation Act 1995* and the *Environment Protection and Biodiversity Conservation Act 1999*. Other priorities for achieving broad conservation goals are the familiar yet critical requirements for sustaining research, education and funding, but added to these is the need for more public debate, engagement of the whole community, and informed inspiration to show that solutions are possible. We conclude that threatened species legislation is important, but just one act in the biodiversity play.

Key words: endangered species, Threatened Species Conservation Act 1995, Fisheries Management Act 1994; scientific experts; threatened species legislation, populations, communities, threatening processes, vulnerable.

Introduction

As biological diversity continues to diminish across the world, a common response by governments has been to pass legislation to endeavour to halt the decline. This response is understandable. Human activity is responsible for the current biodiversity crisis and, as most modern societies are founded on the rule of law, appropriate legislation is essential to rein in our worst excesses and allow the variety of life to be retained. But, does such legislation work? The contributors to this forum agreed that threatened species legislation is important, but diverged in their evaluations of its effectiveness, of the consistency of its administration, and where such legislation fits into the larger picture of biodiversity conservation. Here, we identify just some of the themes arising from the forum and, adapting the concept that species can be viewed as players in the ecological theatre (Hutchinson 1965), put the case that threatened species legislation is but one act in the play.

The forum falls into four parts, with the fourth being the plenary session. In checking the proofs of the plenary, one of the participants (Martin Denny) sent the following message: "I think this session is the most important from the whole meeting. A lot of people speaking honestly, and giving good suggestions." As President of the Ecological Consultants' Association of NSW, Denny represents the view, as he did so adeptly in his paper, that there are many more interested parties than may be apparent by just reading the views of one of the more vocal protagonists. When these other views are neglected, the capacity of both Commonwealth and State threatened species legislation is thereby diminished, as has been crisply demonstrated by Lambert's (2004) pointed contribution. More importantly, these views may well be struggling to find a voice, hence the importance of the public forum on threatened species legislation held by the Royal Zoological Society of NSW.

Among the seemingly wild suggestions in the plenary session was the view put forward by Liz Denny: "Sitting here today I've realised that we have done everything back to front. In fact, everything should be listed, and the game

Pp 180 - 192 in *Threatened species legislation: is it just an Act?* 2004, edited by Pat Hutchings, Daniel Lunney and Chris Dickman. Royal Zoological Society of New South Wales, Mosman, NSW, Australia.

should be to get it off the list." Zany as it may seem, that was in fact among the directions to the Scientific Committee under the *Endangered Fauna (Interim Protection) Act 1991*. Here the Committee was required not only to prepare a list of endangered fauna and be prepared to justify it, but also to be able to justify leaving a species off the list. This meant that it was necessary to prepare the first ever State-wide list of fauna and then to employ a system that was robust enough to be even in its ranking procedures across all vertebrate species (Lunney *et al.* 2004). This unusual requirement is not part of other threatened species acts, including the current NSW *Threatened Species Conservation Act 1995*. In one sense, Liz Denny's proposal is an extreme application of the precautionary principle that is examined in detail in Whelan *et al.* (2004). However, they take the view that the listing process could be precautionary, but a significant amount of information is needed before a population, species or community can be listed, so nominations of taxa for which little information is available are likely to be unsuccessful. Consequently, they note that a precautionary approach to listing would be to use the IUCN classification "Data Deficient" as a category, especially if it were designed to trigger a process of ecological study to gather relevant data, and a Species Impact Statement in cases where a data deficient taxon was likely to be affected by a specific development proposal. This fragment of the debate reveals that in considering threatened species legislation, the precautionary principle has a distinct place. What it also exposes is the problem of how to deal with a lack of information. One approach is to utilise experts, but at this juncture Burgman (2004) jolts everyone who is complacent about the presumed ability of experts.

As Burgman (2004) points out, experts are often the only available source of knowledge to support decisions that would otherwise be made in a vacuum. Group opinions are likely to be better than individual opinions, says Burgman, in that they provide an efficient and convenient way of compiling and aggregating a broad range of knowledge. But, Burgman cautions, there may be a great deal to be learned from other technical disciplines and from the psychology and sociology of risk perception and decision theory. Fortunately, Burgman proposed some positive steps to address the limitations of the expert system, such as noting that performance is enhanced when experts possess appropriate models and are trained to translate subjective assessments into numerical estimates, and that experts perform better when they use a consistent format to communicate, when they are trained to provide unbiased estimates, and when their judgement is reinforced by immediate, unambiguous feedback.

Another of the far-reaching insights from the papers in this forum is that by Farrier and Whelan (2004). Like Burgman's startling contribution, it bridges the disciplines to portray a problem in a new and lucid way. Consider the penultimate paragraph in their paper. It is an example of a superb mix of practical and theoretical studies. It represents a fine example of intellectual adaptive management and also lays bare one of the greatest challenges to the conservation of threatened species. The paragraph runs as follows: "NSW provisions relating to the protection of threatened species habitat betray their roots in a system originally designed to regulate hunting. At a symbolic level, heavy reliance appears to be placed on command regulation triggered directly by threatened species listing, and backed up by the threat of prosecution. In practice, however, this is significantly moderated by the operation of an *ad hoc* approvals process dominated by socio-economic considerations, with conservation outcomes a hoped-for side-effect of ameliorating provisions rather than a primary objective. Decision-making processes are driven by individual development proposals rather than strategic thought, and assessment of cumulative impact is a significant casualty. The government agency with expertise in threatened species conservation and scientific research is marginalised."

Farrier and Whelan (2004) proceed to identify the need for a more effective approach to planning, and to that we can add that all the insights from all the papers and the plenary comments offer an even wider scope for change. This need to broaden our views and come up with new ideas begins with the foreword by Bob Beeton (2004). He felt impelled to add a new dimension as he declared that writing a foreword is "a rare opportunity to be slightly outrageous and perhaps provocative without the normal discipline of modern publication." He was provoked by the title of the forum, which he says, invites such an approach. His foreword draws on his experience of chairing the Australian Threatened Species Scientific Committee whose task is to recommend listings and whose decisions are expected to show "*Solomonic*" judgement. Mark Burgman has yet another source of data on which he can exercise his growing expertise as an iconoclast, especially when some judgements are regarded as "*Solomonic*". Australia would be a poorer place without the skilled heckler, the critic or the expert. Consequently we are lucky to have such a diversity of minds with the willingness to speak, be criticised and then respond by improved performance. While this is the very essence of education philosophy, it is not necessarily always appreciated by those who feel threatened by threatened species legislation.

The debates in the NSW parliament, such as the Threatened Species Conservation amendment bill, as reported in the NSW Legislative Assembly Hansard of 25 September 2002, reveal this conflict and confusion (Appendix 1). This appendix is included here to display the context in which those concerned with threatened species work. It also shows the difficulty of distinguishing a threatened species, with the Green and Golden Bell Frog *Litoria aurea* gaining a mention in the parliamentary debate on this matter. Herein lies the difficulty of managing the rapid change of status of a species, as shown by the Southern Corroboree Frog *Pseudophryne corroboree* as displayed on the cover of this book. Most germane to this forum in this book is the role of scientific committees, the difficulties of implementation of threatened species legislation in the field, such as in rural and regional NSW, and the contest over the importance of conserving threatened species in relation to economic matters, including the question of "who pays?" To us, the key importance of communication of ideas surfaces yet again, and the relativeness newness of threatened species legislation demands that the subject has a thorough public hearing. For those who have been living and breathing threatened species legislation for the last decade or so, it may seem that the subject has been perpetually on the political and scientific agenda. However, the documentation of the debate says otherwise. Wilson (1986), representing the then Australian National Parks and Wildlife Service, stated that it is the State and Territory governments which have the first responsibility for endangered species and that the Commonwealth is involved as the national government through its controls over overseas trade and certain land and marine areas. Wilson noted that some have suggested that the Federal Government should have a greater involvement in the protection and management of endangered species, such as along the lines of the American model. Wilson then comments that "such an all encompassing piece of legislation is unlikely to be passed by the federal government in Australia." It is at this point that the thorough and much-needed paper by Jarman and Brock (2004) becomes so applicable. They have catalogued the sequence of the changes to the legislation in NSW from the 19th century, and it becomes clear that threatened species legislation is both new and only part of a suite of legislation to conserve biodiversity. They consciously avoid the question of effectiveness, but if the State of the Environment reports reflect the state of play, then it can be stated for Australia that "during the 1990s, Australia's biodiversity has experienced continued degradation and decline" (p 7, Williams *et al.* 2001), and for NSW that the "loss of biodiversity is ongoing" (p 191, EPA 2003). We have, as a society, yet to formulate ways of reversing such losses. Any such approaches will include threatened species legislation, but the case we are putting in this chapter is that this is but one act in the play.

Conflict and confusion engender fear and lack of acceptance. For those keen on conserving biodiversity, the need to communicate the science of the subject, and its consequences, has been amply demonstrated by the strong language in the debate as reported in Hansard (Appendix 1). Such debates provoked the Council of the Royal Zoological Society of NSW to run its 2003 forum on this theme. We can now hope that the outcome of this forum will assist those in parliament, and those they represent, to see that there is also an active debate in the scientific and legal world on this matter, and that there is an active public and professional interest in the outcomes.

The scope of threatened species legislation

The term 'threatened species legislation' implies that acts and by-laws are concerned with species and nothing else and, in some older legislation and current jurisdictions, this is certainly true. In New South Wales, for example, the now-superseded *Endangered Fauna (Interim Protection) Act 1991* allowed for the listing and protection only of terrestrial vertebrates, while in Tasmania the *Threatened Species Protection Act 1995* allows for the listing only of recognised species and sub-species. The emphasis on species is not surprising. The conservation efforts of the World Conservation Union are based on identifying and listing species at risk of global extinction (IUCN 2003), and powerful arguments can be made that species are the fundamental 'units' of conservation because they are intuitively obvious and easy to manage (Caughley and Gunn 1996). In addition, news stories about iconic but declining species such as the Koala *Phascolarctos cinereus*, Greater Bilby *Macrotis lagotis* or Green and Golden Bell frog *Litoria aurea* help to keep conservation issues before the public, and maintain support for sometimes costly programs of population restoration.

On its own, however, a focus on threatened species is not enough to achieve the broader goal of biodiversity conservation. In her contribution, Hutchings (2004) points out that invertebrates are overwhelmingly dominant and crucially important in the functioning of both aquatic and terrestrial systems, but our incomplete knowledge of most species means that few are likely to find their way onto threatened species lists. Clearly, the legislative process will be inefficient if it focuses on invertebrates at the species level; this will be true also for many poorly-known and cryptic species of fungi, non-vascular plants and microbes, as well as for hybrids and other life forms that are difficult to categorise. One way out of this dilemma is to conserve communities of invertebrates or their habitats, or to tackle the processes that threaten them. Despite the problems of species-level conservation for invertebrates, Hutchings (2004) argues the case that small numbers of conspicuous, charismatic species should be listed and targeted for conservation because they can raise community awareness. Such species might include the Bathurst copper butterfly *Paralucia spinifera* (Dexter and Kitching 1991), Cumberland Plain land snail *Meridolum corneovirens* (Clark 2004) and Richmond birdwing butterfly *Ornithoptera richmondia* (Sands *et al.* 1997); at present the two former species are listed on the *NSW Threatened Species Conservation Act 1995*, but the spectacular *O. richmondia* remains unlisted.

Although we can extend legislation to incorporate invertebrates and other poorly-known taxa, there is another group of species that will not be covered under any threatened species paradigm. These are the common, widespread and ostensibly non-threatened species. Should common species warrant conservation attention? Lunney *et al.* (2004) comment that reliable identification of threatened species is often fraught with difficulty, and use the species of vertebrates *not* listed in the schedules of the *NSW TSC Act 1995* – the neglected 74% - to make their case. Apart from uncertainties in distinguishing some species as threatened or not, Lunney *et al.* (2004) note that non-listed species are of little interest to planners, managers and others involved in land use decisions because there is no legal imperative to do so; consequently there is little incentive to work on them. With declining research effort on non-threatened species, and so little monitoring on any species, we then have little capacity to recognise when formerly-common species are in trouble or threatened species have recovered. There may be a case for distinguishing species that are close to being threatened, as is done by the World Conservation Union (IUCN 2001), or adding the IUCN category of 'data deficient' (Whelan *et al.* 2004), but there is little sense in making lists of threatened species all-inclusive. Lunney *et al.* (2004) propose broadening the conservation agenda by constructing a database that allows dynamic tracking of all species, so that assessments of status can be made effectively and swiftly as new knowledge is obtained.

In recent years some of the deficiencies of maintaining a strict focus on threatened species have been recognised, and legislation has been broadened to allow listing and protection of other entities (Adam 2004). In New South Wales, for example, the *Threatened Species Conservation Act 1995* (*TSC Act*) took the innovative step of providing for the listing of endangered populations and endangered ecological communities, as well as threatened species; the importance of key threatening processes was also recognised (Auld and Tozer 2004; Baker 2004; Leys 2004). Also, for the first time, it became possible to list taxa other than the familiar vertebrates and vascular plants. As a consequence, six species of fungi, 10 insects, four other species of invertebrates and communities of lichens and fungi have taken their place in the schedules of the *TSC Act 1995*. Similarly broad legislation, the *Fisheries Management Act 1994*, covers marine and many freshwater taxa, as well as endangered communities and key threatening processes. Jarman and Brock (2004) provide a fascinating account of the changing scientific, social and economic conditions that have driven conservation legislation in New South Wales, and note how current laws have increased in complexity and moved to incorporate broad conservation objectives. Not only this, conservation legislation is integrated across local, State and Commonwealth levels of government, and international agreements such as RAMSAR have been developed to recognise that conservation initiatives do not stop at national borders. However, the links and obligations across State, Commonwealth and international boundaries are far from even (Tsamenyi *et al.* 2003).

Threatened species legislation in some jurisdictions has moved much beyond its historically narrow focus on identifying and protecting just the most obvious species. The ability to list populations, ecological communities and threatening processes is a sign of greater sophistication in our thinking about conservation, and also reflects the increasing effectiveness of conservation scientists in making their case to both the media and politicians (Thomas and Salwasser 1989; Burgess 2001). But, is this enough? We are not aware of any legislation that yet addresses the specific needs of metapopulations, hybrids, common but declining species, or the longer term effects of disruption on ecological processes such as dispersal, fire, disturbance, pollination, facilitation or mutualism. Calls to incorporate such concepts into legislation have been made before (*e.g.* Rohlf 1991), but they remain unheeded and serve as challenges for the future. We now turn to the equally important issues of how lists of threatened species and other entities are compiled, and what happens when a listing is made.

The listing process

In most jurisdictions, the listing of a species as threatened will confer recognition of status and trigger some kind of planning process that is intended to manage and oversee the species' recovery. It is important to get this step right to maintain both the species itself and credibility in the process. At several levels, procedures for listing are remarkably similar. At the international level, for example, the World Conservation Union's red lists of threatened species are maintained by Species Specialist Groups operating under the Species Survival Commission. The specialist groups may be concerned with monitoring the status of particular species, such as the Malayan tapir *Tapirus indicus*, or with large groups of species, such as the Australasian Monotreme and Marsupial Specialist Group or the Mollusc Specialist Group. These groups meet with different frequencies and have memberships of between two and over 100 people, but all comprise specialist researchers who know their taxa from first hand experience, and all make recommendations based on adherence to criteria specified by the World Conservation Union. Many specialist groups also take advice from external sources and consider public submissions, thus ensuring that all readily available information is used to decide the status of particular species. In Australia, the Commonwealth and all States and Territories have scientific committees or panels that evaluate species' status within their political boundaries, and all are able (or required) to consider nominations from any person or organisation. The committees either make recommendations for listing to the relevant minister or, in the case of the NSW Scientific Committee, have the final say in what gets listed and what does not. Thus at all levels, there is opportunity for both specialist expertise and broader public input into the listing process. Does this mean that we can have full confidence in the integrity of lists of threatened species and other entities?

In his paper, Adam (2004) points out that some of the first problems besetting the listing process are how to define the entities that can be listed and how to interpret the criteria that would place them on the schedules of the relevant legislation. Using the *TSC Act 1995*, Adam shows that even such familiar terms and concepts as 'indigenous', 'species', 'populations' and 'ecological communities' can mean different things to biologists, lawyers and others. He notes that this places an obligation on the Scientific Committee to make its determinations as clear and unambiguous as possible, although the certainty sought by critics is not realistic and hence unlikely to be attained.

Assuming that issues of definition can be overcome, further obstacles to creating a reliable list of threatened taxa can arise from logistical constraints and a lack of knowledge. In compiling the schedules of threatened vertebrates for the *Endangered Fauna (Interim Protection) Act 1991*, Lunney *et al.* (2004) describe how just one month was allocated for the review of all vertebrates in New South Wales and the identification of species that should be listed. One month! Although there was a longer period of public review after the production of the initial list, the logistical constraints facing the three-person Scientific Committee charged with compiling the list were formidable; the approach taken to deal with this problem has been outlined by Lunney *et al.* (2000, 2004). We do not know the timeframes allowed for the production of similar lists under other pieces of legislation, as this is seldom spelled out. However, if legislators believe that reliable syntheses of large and diverse biotas can be achieved in a matter of just days or weeks, there is a clear disconnect between optimism and reality. When lists of threatened taxa and other entities have been compiled, there is a requirement under some current legislation, such as the *TSC Act 1995* and *Fisheries Management Act 1994*, to review the lists at regular intervals. This is a reasonable requirement, and should help ensure that lists are based on the most up-to-date knowledge from new research and surveys as these are carried out. However, it can also be a logistical impediment. For example, section 17 of the *TSC Act 1995* required that the Scientific Committee keep lists in the schedules under review, and at least bi-annually, to determine whether any changes were necessary. There is a small likelihood that new knowledge would accumulate every six months on all the entities listed in the schedules, and carrying out continuous reviews would be a poor use of time. Dickman (2004) outlines the procedures adopted by the Scientific Committee to meet this statutory requirement, particularly the engagement of consultants to review 'problematic' groups of species, or communities, to provide advice on their status.

Even when time and other resources are available to compile and maintain lists of threatened species, we must often grapple with the further problem that knowledge of many taxa is limited. In a short but provocative essay, Diamond (1987) proposed that lists of threatened species are misleading because they make us think that even the most poorly-known taxa on the lists remain extant. However, due to the difficulty of searching all possible habitats, especially for small or cryptic species, it is usually difficult to prove that extinction has occurred, and so species often stay listed as threatened due to inertia. Would it not be more reasonable to assume such species to be extinct until proven extant? The blue-grey mouse *Pseudomys glaucus* provides a good example. This species is listed as endangered (*TSC Act 1995*) or critically endangered (IUCN 2003), but as the last individual was seen in 1956 in an area that has since been extensively disturbed (Dickman 1993), a more realistic assessment would be that it is extinct. In the absence of specific knowledge, but armed with the general understanding that habitats are at risk of disruption everywhere, perhaps the onus of proof should shift to confirming that highly threatened species are still hanging on.

Whelan *et al.* (2004) extend this line of thinking, arguing that our incomplete knowledge of the ecology of many threatened species should in itself act as a trigger for application of the precautionary principle. Thus, if a species is little known, why assume that it is secure and leave it off a list of threatened taxa? Species are likely to be poorly-known precisely because they are sparse or occupy small geographical ranges; if listing triggers subsequent ecological research, we will then gain the information needed to make an informed decision about the species' status. Likewise, in assessments of the impact of development, such as the eight-part-test specified in the NSW *Environmental Planning and Assessment Act 1979*, why assume that a development will have no significant effect just because a species is too poorly-known to be sure what the impact will be? It is arguably better to adopt the precautionary approach that there might be a significant impact, despite any proposed mitigation or remediation works, and evaluate the proposal for development as thoroughly as possible before any habitat is modified. As Auld and Tozer (2004) emphasise in their contribution, fragmentation and attrition of remnant habitats undermine the long-term viability of ecological communities and their constituent species, so any proposal for new development must be taken seriously. And, as the impacts of any development are virtually never monitored after proposals have been approved and the development begun, there must be much more onus on evaluating the impacts effectively in the first place. Whelan *et al.* (2004) note that developments will often go ahead even if significant impacts are predicted, because socio-economic and other imperatives are part of the decision-making process. However, if the precautionary principle is used more extensively to identify species that will be affected by developments, these authors argue cogently that decision-makers will be forced, for the first time, to be honest about the impacts they are authorising.

The issue of limited knowledge in the listing process is explored further by Burgman (2004), who questions the uses and limitations of expert judgement in arriving at listing decisions. A worked example of one way the problem has been tackled is in the paper by Lunney *et al.* (2004). That work documents the range in expert opinion that was received when threatened vertebrates were being identified for listing in the schedules of the NSW *Endangered Fauna (Interim Protection) Act 1991*, and shows that variance in expert opinion about species' status is related inversely to how much we know about those species. In listing 26% of the vertebrate fauna of New South Wales as threatened, the Scientific Committee established under the *Endangered Fauna (Interim Protection) Act 1991* steered a middle course between conservatism (just 14% of the State's vertebrates would have been listed if all experts consulted had agreed) and speculation (40% would have been listed if just one expert voted for listing), using its own expertise to adjudicate the responses of other experts. The present Scientific Committee, established under the NSW *Threatened Species Conservation Act 1995*, also relies on advice from external experts, and uses a process of extensive information-gathering, debate and review to arrive at its decisions on listing (Dickman 2004).

The consequences of listing

What happens when a species, population or ecological community is listed as threatened, or a key threatening process is recognised? If there is no statutory recognition of the listing, not much happens at all. For example, the *Red Lists* produced by the World Conservation Union (IUCN 1996, 2003) represent the most comprehensive and quantitatively-assessed data bases of threatened species that are available (Lamoreux *et al.* 2003), but many national governments pay little more than lip service to them. There is no international law that dictates otherwise. However, when there is statutory recognition of threatened species or threatening processes, usually at national, state or local levels, different requirements for administration, research and management come into play. Here, we consider these requirements and how effectively they work.

In all Australian States and Territories, as well as the Commonwealth, the listing of a species as threatened triggers the need to produce a recovery plan. The timeframes specified for writing recovery plans differ between jurisdictions, and plans vary greatly in the emphasis they place on biological review, research, monitoring and active management (Dickman 1996). In New South Wales, for example, the recovery plan for Coxen's Fig-parrot *Cyclopsitta diophthalma coxeni* calls for an extensive program of captive breeding, field survey, raising of community awareness and habitat assessment and restoration, whereas that for the Painted Burrowing Frog *Neobatrachus pictus* recommends only surveys for extant populations and engaging the community (NPWS 2000, 2002). Although all recovery plans specify biodiversity conservation as their ultimate goal, it is not surprising that they recommend different tactics to achieve it. Species with historically small but stable populations may be threatened by random events or human incursions into their habitats and need only to be monitored, whereas declining species must usually be actively managed to reverse the downward trend in their populations (Caughley 1994). There are also many different methods in the 'ecological toolbox' (Norris 2004), and their effectiveness in achieving species recovery depends on the nature of the threats faced by particular species as well as on the philosophy of the recovery planners (Carroll *et al.* 1996; Green 2002; Norris 2004). Sometimes several taxonomically or geographically related species may be grouped together for management; threat abatement plans written to reduce the impacts of key threatening processes will potentially also assist suites of threatened and non-threatened taxa.

Analyses of the effectiveness of recovery and threat abatement planning are scarce. However, there are general concerns that plans take too long to be written and approved (Tear *et al.* 1995), that cute and glamorous species get both the earliest attention and most resources (Restani and Marzluff 2002), and that recovery planning is not always cost effective. With respect to cost, perusal of recently approved and draft recovery plans in New South Wales suggests that the combined funding needed to achieve broadscale species recovery is in the order of several million dollars each year; for threatened Australian birds alone, almost $28 million was dedicated to recovery between 1993 and 2000 (Garnett *et al.* 2003*a*). With respect to the effectiveness of recovery plans in achieving their objectives, Silver (2003) commented that only 30 of 1262 species listed as threatened on the US Endangered Species Act had been removed from the list since 1973; of these, seven species were removed because they became extinct. A selective review by Reeves (1999) suggested that no Australian-based projects had successfully brought species back from the brink, but intensive management has in fact been successful for several threatened species of vertebrates (Kinnear *et al.* 2002; Garnett *et al.* 2003*b*), invertebrates (New 1997; Sands 1999) and plants (Johnston and Dollary 1998). Although no comprehensive analyses have been undertaken to evaluate the current effectiveness of recovery planning in Australia, case studies and perspectives provided by forum participants suggest that both cautious optimism, and more work, are needed.

Perspectives from case studies

In New South Wales, the *Threatened Species Conservation Act 1995* is a major legislative instrument for the conservation of terrestrial biota, and primary responsibility for its administration lies with the Department of Environment and Conservation. The *Fisheries Management Act 1994* mandates the conservation of the State's aquatic biota, and its administration lies with NSW Fisheries.

Papers by Baker (2004) and Shields (2004) make the case that recognising and listing threatened taxa are fundamentally different processes from those needed to prevent extinction and promote recovery. Both are concerned that we fail collectively to clearly articulate this difference, and question whether the bottom-line objective – conservation of biodiversity – is being achieved. Baker (2004) applauds the provision in the TSC Act that allows the listing of endangered populations, but laments our inability to monitor or manage, in any way, the majority of the endangered populations that have been listed so far. Shields (2004) elaborates on this issue, and describes a series of steps to ensure effective species recovery and threat abatement. These include articulating a clear vision to guide environmental legislation, defining key actions for success and, crucially, ensuring that they are implemented. Shields also emphasises the importance of gaining community support for environmental legislation, the need for improved resourcing, less bureaucracy, and the need to adopt conservation outcomes that are measurable. His challenging conclusion is that conservation of biodiversity cannot be achieved in piecemeal fashion by making ever-expanding lists of threatened taxa, but rather by managing overall threats so that such lists are of secondary importance in conservation planning.

Case studies of the management of listed entities are provided by Murphy and Nally (2004), Talbot *et al.* (2004) and Auld and Tozer (2004). For example, for those keen on conserving invertebrates hope that their contribution will assist in making the point that landscape planning is not guaranteed to conserve threatened species. However, the striking feature of these studies is that they demonstrate that threatened species legislation can be intelligently applied to difficult problems with good local outcomes. The case studies demonstrate the level of skill needed, the time taken and the resources that will be necessary to achieve worthwhile results. More importantly, these case studies show that the scale of the total problem is large if we are to go down the route of recovery planning for an ever-increasing list of entities. Finally, the case studies, covering such diverse entities as sharks, snails and butterflies, and endangered ecological communities from a botanical perspective, give a far greater insight into the complexity of this subject of conservation biology than just the particular studies reported. Being able to draw on such skilled studies is a great asset in this broader debate.

Community perspectives

Several contributors to the forum highlighted the importance of community involvement and ownership in conserving biodiversity, both in their remarks on the day itself (see plenary discussion) and in the written papers presented here. This issue has also been recognised at previous forums of the Royal Zoological Society (Bennett *et al.* 1995; Dickman and Lunney 2001; Eby and Lunney 2002; Lunney and Dickman 2002; Lunney *et al.* 2002). In her contribution, Lambert (2004) re-emphasises the role of the community and commends the potential value of State and Commonwealth legislation in allowing people to get involved in a hands-on way in the conservation of threatened species and their habitats. However, she also sounds a sobering note of caution, commenting that individual and community interests in the environment are often subverted in the courts or by ministerial discretion to favour yet more development. At North Head, home to endangered populations of Long-Nosed Bandicoots *Perameles nasuta* and Little Penguins *Eudyptula minor* as well as remnants of the endangered Eastern Suburbs *Banksia* scrub community, such decisions have already eroded habitat and may increase traffic flows, putting the bandicoot population, at least, at increased risk (Banks 2004). If local communities become disillusioned, local governments will also face increasingly uphill battles to carry out their statutory obligations under State and Commonwealth legislation, and biodiversity conservation will suffer. To counter these trends and offset the bullying tactics of the pro-development lobby using the existing legislation, Lambert (2004) urges stronger collaborations between concerned scientists, local people and local government.

Consultant perspectives

Although local communities, councils, government agencies, scientists, lawyers and others may all deal with threatened species legislation on a regular basis, ecological consultants are left to hold the pointy end of the stick more than most. In New South Wales, consultants are often called to judge whether a proposed development is likely to have a significant impact under the terms of the *Environmental Planning and Assessment Act 1979*, or to carry out the much more detailed requirements of a Species Impact Statement as specified under the *Threatened Species Conservation Act 1995*. Writing as an experienced consultant and President of the Ecological Consultants Association, of New South Wales, Denny (2004) highlights some of the problems encountered in interpreting the legislation and in keeping up to date with the expanding lists of threatened taxa that must be considered. Denny (2004) also echoes the growing concern that threatened species legislation is narrowing the conservation focus, inappropriately, on just a small subset of the overall biota; that is, the species, populations and ecological communities that are listed. Common species, including the 'neglected 74%' of non-threatened vertebrates in New South Wales, simply do not rate in assessments of impact, despite the fact that many are likely to have ecologically important roles in the natural system or that many could be listed as threatened if we had up-to-date information on their status (Dickman and Steeves 2004; Lunney *et al.* 2004). For Denny (2004), and probably many others concerned with the conservation of biodiversity generally, the limited focus of threatened species legislation is therefore frustrating and unsatisfactory.

Landowner perspectives

The issue of landowner responses was raised in several presentations and also in the plenary discussion at the forum. It has become increasingly clear in recent years that conservation efforts cannot be confined to protected areas such as national parks and reserves because these areas represent just fractions of the entire landscape. Moreover, they contain only a small subset of the overall suite of habitats and land systems, and sometimes represent just the unproductive areas that could not be used for agriculture (Hobbs 1994). Hence, much recent effort has gone into engaging private landowners and into ensuring that the matrix of production and non-production land is managed to the best effect (Bennett *et al.* 1995; Craig *et al.* 2000). Given the importance of private land to the conservation effort, what are the incentives for landowners to protect threatened species and what are the constraints on land development?

Among the incentives are the values placed on biodiversity by people on the land themselves, and their desire to see it maintained. Such incentives are extremely important, but of course, they do not pay the bills. Financial or assistance-based incentives include joint management agreements, voluntary conservation agreements and other kinds of covenants where a landholder sets aside an area of land in return for tax breaks, help with management or some other compensation for lost opportunity. These kinds of compensation are outlined in detail by Farrier and Whelan (2004) and Shields (2004). Such incentive schemes have yet to achieve any kind of profile in Australia, but they are being contemplated increasingly in the United States (Silver 2003).

In contrast to the provision of carrots, sticks are also used to constrain over-development of land. It is illegal to clear native vegetation without a permit in most Australian States, and the listing of species, populations and communities as threatened is also intended to provide legislative protection. But these protective laws are often flouted. In New South Wales, Bartel (2003) notes that few cases of illegal land clearance have been prosecuted since 1998, that monitoring and enforcement are insufficient, and that political will to change this situation is weak due to stakeholder influence. Farrier and Whelan (2004) comment further that command regulation in an environmental context has traditionally been used to direct developments towards approval processes rather than to outlaw them altogether. As approvals are usually *ad hoc* and driven by individual development proposals, habitats, communities and areas generally of high conservation value get whittled away with no consideration of cumulative impacts.

Given the pros and cons of extending the operation of threatened species legislation to private land, what do landowners do when a species, population or ecological community that occurs on their property is listed? Unfortunately, we have little documentation of this in Australia, but a recent study from the United States by Brook *et al.* (2003) is instructive. Following listing of Preble's Meadow Jumping Mouse *Zapus hudsonius preblei* as threatened under the Endangered Species Act, Brook *et al.* (2003) mailed a questionnaire to landowners on whose property the mouse occurred. They found that 56% of respondents would not allow any biological survey of their property, hence precluding collection of data needed for conservation. In terms of hectares owned, 25% of respondents reported managing their land to improve habitat for the mouse, whereas 26% said that their management had sought to harm it. Thus, the listing of Preble's meadow jumping mouse did little to increase its prospects of survival on private land. Brook *et al.* (2003) suggested that mutually better outcomes would be obtained by alleviating the economic concerns of landowners about listings, increasing the use of social networks and collaborative processes to improve communication, and providing institutional assurances that landowner interests would not be harmed by managing their land to conserve threatened species. There are lessons to be learnt for conservation in Australia.

Alternatives / additions to the current legislation

We must now ask what other options might be available for achieving broad conservation goals, and where threatened species legislation fits into the bigger picture. One of the themes to emerge from the forum papers was that listing and saving species (or populations or ecological communities) is very costly if this is done on a one-by-one basis. The costs of species-based recovery programs, when summed, are clearly prohibitive, and this prompted several authors at the forum (e.g. Baker 2004; Hutchings 2004; Lunney *et al.* 2004; Shields 2004; Whelan *et al.* 2004) and elsewhere to question what can be done (Wilcove and Chen 1998; Possingham *et al.* 2002; Williams *et al.* 2001). As this issue has been canvassed broadly in other forums (Beattie and Ehrlich 2001; Archer 2002; Foran and Poldy 2002; Lunney and Dickman 2002; Archer and Beale 2004), we present only some brief and selective comments here.

Priority lists

If there are insufficient funds to recover all listed taxa, one solution is to create a list to rank the priorities for the allocation of resources. At present, there is a tendency for the most endangered species to be awarded most attention and funding for recovery (Garnett *et al.* 2003*a*), although this *de facto* priority list has been criticised because it sometimes means that money is thrown at lost causes (Possingham *et al.* 2002) or ignores threatening processes (Lunney *et al.* 2003). If species were ranked according to some cost-effectiveness measure, such as rate of population increase or rate of geographical spread per dollar allocated, we may see some spectacular recoveries of not-so-threatened species and smiles on the faces of conservation economists. However, biological data to inform such a measure would be lacking for many species and, because of their reduced ability to respond quickly, many seriously threatened species would rank poorly and thus be consigned to oblivion.

Other possibilities for ranking might be to identify 'zombie' species, *i.e.* the living dead, or taxa that otherwise appear to be unimportant, and remove these from funding formulae. Zombies are species whose time is almost up, and which persist in ever-diminishing numbers. Some might be identified as taxa that will lose all their habitat in the next few years due to climate change; examples would include high altitude specialists such as the Mountain Pygmy-possum *Burramys parvus*, Leadbeater's Possum *Gymnobelideus leadbeateri* (Lindenmayer *et al.* 1991), or the Bramble Cay Melomys *Melomys rubicola*, whose tiny island home in Torres Strait is eroding inexorably into the sea (Limpus and Watts 1983). Other potential zombie species might include those with very small populations that could vanish during a bad year or due to a fire or other catastrophic event; examples would include the Northern Hairy-nosed Wombat *Lasiorhinus krefftii* and Lord Howe Stick Insect *Dryococelus australis*. Of course, these and many other potential zombie species are well known and iconic. In the absence of certainty about the directions of future climate change, writing them off prematurely would be a brave but foolish act, and one that would also be damaging to the collective psyche of conservation biologists. Further, a converse argument can be mounted that the studies of species near mountain tops are giving the clearest early warning of the impending impact of climate change (*e.g.* Williams *et al.* 2003; Winter 2004), and that the Leadbeater's possum has been the most important animal in Australia to challenge the orthodoxy of woodchipping and logging as usual, and the effort to conserve it has led to scanning the international literature on conservation biology, identifying its limitations, then beginning to plug the gaps (*e.g.* Lindenmayer and Franklin 2002).

How about 'unimportant' species? Species are sometimes considered unimportant if they have no obvious role in an ecological community (see Walker 1992, 1995 for discussion on this point), or if they do not engage in diverse interactions with others (Soulé *et al.* 2003). If a species has many close relatives, its loss is arguably of less importance in terms of conserving genetic diversity than that of another species that is the last member of its family or order (Strahan 1986). Strahan's thesis is extraordinary because it of its boldness, its deep grasp of the evolutionary place that each species occupies and its challenge to priorities, or more importantly, our inability to either see them or to articulate them. Strahan is now famous for his book on Australian mammals (Strahan 1995) and it is to his view of mammals that he expounds his thesis. 'Importance' may also be assigned by ranking on scales of cuteness, tourist dollar value, use in drug production, medical research and so on. The obvious difficulty with all such ranking schemes based on importance is that value is in the eye of the beholder, so that agreement across any two or more ranked lists is likely to be low (Dickman *et al.* 2005). In addition, how might we rank such different entities as species, populations and endangered ecological communities? Any system that ranks some species or entities above others is, in a sense, an admission that a component of biodiversity is to be cut adrift from any conservation effort; it is an admission of defeat. Perhaps we should ask not how to slice the conservation budget ever more finely, but ask rather how we can increase the overall allocation.

Beyond species-based conservation: threatening processes

A strong theme to emerge from the forum was that conservation of species, populations and ecological communities could often be achieved most effectively by managing the threats that afflict them. This point was made most cogently by Leys (2004), but echoed by Auld and Tozer (2004), Murphy and Nally (2004) and Talbot *et al.* (2004), among others at the forum. The advantages of effective threat abatement are clear: if a threatening process such as predation by the Red Fox *Vulpes vulpes* can be reduced, many native species will benefit. Threat abatement is often an expensive, on-going and labour-intensive enterprise that sometimes has unexpected consequences (*e.g.* Roemer and Wayne 2003), but there is no doubt that it will continue to be a crucial act in the biodiversity play.

Getting conservation onto the agenda

Five years ago, Recher (1999) made the seemingly apocalyptic prediction that up to 50% of Australia's bird species would be come extinct within the next century. This remark is now bordering on a citation classic. However, it is likely to be eclipsed by the more recent view of one of the world's leading conservation biologists, Peter Raven, from the Missouri Botanic Gardens in the United States, who said on Earthbeat on 4 September 2004: "that if we continue with business as usual, up to two-thirds of the species on Earth will be gone within a hundred years" (http://www.abc.net.au/rn/science/earth/stories/s1191446.htm). In this context, Liz Denny's view that "everything should be listed, and the game should be to get it off the list" can be seen as both a smart remark and very smart biology. In the context of this prediction of mass extinction, threatened species legislation will be dealing with most species, communities and issues of biological conservation. This seems a most uneconomical way to proceed, and makes little use of the many other acts in the play, including, research, education, reserve selection, community involvement, and most importantly, public debate about the future of Australia's biological heritage. Indeed, we need to lift Cassandra's curse - that dire predictions by conservation biologists for the future, though true, are not believed - and shift conservation biology from a discipline of doom and despair to one of informed inspiration (Redford and Sanjayan 2003), from science to political action (Haber 1992), and to engagement of all the community. This forum was convened and edited in this spirit.

Acknowledgements

We are indebted to many colleagues who have contributed to our ideas over the years, to the participants at this forum for their insights and perspectives on threatened species legislation, and to Carol McKechnie for critical comments on drafts of this manuscript.

References

(Note: references that appear in the body of the text with a 2004 date that do not appear below refer to other chapters in this book, and can be found listed in the contents page.)

Archer, M. 2002. Confronting crises in conservation: a talk on the wild side. Pp. 12-52 in *A Zoological Revolution: Using Native Fauna to Assist in its own Survival*, edited by D. Lunney and C. Dickman. Royal Zoological Society of New South Wales, Mosman, and Australian Museum, Sydney, NSW, Australia.

Archer, M. and Beale, B. 2004. *Going Native: Living in the Australian Environment*. Hodder Headline, Sydney.

Banks, P.B. 2004. Population viability analysis in urban wildlife management: modelling management options for Sydney's quarantined bandicoots. Pp. 70-77 in *Urban Wildlife: More than meets the Eye*, edited by D. Lunney and S. Burgin. Royal Zoological Society of New South Wales, Mosman, NSW.

Bartel, R.L. 2003. Compliance and complicity: an assessment of the success of land clearance legislation in New South Wales. *Environmental and Planning Law Journal* **20:** 116-141.

Beattie, A. and Ehrlich, P.R. 2001. *Wild Solutions: How Biodiversity is Money in the Bank*. Melbourne University Press, Melbourne.

Bennett, A., Backhouse, G. and Clark, T. (eds). 1995. *People and Nature Conservation: Perspectives on Private Land Use and Endangered Species Recovery*. Royal Zoological Society of New South Wales, Mosman, NSW.

Brook, A., Zint, M. and de Young, R. 2003. Landowners' responses to an Endangered Species Act listing and implications for encouraging conservation. *Conservation Biology* **17:** 1638-1649.

Burgess, B.B. 2001. *Fate of the Wild: the Endangered Species Act and the Future of Biodiversity*. University of Georgia Press, Athens, USA.

Carroll, R., Augspurger, C., Dobson, A., Franklin, J., Orians, G., Reid, W., Tracy, R., Wilcove, D. and Wilson, J. 1996. Strengthening the use of science in achieving the goals of the Endangered Species Act: an assessment by the Ecological Society of America. *Ecological Applications* **6:** 1-11.

Caughley, G. 1994. Directions in conservation biology. *Journal of Animal Ecology* **63:** 215-244.

Caughley, G. and Gunn, A. 1996. *Conservation Biology in Theory and Practice*. Blackwell Science, Oxford.

Clark, S.A. 2004. Native snails in an urban environment – conservation from the ground up. Pp. 78-81 in *Urban Wildlife: More than meets the Eye*, edited by D. Lunney and S. Burgin. Royal Zoological Society of New South Wales, Mosman, NSW.

Craig, J.L., Mitchell, N. and Saunders, D.A. (eds). 2000. *Nature Conservation 5: Conservation in Production Environments: Managing the Matrix.* Surrey Beatty & Sons, Chipping Norton, NSW.

Dexter, E.M. and Kitching, R.L. 1991. *Nomination for the Register of the National Estate.* Unpublished report, Australian Heritage Commission, Canberra.

Diamond, J.M. 1987. Extant unless proven extinct? Or, extinct unless proven extant? *Conservation Biology* **1:** 77-79.

Dickman, C.R. 1993. The biology and management of native rodents of the arid zone in New South Wales. *NSW National Parks and Wildlife Service Species Management Report* No. **12:** 1-149.

Dickman, C.R. 1996. Incorporating science into recovery planning for threatened species. Pp. 63-73 in *Back from the Brink: Refining the Threatened Species Recovery Process*, edited by S. Stephens and S. Maxwell. Surrey Beatty & Sons, Chipping Norton, NSW.

Dickman, C.R. and Lunney, D. (eds) 2001. *A Symposium on the Dingo*, Royal Zoological Society of New South Wales, Mosman, NSW.

Dickman, C.R., Pimm, S.L. and Cardillo, M. In press. Extinction risk, biodiversity loss, and does it matter? In *Hot Topics in Conservation Biology*, edited by D.W. Macdonald. Oxford University Press, Oxford.

Dickman, C.R. and Steeves, T.E. 2004. Use of habitat by mammals in eastern Australian forests: are common species important in species management? Pp. 761-73 in *Conservation of Australia's Forest Fauna*, second edition, edited by D. Lunney. Royal Zoological Society of New South Wales, Mosman, NSW.

Eby, P. and Lunney, D. (eds). 2002. *Managing the Grey-headed Flying-fox as a Threatened Species in New South Wales.* Royal Zoological Society of New South Wales, Mosman, NSW.

EPA. 2003. *New South Wales State of the Environment 2003.* Department of Environment and Conservation, Sydney, NSW.

Foran, B. and Poldy, F. 2002. *Future Dilemmas: Options to 2050 for Australia's Population, Technology, Resources and Environment.* Available as a 61-page summary, '*Dilemmas Distilled*', or as the full 337-page report (www.cse.csiro.au/futuredilemmas).

Garnett, S.T., Crowley, G.M. and Balmford, A. 2003a. The costs and effectiveness of funding the conservation of Australian threatened birds. *BioScience* **53:** 658-665.

Garnett, S.T., Crowley, G.M. and Stattersfield, A.J. 2003b. Changes in the conservation status of Australian birds resulting from differences in taxonomy, knowledge and the definitions of threat. *Biological Conservation* **113:** 269-276.

Green, R.E. 2002. Diagnosing causes of population declines and selecting remedial actions. Pp. 139-156 in *Conserving Bird Biodiversity*, edited by K. Norris and D.J. Pain. Cambridge University Press, Cambridge.

Haber, W. 1992. On transfer of scientific information into political action; experiences from the German Federal Council of Environmental Advisors. *INTECOL Bulletin* **20:** 3-13.

Hobbs, R.J. 1994. Effects of landscape fragmentation on ecosystem processes in the Western Australian wheatbelt. *Biological Conservation* **64:** 193-201.

Hutchinson, G.E. 1965. *The Ecological Theatre and the Evolutionary Play.* Yale University Press, New Haven, Connecticut.

IUCN. 1996. *1996 IUCN Red List of Threatened Animals.* IUCN, Gland, Switzerland.

IUCN. 2001. *IUCN Red List Categories and Criteria. Version 3.1.* IUCN Species Survival Commission, Gland, Switzerland.

IUCN. 2003. *2003 IUCN Red List of Threatened Species.* IUCN Species Survival Commission, Gland, Switzerland.

Johnston, B. and Dollary, C. 1998. Survey Report – *Grevillea kennedyana.* Unpublished report to NSW National Parks and Wildlife Service, Hurstville, NSW.

Kinnear, J.E., Sumner, N.R. and Onus, M.L. 2002. The red fox in Australia – an exotic predator turned biocontrol agent. *Biological Conservation* **108:** 335-359.

Lamoreux, J., Akçakaya, H.R., Bennun, L., Collar, N.J., Boitani, L., Brackett, D., Bräutigam, A., Brooks, T.M., da Fonseca, G.A.B., Mittermeier, R.A., Rylands, A.B., Gärdenfors, U., Hilton-Taylor, C., Mace, G., Stein, B.A. and Stuart, S. 2003. Value of the IUCN Red List. *Trends in Ecology and Evolution* **18:** 214-215.

Limpus, C.J. and Watts, C.H.S. 1983. *Melomys rubicola*, an endangered murid rodent endemic to the Great Barrier Reef of Queensland. *Australian Mammalogy* **6:** 77-79.

Lindenmayer, D.B. and Franklin, J. F. (eds). 2002. *Conserving Forest Biodiversity: a comprehensive, multiscaled approach.* Island Press, Washington, D.C., USA.

Lindenmayer, D.B., Nix, H.A., McMahon, J.P., Hutchinson, M.F. and Tanton, M.T. 1991. The conservation of Leadbeater's possum, *Gymnobelideus leadbeateri* (McCoy): a case study of the use of bioclimatic modelling. *Journal of Biogeography* **18:** 371-383.

Lunney, D. and Dickman, C. (eds). 2002. *A Zoological Revolution: Using Native Fauna to Assist in its own Survival.* Royal Zoological Society of New South Wales, Mosman, and Australian Museum, Sydney, NSW, Australia.

Lunney, D., Dickman, C. and Burgin, S. (eds). 2002. *A Clash of Paradigms: Community and Research-based Conservation.* Royal Zoological Society of New South Wales, Mosman, NSW.

Lunney, D., Matthews, A., Stein, J.A. and Lunney, H.W.M. 2003. Australian bat research: the limitations of *The Action Plan for Australian Bats* in determining the direction of research. *Pacific Conservation Biology* **8:** 255-260.

New, T.R. 1997. *Butterfly Conservation*, 2nd edition. Oxford University Press, Oxford.

Norris, K. 2004. Managing threatened species: the ecological toolbox, evolutionary theory and declining-population paradigm. *Journal of Applied Ecology* **41:** 413-426.

NPWS. 2000. Painted Burrowing Frog (*Neobatrachus pictus*) Recovery Plan. NSW National Parks and Wildlife Service, Hurstville, NSW.

NPWS. 2002. Recovery Plan for the Coxen's Fig-parrot *Cyclopsitta diophthalma coxeni* (Gould). NSW National Parks and Wildlife Service, Hurstville, NSW.

Possingham, H.P., Andelman, S.J., Burgman, M.A., Medellin, R.A., Master, L.L. and Keith, D.A. 2002. Limits to the use of threatened species lists. *Trends in Ecology and Evolution* **17:** 503-507.

Recher, H.F. 1999. The state of Australia's avifauna: a personal opinion and prediction for the new millennium. *Australian Zoologist* **31:** 11-27.

Redford, K. and Sanjayan, M.A. 2003. Retiring Cassandra. *Conservation Biology* **17:** 1473-1474.

Reeves, G.W. 1999. *Midterm Review of the Natural Heritage Trust: Bushcare Program.* CSIRO Centre for International Economics, Canberra.

Restani, M. and Marzluff, J.M. 2002. Funding extinction: biological needs and political realities in the allocation of resources to endangered species recovery. *BioScience* **52:** 169-177.

Roemer, G.W. and Wayne, R.K. 2003. Conservation in conflict: the tale of two endangered species. *Conservation Biology* **17:** 1251-1260.

Rohlf, D.J. 1991. Six biological reasons why the Endangered Species Act doesn't work – and what to do about it. *Conservation Biology* **5:** 273-282.

Sands, D.P.A. 1999. Conservation status of Lepidoptera: assessment, threatening processes and recovery actions. Pp. 382-387 in *The Other 99%: the Conservation and Biodiversity of Invertebrates*, edited by W. Ponder and D. Lunney. Royal Zoological Society of New South Wales, Mosman, NSW.

Sands, D.P.A., Scott, S.E. and Moffatt, R. 1997. The threatened Richmond birdwing butterfly (*Ornithoptera richmondia* (Gray)): a community conservation project. *Memoirs of the Museum of Victoria* **56:** 449-453.

Silver, S. 2003. Attacking the Act. *Frontiers in Ecology and the Environment* **1:** 59.

Soulé, M.E., Estes, J.A., Berger, J. and del Rio, C.M. 2003. Ecological effectiveness: conservation goals for interactive species. *Conservation Biology* **17:** 1238-1250.

Strahan, R. 1986. Why save endangered species and which ones do we save? Pp. 2-17 in *Endangered Species: Social, Scientific, Economic and Legal Aspects in Australia and the South Pacific*, edited by S. Burgin. Total Environment Centre, Sydney.

Strahan, R. (ed.). 1995. *The Mammals of Australia*. Reed, Chatswood, NSW.

Tear, T.H., Scott, J.M., Hayward, P.H. and Griffith, B. 1995. Recovery plans and the Endangered Species Act: are criticisms supported by data? *Conservation Biology* **9:** 182-195.

Thomas, J.W. and Salwasser, H. 1989. Bringing conservation biology into a position of influence in natural resource management. *Conservation Biology* **3:** 123-127.

Tsamenyi, M., Rose, G. and Castle, A. 2003 International marine conservation law and its implementation in Australia. Pp. 1-17 in *Conserving Marine Environments. Out of sight, out of mind*, edited by P. Hutchings and D. Lunney. Royal Zoological Society of New South Wales, Mosman, NSW.

Walker, B.H. 1992. Biological diversity and ecological redundancy. *Conservation Biology* **6:** 18-23.

Walker, B.H. 1995. Conserving biological diversity through ecosystem resilience. *Conservation Biology* **9:** 747-752.

Wilcove, D.S. and Chen, L.Y. 1998. Management costs for endangered species. *Conservation Biology* **12:** 1405-1407.

Williams, J., Read, C., Norton, A., Dovers, S., Burgman, M., Proctor, W. and Anderson, H. 2001. *Biodiversity, Australia State of the Environment Report 2001 (Theme Report)*. CSIRO publishing on behalf of the Department of the Environment and Heritage, Canberra, Australia.

Williams, S.E., Bolitho, E.E. and Fox, S. 2003. Climatic change in Australian tropical rainforests: an impending environmental catastrophe. *Proceedings of the Royal Society of London B* **270**:1887-92.

Wilson, G. R. 1986. Endangered species – an overview. Pp 18-25 in *Endangered Species: Social, Scientific, Economic and Legal Aspects in Australia and the South Pacific*, edited by S. Burgin. Total Environment Centre, Sydney.

Winter, J. W. 2004. Forest mammals of northern Queensland: is their conservation status improving? Pp. 435-451 in *Conservation of Australia's Forest Fauna* (second edition), edited by D. Lunney. Royal Zoological Society of NSW, Mosman, NSW, Australia.

APPENDIX I

Appendix 1.

In the Threatened Species Conservation amendment bill, as reported in the NSW Legislative Assembly Hansard of 25 September 2002 (http://www.parliament.nsw.gov.au/prod/parlment/hansart.nsf/V3Key/LA20020925074), the following extracts show some of the community divisions and strength of feeling on the matter.

"Mr Slack-Smith: In the United States of America if a species is important to the community and to the environment, the community contributes towards the preservation of that species. That is not happening in New South Wales. Two per cent of the population is bearing 98 per cent of the costs. That must be changed. The New South Wales community should be responsible for all endangered species. If the community accepts that responsibility, many species will be taken off the threatened species list. Treasury cannot afford to retain some of the endangered species that have been placed on that list by the Scientific Committee. It all goes back to dollars and cents. If the Scientific Committee wants to retain threatened species on that list the whole New South Wales community should contribute towards the preservation of those species. At the moment 98 per cent of those costs are being borne by land-holders. Let us look seriously at making the entire community the custodians of all threatened species. Primary producers should be paid if they are required to preserve some of these threatened species. Thirty years ago farmers in the United States of America used to shoot, bury and burn any threatened species on their properties and say nothing about it. That is what will happen in New South Wales if this legislation is enacted."

"Mr Armstrong: (Lachlan) [10.33 p.m.]: I listened with great interest to the contributions of honourable members to debate on this bill. Some good and cogent points were made by a number of honourable members who contributed to this debate. We are debating a complex subject-a subject that the scientific community and the broader community do not fully understand. Everyone can make a positive contribution to an issue that will evolve over the years. What is a threatened species? Can a threatened species be identified? It is a matter of personal opinion, preference and management. I am in favour of preserving genuine and unique threatened species, but we have to determine what constitutes a unique species."

"**Mr Armstrong:** ... the superb parrot is a beautiful bird. Signs on Trunk Road 56, which leads to Cowra and Yass, and on Main Road 248, which leads to Young and Crookwell, reflect the fact that Boorowa is the home of the superb parrot. The mayor is proud of that fact, but he is not nearly as proud as the local member, the honourable member for Burrinjuck. Two years ago this Government allocated $1 million to lay bitumen on the road from Young to Boorowa-a welcome amount of money to complete a road that is used by school buses almost every day. The budget blew out by $160,000 because it took nine months for a decision to be made about the relocation of 10 nests of the superb parrot. The Roads and Traffic Authority, the National Parks and Wildlife Service, and the local council were involved in the relocation of the superb parrots."

"**Mr Whelan:** Do you think they should have been killed?"

"**Mr Armstrong:** No, they should have been relocated. I am simply saying that we have to work out a methodology before we start to lay down the law. I am very much in favour of such legislation, but we need to get our priorities right. We need to work out the mechanics before we inflict this sort of legislation on local government, farmers, State Government instrumentalities and the broader community. Who paid for the relocation of the parrots? The people of Young. The funding shortfall to complete the road had to come out of ratepayers' funds because the Government did not advise properly. There is an inequity in the management of the process. This legislation does not provide for the proper process, and therefore the end objective cannot be achieved. We are starting the proposition in the middle of the management exercise."

"**Mr Oakeshott:** (Port Macquarie) [10.48 p.m.]: In the recent past the farming community on the mid North Coast and in the electorate of Port Macquarie has been hit by a whole range of regulations. In addition to the threatened species legislation, there has been an obvious and public debate about the native vegetation plans of management and the acid sulfate soil plans of management that are currently the subject of consultation." "New South Wales farmers in general are at the crossroads." "Sadly, the research of some Independents has shown that even Liberal Party and National Party members should hang their heads in shame in this regard."

"**Mr Fraser:** You should talk about shame; you walked away from the National Party."

"**Mr Oakeshott:** And with very good reason. You guys are selling your constituents down the drain, and this debate is a very good example of that."

"**The Chairman:** Order! If the honourable member for Coffs Harbour does not cease interjecting, I will call him to order."

"**Mr D. L. Page** (Ballina) [10.58 p.m.]: Since I have been challenged by the honourable member for Port Macquarie in relation to native vegetation management, I would like to clarify the position. As I said when the native vegetation legislation went through Parliament and again more recently in a matter of public importance, we support responsible management of native vegetation. If the honourable member for Port Macquarie had any brains he would too."

"**Mr Debus:** (Blue Mountains-Attorney General, Minister for the Environment, Minister for Emergency Services, and Minister Assisting the Premier on the Arts) [11.08 p.m.]: Australia is home to more than one million species of fauna and flora, many of them found nowhere else in the world. Changes to the landscape and loss of native habitat resulting from human activity have put many of these unique species at risk. Over the past 200 years many species have become extinct-they are gone for ever. For other species of plants and animals whose very survival is threatened, a range of management and conservation measures are in place.

The New South Wales Government's mechanism for conservation of threatened species is the Threatened Species Conservation Act-landmark legislation that was passed seven years ago, in 1995. However, one would hardly guess that from the contributions of members opposite tonight. The threatened species conservation legislation was passed seven years ago. The Act aims to secure the recovery of plant and animal species that are in imminent danger of becoming extinct. It integrates species impact assessment into the concurrence provisions of the Environmental Planning and Assessment Act. That means that almost all actions that are likely to have a significant effect on threatened species do not require separate licences under the Act. That is worth reiterating.

The amendments contained in this bill will enhance the biodiversity conservation objectives of the Threatened Species Conservation Act and they will, indeed, safeguard the independence of the Scientific Committee. I point out that this bill responds to the recommendations of the joint parliamentary committee and to the enactment of the new Commonwealth threatened species legislation. That is, it responds to the unanimous recommendations of the joint select committee which considered legislation that was passed in 1995, and the Opposition is now lining up to pretend that something entirely new is happening. In fact, the present bill amounts to finetuning of a bill that has been in place since 1995. In no respect does it make life more difficult for farmers."

APPENDIX I

"Almost every concurrence request is approved but, importantly, only after the proponent has modified the development to ensure that threatened species are, wherever possible, protected. That is to say, this bill ensures that ecologically sustainable development is carried out. The Government cannot support the establishment and function of a social and economic committee as has been proposed by the Opposition. The amendments moved by the Opposition would subject a factual decision as to whether a species is threatened with extinction, and therefore listed or not listed under the Act, to political considerations, and that is not appropriate."

"**Mr Fraser:** But compensation is what is wanted. You said you would pay compensation."

"**Mr Debus:** ... it is simply a contradiction to suggest that a Scientific Committee cannot make a decision on a scientific basis and that instead it has to take into account political considerations."

"The Government will support the establishment of a social and economic committee on the understanding that that committee will be established through administrative processes rather than by statutory means." "In carrying out that role it is not intended that the committee should compromise the independence of the Scientific Committee in its scientific assessment of the conservation status of a species, a population or an ecological community, or its deliberations regarding threatening processes."

"**Mr Stoner:** This is madness. You have sold out rural and regional New South Wales."

"**Mr Debus:** I must say the Opposition is more than usually confused tonight."

"**Mr Debus:** For those who have ears to hear, I have demonstrated that the Government is serious about taking into account the social and economic effects of its threatened species legislation."

"**Mr Fraser:** (Coffs Harbour) [11.27 p.m.]: " The Minister went on about science and about nobody understanding science. In the endangered species listing in this bill is a frog whose common name is the green and golden bell frog. If the Minister goes to the Parliamentary Library and gets the Australian book of frogs, produced by the Australian Museum-which has members on the Scientific Committee-he will wee [*sic*] that in that book the Australian Museum has listed the frog as being abundant up and down the east coast. Under this legislation it is listed as an endangered or threatened species.

I ask the Minister, "Why not have checks and balances?" Emotion, as well as science, is involved in these issues."

"Mr Debus: (Blue Mountains-Attorney General, Minister for the Environment, Minister for Emergency Services, and Minister Assisting the Premier on the Arts) [11.32 p.m.]: I move: That the report be now adopted.

The House divided. Ayes, 46, Noes, 31.